Frontispiece **Jack Simmons, June 1955** (Photograph courtesy of the Leicester Mercury Newspaper Ltd)

The Impact of the Railway on Society in Britain

Essays in honour of Jack Simmons

Edited by
A.K.B. EVANS AND J.V. GOUGH

ASHGATE

Published by
Ashgate Publishing Limited
Gower House
Croft Road
Aldershot
Hants GU11 3HR
England

Ashgate Publishing Company
Suite 420
101 Cherry Street
Burlington, VT 05401-4405
USA

Ashgate website: http://www.ashgate.com

British Library Cataloguing in Publication Data
The impact of the railway on society in Britain : essays in
    honour of Jack Simmons
    1.Railroads - Great Britain - History 2.Railroads - Social
    aspects - Great Britain 3.Railroads - Great Britain -
    Historiography
    I.Evans, A. K. B. II.Simmons, Jack, 1915-2000
    385'.0941

Library of Congress Cataloging-in-Publication Data
The impact of the railway on society in Britain : essays in honour of Jack Simmons /
edited by A.K.B. Evans and J.V. Gough
    p. cm.
    Includes bibliographical references and index.
    ISBN 0-7546-0949-9 (alk. paper)
    1. Railroads--Great Britain--History. 2. Great Britain--Social conditions. I. Simmons,
Jack, 1915-2000 II. Evans, A. K. B.

HE3018 .I45 2002
385'.0941--dc21

2002024895

ISBN 0 7546 0949 9

Typeset in Times by N^{2}productions
Printed and bound in Great Britain by MPG Books Ltd., Bodmin.

Contents

List of Illustrations		vii
List of Contributors		ix
Foreword *Derek H. Aldcroft*		xiii
Editors' Preface		xv
Abbreviations		xvii
Three Tributes *Alan Everitt, J. Mordaunt Crook, Dame Margaret Weston*		xix
1	Jack Simmons: the Making of an Historian *Michael Robbins*	1

Section I   The Railway: Origins and Working

2	Pre-locomotive Railways of Leicestershire and South Derbyshire *Marilyn Palmer and Peter Neaverson*	11
3	The Transport Geography of the Wigan Coalfield: the Canal and Railway Contributions *David Turnock*	33
4	Rolling Stock, the Railway User, and Competition *Michael Harris*	47
5	A Note on Midland Railway Operating Documents *John Gough*	61
6	Financing the Bagdadbahn: Barings, the City, and the Foreign Office, 1902–3 *P.L. Cottrell*	77

Section II   Spirit, Mind and Eye

7	The 'Broad Gauge' and the 'Narrow Gauge': Railways and Religion in Victorian England *R.C. Richardson*	101

8   Railways, their Builders, and the Environment   117
     Gordon Biddle

9   Ruskin and the Railway   129
     J. Mordaunt Crook

10   Philip Larkin's Railways   135
     Roger Craik

11   BEWARE OF THE TRAINS: Reflections and a Few Footnotes on the
     Railways of Suffolk   151
     Norman Scarfe

12   The Train in the Landscape: Dovey Junction c. 1932   163
     Gwyn Briwnant-Jones

Section III   The Opening Up of Britain

13   The London Railway Suburb 1850–1914   169
     Alan A. Jackson

14   The Railway and Rural Tradition 1840–1940   181
     Alan Everitt

15   Tourism and the Railways in Scotland: the Victorian and Edwardian
     Experience   199
     Alastair J. Durie

16   Railways and the Evolution of Welsh Holiday Resorts   211
     Roy Millward

17   Sir George Samuel Measom (1818–1901), and his Railway Guides   225
     G.H. Martin

Section IV   Heritage and History   241

18   The North Eastern Railway Museum, York – 'the germ of a truly
     National Railway Museum'   243
     Dieter Hopkin

19   Transport Museums and the Public Appreciation of the Past   259
     Colin Divall and Andrew Scott

20   Writing the History of British Railways   269
     Terry Gourvish

21   'Bibbling' the Railways   279
     George Ottley

Appendix   Jack Simmons: a Bibliography of his Published Writings   283
           Diana Dixon and Robert Peberdy

List of Sponsors   307

Index   309

List of Illustrations

Frontispiece   Jack Simmons

Figure 2.1   Sketch map showing the western end of the Charnwood
                Forest Canal built by the Leicester Navigation Company
                (redrawn from Christopher Staveley's plan, 1791)   15

Figure 2.2   Sketch map showing the proposed branches of the Ashby
                Canal to the limestone quarries and collieries at Breedon
                Cloud Hill, Staunton Harold, Ticknall, and Lount. The system
                of railways eventually constructed is also shown (redrawn
                from Robert Whitworth's survey plan of 1792, LLRRO
                DE421/4/22)   19

Plate 2.1   A section of the Ashby Canal railway, Ticknall branch,
                showing the plate rails with points and crossover sections
                (Science Museum/Science & Society Picture Library)   20

Figure 2.3   A typical layout of 13 lime-kilns in the Ticknall limeyards
                served by railway from the Ashby Canal, based on the 1881
                OS 25-inch map   21

Plate 2.2   The bridge c.1802 carrying the Ashby Canal railway over the
                main street in Ticknall (authors)   22

Plate 2.3   An excavated stretch of the stone blocks supporting the
                21-inch gauge railway in Margaret's Close limeyards at
                Ticknall (authors)   23

Figure 2.4   The railway branches at Dimminsdale and Heath End, from
                a site survey in 1994   24

Figure 2.5   The railways connecting ironworks and collieries on Ashby
                Woulds to the northern end of the Ashby Canal, showing
                their possible full extent between 1800 and 1840   27

Plate 2.4   Sections of the Belvoir Castle railway track of c.1815 as
                reconstructed by the National Railway Museum. The upper
                fish-bellied edge-rails include a point system on the right,
                while the lower rails show a crossover (Science Museum/
                Science & Society Picture Library)   28

Plate 2.5   A load of peat being transported on the Belvoir Castle
                railway: the castle may be seen in the background (Science
                Museum/Science & Society Picture Library)   29

Figure 3.1   The Wigan Coalfield: cumulative map of mines, canals, and
                railways   34

Figure 5.1   A page from an early Midland Railway Working Time-Table   66
Figure 5.2   A page from a Midland Railway Sectional Appendix, giving
              information to drivers about the approach to London   73
Figure 5.3   A Midland Railway notice to staff   75
Figure 11.1   Peter Bruff's timber viaduct at Cattawade on his main
              Colchester–Norwich line of the Eastern Union Railway, 1846
              (*Illustrated London News*, 1846 / Hugh Moffat)   155
Plate 11.2   Worlingworth station, on the Mid-Suffolk Light Railway,
              opened 1904, closed 1952 (Jack Simmons, 1953)   156
Plate 11.3   Capel station, on the Hadleigh branch line, opened 1849,
              demolished 1973 (Jack Simmons, 1953)   157
Figure 11.4   Reg Carter's drawing in the Sorrows of Southwold postcard
              series No. 2 (Southwold Museum)   159
Figure 11.5   Reg Carter's drawing in the Sorrows of Southwold postcard
              series No. 1 (Southwold Museum)   160
Plate 12.1   Dovey Junction c.1932. Painting by Gwyn Briwnant-Jones,
              which appears in colour on the jacket. Reproduction by
              courtesy of the National Library of Wales   164
Plate 17.1   The multi-coloured paper cover of a one-shilling Measom
              guide (1856)   229
Plate 17.2   The maroon and gilt glazed cloth cover of a Measom guide
              (1856)   230
Figure 17.1   The Crystal Palace: the transept and lily pond (*The Official
              Illustrated Guide to the London and South-Western, North
              and South Devon, Cornwall, and West Cornwall Railways*,
              Griffin & Co., London, 1864)   231
Figure 17.2   Mayall's portrait of Measom (*Guide to the London and
              South-Western*, 1864)   233
Figure 17.3   Horne the house-painter, at home and away (*Guide to the
              London and South-Western*, 1864)   236
Figure 17.4   An interior view of Wells Cathedral (*Guide to the London
              and South-Western*, 1864)   239
Figure 17.5   Man and nature: the sea wall and railway at Dawlish (*Guide
              to the London and South-Western*, 1864)   240
Plate 18.1   The Small Exhibits Section of the Museum, with E.M.
              Bywell (Curator), J. Dixon (Museum Sub-Committee) and
              Mr Lister (Museum Custodian), 18 July 1947 (National
              Railway Museum – 37/00)   249
Plate 18.2   The first public visitors to the Queen Street Large Exhibits
              Section following reopening after the Second World War,
              18 July 1947 (National Railway Museum – 36/00)   254
Plate 18.3   The entrance to the Queen Street Museum showing
              Stephenson's Hetton Colliery locomotive and admission
              prices c.1960 (National Railway Museum — 894/89)   256

List of Contributors

DEREK H. ALDCROFT is currently University Fellow in the Department of Economic and Social History, University of Leicester. Formerly Research Professor in Economic History at Manchester Metropolitan University (1994–2001), Professor and Head of Department of Economic History at the universities of Sydney (1973–6) and Leicester (1976–94). At Leicester University in the later 1960s and early 1970s he had frequent contact with Jack Simmons regarding mutual interests in transport history. He later became one of the editors of *The Journal of Transport History*. He has written extensively on transport history and economic growth and is listed in *Who's Who in Economics* (Edward Elgar).

GORDON BIDDLE, a surveyor by profession, has concentrated since retirement on his long-standing interest in transport history, particularly railways and inland waterways, on which he has written extensively. Among his books is *The Railway Surveyors*, and he was co-editor with Jack Simmons of *The Oxford Companion to British Railway History*. He has made a special study of historic railway structures: some 2 000 sites of importance in the railway heritage appear in *Britain's Historic Railway Buildings: an Oxford Gazetteer of structures and sites* (Oxford, Spring 2003). He is a Vice-President of the Railway & Canal Historical Society.

GWYN BRIWNANT-JONES (born Machynlleth, Montgomeryshire) taught in grammar and comprehensive schools in Wales before embarking on a second career devoted to researching into Welsh Railways; publications and radio and television broadcasts followed.

PHILIP L. COTTRELL is Professor of Economic and Social History, University of Leicester. He was fortunate to benefit both from Jack Simmons's particular guidance and from the general provision that Jack made at Leicester through developing (and donating many volumes to) a special transport history collection within the University Library and by establishing *The Journal of Transport History*.

ROGER CRAIK was born in Leicester, where his father taught English at the University. He is an Associate Professor of English at Kent State University, Ohio, and specializes in twentieth-century poetry, especially that of Philip Larkin.

JOE MORDAUNT CROOK is Professor Emeritus of Architectural History and former Director of the Victorian Studies Centre, University of London, at Royal

Holloway and Bedford New College, and has published widely in these fields. He is a Fellow of the British Academy. His first post as a university lecturer was in Jack Simmons's Department of History, University of Leicester (see below p. xxi).

COLIN DIVALL is Professor of Railway Studies in the Institute of Railway Studies, a joint initiative of the National Railway Museum and the University of York. He is the author, with Andrew Scott, of *Moving On: Making Histories in Transport Museums* (Leicester University Press, 2001) and is on the editorial board of *The Journal of Transport History*.

DIANA DIXON studied history at Leicester University and subsequently worked as Assistant Bibliographer in the Victorian Studies Centre, Leicester. She then lectured at the College of Librarianship, Wales, and Loughborough University. Her research interests include the bibliography and history of English provincial newspapers and nineteenth-century children's magazines.

ALASTAIR DURIE, Senior Lecturer in Economic and Social History at the University of Glasgow, was originally a textile historian interested in the growth of the Scottish linen industry in the eighteenth century. But in recent years, due in no small part to Jack Simmons's encouragement, his interests have shifted into how tourism developed in Scotland. His study, *Scotland for the Holidays? Tourism in Scotland, c.1780–1939* (Tuckwell Press, East Linton, 2002), will appear shortly.

ALAN EVERITT was Hatton Professor and Head of the Department of English Local History in the University of Leicester until his retirement; he is also a Fellow of the British Academy. He has written widely on traditional society and is interested in the survival of rural custom. At present he is engaged on the history of common land in England since the Norman Conquest.

JOHN GOUGH was a member of the Department of English of Leicester University. He has always been interested in the history and operation of railways and has written extensively in both areas. He is Deputy Chairman of the Railway History Heritage Committee and has held office in a number of other bodies of railway interest.

TERRY GOURVISH has been Director of the Business History Unit at the London School of Economics since 1989. His writings on railways include: *Mark Huish and the London & North Western Railway* (Leicester University Press, 1972); *Railways and the British Economy* 1830–1914 (Macmillan, 1980); *British Railways 1948–73: A Business History* (Cambridge University Press, 1986), and *British Rail 1974–97: From Integration to Privatisation* (Oxford University Press, 2002). He is currently writing an official history of the Channel Tunnel.

MICHAEL HARRIS (1945–2001), after reading politics and sociology at the University of York, spent almost all his working life with railways; involved in their operation on BR, with the private equipment manufacturers, and with railway publishing. He is the author of a dozen books on railways, specialising in rolling stock on which he was an authority. For many years he edited a popular monthly magazine, was one of the founders of Transport 2000, and took a special interest in heritage affairs.

DIETER HOPKIN has been Head of Library and Archive Collections at the National Railway Museum since 1994; before this he was Curator of Collections from 1989. He is responsible for all the NRM's two-dimensional collections ranging from printed works on paper and archives to photographs and works of art. He has written on the historical development of railway museums and museology, aspects of the NRM's collections and the iconography of early railways.

ALAN A. JACKSON is a retired Treasury civil servant who, since the early 1960s, has written some 15 books on railway and modern social and local history. He is a past President of the Railway & Canal Historical Society, and was the first President of the London Underground Railway Society.

GEOFFREY MARTIN is Research Professor of History in the University of Essex. He was formerly Keeper of Public Records (1982–8) and Professor of History in the University of Leicester (1973–82), where he was a colleague of Jack Simmons from 1952.

ROY MILLWARD was Lecturer and later Reader in Historical Geography at Leicester University 1947–82. He has published in the field of British landscape history including, in collaboration with A.H.W. Robinson, two books on the English–Welsh borderland and one on the landscape history of North Wales. He contributed a study of Lancashire to W.G. Hoskins's *Making of the English Landscape* series and for the later volumes acted as Assistant Editor.

PETER NEAVERSON trained as a physicist and has worked in engineering and electronics. He has been an active industrial archaeologist for the past 25 years and is an Honorary Visiting Fellow in the School of Archaeology and Ancient History at Leicester University. He has a particular interest in the history and archaeology of railways. He and Marilyn Palmer have jointly edited *Industrial Archaeology Review* for nearly 18 years and published books and articles. These include *Industrial Landscapes of the East Midlands* (1992), *Industry in the Landscape: 1700–1900* (1994) and *Industrial Archaeology: Principles and Practice* (1998).

GEORGE OTTLEY compiled the standard bibliography of British railway history. He spent his working life as a librarian, first at University College London, 1934–7, then in the British Museum Library, 1937–67, and from 1967 to retirement in 1982 in Leicester University Library. The three volumes of the Bibliography describe 20 000 monographs (the work is continued annually in the *Journal of the Railway & Canal Historical Society*).

MARILYN PALMER is Professor of Industrial Archaeology and Head of the School of Archaeology and Ancient History at Leicester University. She serves on committees concerned with industrial archaeology for English Heritage, the Council for British Archaeology and The National Trust, and is a past President of the

Association for Industrial Archaeology. She has worked to ensure the acceptance of industrial archaeology as an academic discipline and to define a methodological framework for the study of industrial structures and landscapes within an archaeological context.

ROBERT PEBERDY studied history at Merton College, Oxford, and later undertook research on mediæval urban history in the Department of English Local History at Leicester University. He is an Assistant Editor of the Victoria History of Oxfordshire and has given many evening classes and day-school lectures on local and mediæval history for the Oxford University Department for Continuing Education.

ROGER RICHARDSON, now Professor of History and Director of International Relations at King Alfred's College, Winchester, was an undergraduate in Jack Simmons's Department of History, University of Leicester. He is the author of many books and articles, most recently *The Study of History: A Bibliographical Guide* (2nd edn, Manchester University Press, 2000) and *The Changing Face of English Local History* (Ashgate, 2000).

MICHAEL ROBBINS was a board member (managing director, railways), London Transport. He was co-founder, with Jack Simmons, of *The Journal of Transport History*. He has been president of the Institute of Transport; chairman of the governors, Museum of London; president, Society of Antiquaries.

NORMAN SCARFE was a lecturer in Jack Simmons's History Department, Leicester University, 1949–63. He co-founded the Suffolk Records Society in 1958. He is Founder-Chairman of the Suffolk Book League and Quondam Chairman, Board of the Centre of East Anglian Studies, University of East Anglia. His publications include three Shell County Guides, books on Suffolk, and, most recently, three on French travellers in eighteenth-century Britain (which particularly interested Jack Simmons).

ANDREW SCOTT is Head of Museum, National Railway Museum, and was previously Director of the London Transport Museum. He is a member of the Railway Heritage Committee. He is the co-author, with Colin Divall, of *Moving On: Making Histories in Transport Museums* (Leicester University Press, 2001), the first analysis of the role of transport museums for many years.

DAVID TURNOCK is Reader in the Department of Geography, University of Leicester. He has a long-standing interest in the economic and social impact of railways, particularly in industrial regions and especially in his home area of Central Lancashire.

DAME MARGARET WESTON was Director of the Science Museum 1973–86, when Jack Simmons was a member of its Advisory Council (see below p. xxiii). She is President of the Association of Independent Railways and Preservation Societies.

Foreword

Derek H. Aldcroft

A Festschrift on railways in honour of Jack Simmons is a fitting tribute to one who did so much to establish railway history as a serious field of study. One should bear in mind, however, that railway history was only one of Jack Simmons's many academic interests. In the later years of his writing career it became his major preoccupation, and his work in this field alone would mark him out as a scholar and historian of distinction. But Jack had many strings to his bow. In his early career and also for much of the period when he was Professor of History in the University of Leicester (1947–75), he was a prolific writer on a wide range of subjects, including industrial archaeology, parish churches, Lakeland poets, local history, transport museums, town and country studies, universities, and, in a wider context, British and Imperial history. All this was achieved while undertaking a very full academic load in the University, in both teaching and administration.

Scholarship alone was by no means Jack's only forte. He had the rare gift of being able to write elegantly, in the words of H. J. Dyos (*Journal of Transport History*, n.s., 3, 1975–6) a 'smoothly flowing historical narrative', and to bring alive his subject matter. This stemmed from his sympathetic mentality towards both things and people, which was reflected not only in his writing, but also in his humane and caring approach to colleagues and students, so many of whom benefited from his erudition and enthusiasm for history and learning. Academic discussion with Jack Simmons could never fail to stimulate the mind.

Since Jack Simmons retired, university life has undergone a radical transformation, and in many respects it is almost unrecognizable from what it was a generation ago. Some would say that change has not always been for the better, and one suspects that Jack was inclined to agree with this sentiment. Yet notwithstanding the many changes, the contributions in this volume demonstrate that railway history is still thriving and maintaining the standard of excellence set by Jack Simmons.

Editors' Preface

This volume of essays by colleagues, friends, and admirers of Jack Simmons was planned to honour him in his eighty-sixth year. Sadly, he died on 3 September 2000 a few days after his eighty-fifth birthday. The project had been kept from him, as an intended surprise, but it was possible during his last days to tell him about it: he listened intently and was pleased.

The theme of the book was one that he had made his own. It was chosen to reflect his broad view of the study of history: that it embraces all aspects of civilized life, political, intellectual, cultural, aesthetic, scientific and technical, which are all interrelated. This belief, set out as long ago as 1948 in his inaugural lecture as the first Professor of History in the then University College of Leicester, was realized in his own achievement. His interests were wide and deep: local, national and imperial history; literature and the arts; landscape, architecture, towns and villages; museums, archaeology and antiquities. A collection of essays from his many friends in these various fields could have seemed random. For coherence, it was necessary to choose one topic which impinged on all these fields.

The choice of the railway was both apt and inevitable, for railway history is the area of study for which he is best known, and it is one to which he brought a new dimension, as Michael Robbins has pointed out: an all-embracing approach to the effects of the railway on the lives of the communities in which it operated. The four sections of this book look at the impact of the railway on society in Britain in ways that were of particular interest to him. Three of the sections reflect in their titles the divisions he proposed for his intended four-volume work on the railway in England and Wales (which, because of regrettable publishing decisions, was never fully realized). Our fourth section is directly concerned with railway history.

Section one, Origins and Working, is about the railway itself: how rails were used for transport of coal and iron before the advent of steam traction; how railways, canals, and canal-related tramroads transformed a coal industry; what forces influenced the railway companies to improve their rolling stock; how the organization of a railway's daily working was effected by its internal publications; and how politics could affect the financing of railway construction.

The essays in the second section treat aspects of railway study particularly dear to Jack Simmons's heart: Mind and Eye. The coming of the railway not only brought radical changes in transport but had an unprecedented visual impact on the landscape (greater even than the Romans had achieved) and gave a new direction to habits of mind. Its presence and the treasury of images and metaphors it provided influenced painters and cartoonists, prose writers and poets, and also religious organization and symbolism – a new field of study.

The third section deals with the widening of horizons resulting from the easy and relatively cheap mobility the railway afforded. It affected the pattern of suburban development (here examined in the major city of London), enlarging not only the area of possible residence but also the range of leisure activities. The countryside was brought within reach for many who would not otherwise have ventured beyond their own locality, enriching their periods of leisure and introducing them to rural life and its customs. This change in the relationship between town and country encouraged both the development of tourism, which was to become one of the great industries of the twentieth century, and the serious study of regional history. A whole new area of publishing was prompted by the new mobility: one element of this was the provision of guidebooks for the traveller.

The last section, Heritage and History, looks at two ways of approaching railway history: through artefacts and by the written word. The essays on museums investigate the emergence of collections of railway memorabilia and discuss the part played by museums of transport in the public understanding of its past development and appreciation of its social and economic importance. The essays on railway literature explain the challenges and opportunities involved in writing the history of railways up to recent times; and how the steadily increasing numbers of publications about railways have created a demand for bibliographical guidance, essential to researchers (the 'bibbling' of the railways).

These four groups of essays are framed by appreciations of the man to whose memory the volume is dedicated. Michael Robbins, his oldest friend and collaborator in joint enterprises, looks at the development of his historical interests from his school and undergraduate years. The Foreword and the Tributes commemorate his warm personality and some of the many and diverse fields of his intellectual and cultural activity. At the end of the book, the bibliography of his published work forms a permanent record of the great contribution he made to historical studies in his long and productive scholarly life.

Leicester   A.K.B. Evans
                                                                             J.V. Gough

Abbreviations

BR	British Railways; British Rail
DNB	*Dictionary of National Biography*
LLRRO	Leicester, Leicestershire, and Rutland Record Office
NRM	National Railway Museum, York
Oxf. Comp.	*Oxford Companion to British Railway History*, ed. Jack Simmons and Gordon Biddle, 1997
PRO	Public Record Office, Kew, London
RO	Record Office
TLAHS	*Transactions of the Leicestershire Archaeological and Historical Society*
TLAS	*Transactions of the Leicestershire Archaeological Society*
VCH	*Victoria History of the Counties of England*

Three Tributes

In December 2000 the University of Leicester held a celebration of Jack Simmons's life and work. Three of the addresses given on that occasion which are particularly relevant to this collection of essays are presented here.

Jack Simmons and English Local History

Alan Everitt

Jack Simmons is always thought of first as a railway historian. Yet in some ways that does less than justice to his reputation. His ever-questing mind ranged far beyond the development of railways for their own sake. It was their entire impact upon society, and the human and local response to their expansion, that dominated his most original work. For him, I used to think, as for me, the study of history was itself like a journey into the past: a journey in which one discovers that places are like people: every one has a life and personality of its own. He never forgot that we live in one of the world's oldest, most varied, and most richly documented countries. The way cities, towns, and villages reacted to the coming of the railway was moulded by their past as well as the present. He remarked, 'The past always lives on in the present' and that approach to it, as it survives in the world around us, remained one of his deepest convictions.

We both also believed that the English language had given birth to the world's greatest and most diverse literature. His own omnivorous reading greatly enriched his understanding of history, an understanding that deepened as he grew older. It enabled him to see beyond the formal evidence into that human world about which it is so often silent. It gave him an abiding sense of the complexity of human relationships, which we all recognize in our daily lives yet too frequently forget in turning to the past.

These were some of the convictions that made Jack so forceful a supporter of the establishment of English local history at Leicester. In that respect his name must be joined with those of its successive founding fathers, W.G. Hoskins and H.P.R. Finberg, and with the support of its original proponent, Principal Attenborough, himself a notable scholar and photographer of the countryside. Beyond him we must trace a further link to the name of Thomas Hatton, whose original benefaction of 2 000 topographical works to the University College in 1921, even before its doors were opened, made the foundation of the department practicable.

Though Hoskins, Finberg, and Jack did not always see eye to eye, on two fundamental principles they were united. First, they believed in the necessity of breaking away from the purely antiquarian tradition of local historiography in establishing the new department. Secondly, they believed it was essential to cover the whole of English society, on a comparative basis, and not limit the new venture to Leicestershire. It attracted the interest of a number of leading historians elsewhere, especially R.H. Tawney and F.M. Stenton. That it happened at all at Leicester, however, then the smallest and poorest of English university colleges, probably owed more to the energy and vision of Jack himself than any of us now know.

His own contribution to the study of local history demonstrates how broad that vision was. First, the great series of county history reprints: the re-publication of Nichols's *Leicestershire* was followed by new editions of comparable works for many other parts of the country, Gloucestershire, Hertfordshire, Nottinghamshire, Dorset, Cornwall, Kent, Cumberland, and Westmorland among them. The whole scheme entailed a truly daunting task of organization on Jack's part, coordinating the activities not only of printers and publishers but of county librarians, local editors, leading bookshops, and influential figures in the localities.

That it proved to be such a success in shire after shire was very largely due to Jack's unflagging energy, his attention to detail, and (one must add) his personal kindness. When I had finished my own 'Introduction' to Edward Hasted's *History and Topographical Survey of Kent*, it happened I was unable to visit him at Blackheath to discuss it. He had been through my typescript in detail and immediately suggested making the journey himself. We went through it together, paragraph by paragraph, in the refreshment room at Rugby Station! Six years later, eight of the county 'Introductions' were republished as a book, prefaced by a splendid survey by Jack on 'The Writing of English County History'. Its 20 pages are a striking testimony to the breadth of his own grasp of the subject and the perception of his judgement, every word straight to the point.

Finally, I must single out one work from the extraordinary efflorescence of his later years, *The Railway in Town and Country* (1986), written at the age of 70. It seems to me a far more original and wide-ranging book than its modest title suggests. Its subject is not so much the development of the system, which he had described in an earlier volume, as its *impact* upon the varied types of community that make up the whole realm: London and its suburbs, the leading provincial cities, major towns, railway towns, ports and docks, watering places, market centres, and the countryside itself. It is the most richly textured of all Jack's books. A mere glance at its 1 616 notes will convince anyone of the astonishing range of learning behind it.

The great challenge that confronted Jack as he began that work, and from my own small part in it I know deepened as he proceeded, was that in so few places had this great subject ever been seriously investigated at the local level. In some town histories, as at Bolton and Hitchin, it had simply been ignored. In others, facile assumptions about automatic expansion had been made without any real evidence to support them. The facts, as Jack discovered them in place after place, every one of them chosen with his unerring eye for the luminous example, proved to be far more diverse, and in his hands far more enlightening. Who but he, in the space of a single page, could have made one see that Surbiton (for example) was not merely one of the joke-suburbs of London, the epitome of metropolitan dowdiness, but actually a most interesting place, still well worth visiting as the very first railway-suburb in Europe – perhaps in the world? Who but he would have lit on Dulverton in Somerset, and Melbourne in Derbyshire to illuminate the impact of the railway on two contrasting village communities, and their own very distinctive contribution to the evolution of the system itself?

So the reader is led on with beguiling fascination from one place to another, and learns something fresh every time about the diversity of the local response to the coming of the railway. Every part of the country, every city and town and village, even individual farmsteads in some counties, rose to the new challenge and the new opportunity it offered in its own way. For its growth or stagnation depended not only on the revolution in transport but on its own individual life: its energies, resources, and circumstances, its legacy from the past, and above all perhaps the character and initiative of its people. Yet the pattern of diversity that emerges is a far from meaningless pattern: the detailed threads are all woven into one great theme in the history of the entire country. In each chapter the plea is made for more detailed local study, and the way forward is clearly indicated for other scholars. *The Railway in Town and Country* is one of those rare books that seem to open a new world. It is a triumphant achievement.

Jack Simmons and Victorian Studies

J. Mordaunt Crook

October 1963: I was 26 and Jack was 48. He was a Pro-Vice-Chancellor, and I was an Assistant Lecturer. And there he was, at the other end of the seminar table – plump and pink and mischievous – fizzing with ideas like some high-explosive firework. Together we were supposed to be teaching a new Special Subject. In fact, I was learning and he was teaching, and the students formed an appreciative audience. And how we enjoyed ourselves! The subject was Victorian England, and most unusually its set books covered the whole spectrum of Victorian thinking in politics, theology, philosophy, economics, literature. There were novels, and speeches, and letters, and essays. And even that wasn't enough: the conversation quickly got round to architecture, and topography, and aesthetics. Officially, we divided the syllabus between us. He took Gladstone, and John Stuart Mill, and Cardinal Newman, and Trollope. I was left with Dickens and George Eliot, Matthew Arnold and Walter Bagehot. Quite enough to be going on with: I was already lecturing on the Tudors and the Regency.

The use of literature as historical evidence certainly appealed to me. But as a product of the Oxford History School, nursed on a diet of constitutional documents, I had only a hazy idea of how to handle such material. Soon enough, I got the hang of it, using the texts to explain Victorian attitudes: the purpose of education, the nature of liberalism, the psychology of class, the status of women. Pretty novel in the teaching of History in 1963 but it had all been mulled over by Jack almost before I was born. As early as 1944 – in a review in *Time and Tide* – he was already toying with the idea of 'the Victorian mind', the first use of that specific phrase, incidentally, that I have found.

In a way, Jack understood the Victorians because he was himself a Victorian, born just as the Great War swept away the world the Victorians took for granted. He shared their energy, their commitment, their patriotism and pride, their obsession with detail, their feeling for the music of words, their reticence, their crushing sense of duty, their sentiment, their instinct for the absurd. And I think he saw something of himself in the figure of one particular late Victorian, Cecil Torr. Torr was a bachelor-gentleman-scholar who retired to the tiny village of Wreyland, near Dartmoor, recording his impressions as he watched the world go by in all its endearing absurdity. 'His intelligence', Jack noted in the introduction to *Small Talk at Wreyland* (1980),

> was a highly sceptical one... deepened and enriched by his prolonged study of history... He [took] a quiet pleasure in pointing to ludicrous mistakes made by the eminent; but... deeper still, governing even his scepticism, lay his charity and an ultimate fairness of judgement... [For] scepticism [can become] an occupational disease among historians... if it passes into cynicism... [and there is a world of] difference between the historically minded sceptic and the cynic... For judgement is what matters, the relentless pursuit of truth, exact and prosaic... [Torr] had no patience with modish theoretical fads in teaching... His reading was catholic... [but his] imagination was yoked to a fastidious exactitude, and both were played upon by a delicious sense of the comic.

Jack was describing a man born in the mid-Victorian period but he could almost have been writing his own historical credo.

Those were the attitudes – Victorian in origin – which Jack brought to Victorian studies at Leicester. From these stemmed the undergraduate Victorian Special Subject, the postgraduate Victorian Studies Group, and the research-oriented Victorian Studies Centre; most of all, they were the origin of the Victorian Library, the jewel in the crown of Leicester University Press. Jack had sufficient confidence in me to make me responsible for one of its first volumes, Eastlake's *History of the Gothic Revival*. But the whole programme – the format, the apparatus, the choice of books – was very much under Jack's control. The Victorian Library, in its scholarship and range, as well as in the superb quality of its production, was the product of Jack's extraordinarily eclectic vision – a Victorian vision, of which I am irresistibly reminded whenever I board the train at St Pancras Station.

I used to be puzzled sometimes by Jack's eager decision to leave Oxford behind – to swap Christ Church and All Souls for Leicester. I think I only really understood his motivation when I came to examine his diaries – those bulging red-backed volumes we all remember, jam-packed with appointments and memos. They are the record of an administrator extraordinary: committees in Leicester and London; committees at every level, national, local, departmental; so many committees, one suspects the scholar ultimately became submerged by the committee-man. I think he always had a hankering after a bigger world, the world of imperial affairs occupied by his predecessors in the Beit Lectureship at Oxford, Lionel Curtis and Reginald Coupland. Jack saw himself, obviously, as a historian, explaining the past to the present; but he also felt himself to be, ultimately, a public servant, in a great Victorian tradition; almost a professorial proconsul, pushing back the frontiers of ignorance; recording day by day, year by year, his struggles with incompetence, and stupidity, and greed. Those endless diaries, recording a life of truly heroic endeavour, will give him some claim – academically speaking – to be regarded as the last of the Victorians.

Jack and Museums

Dame Margaret Weston

Jack Simmons had a twenty-one year long association with the three museums I knew best: the Science Museum and its National Railway Museum and National Museum of Photography, Film, and Television. I was fortunate enough to be the Science Museum's Director from January 1973 to March 1986, a period central to the story. Professor Simmons, as we then knew him, joined the Museum's Advisory Council four years earlier in 1969 when Sir David Follett was intent on preserving the Clapham and York railway collections, planning to combine them into a new National Railway Museum. Discussions as to where this new museum should be located delayed matters. So it was on my first day as Director that it was announced that the NRM would go ahead and be in York. It would open in September 1975. There were a few other things on the go too; there was approval for an extension to be built in the Science Museum's East Court and the offer of the Wellcome Collection, which if accepted would add Medicine to the Museum's fields of Science and Engineering. Also there was the requirement to set up and administer a Fund, working countrywide to ensure the preservation of Scientific and Technological material. This Fund is still doing valuable work today and is now known as PRISM.

Jack decided that he was ready to be involved in this in addition to becoming a committee member of the new National Railway Museum, which was to be headed by Dr John Coiley. I decided to have a small committee advising on the operation and management of the Fund. I asked Jack to chair it and I also took part in its work. Later, in 1981, he took over as Chairman of the NRM Committee. Then at the end of his term in 1984 he said that he did not think it right to remain on that committee but was very willing to serve on the Committee of the National Museum of Photography, Film, and Television which we had opened in Bradford in 1983. This time the museum had been my idea, but I would like to pay tribute to the Science Museum Advisory Council who after I presented it to them said, 'If this is one of the things you want to do, and you think you can, then certainly you should go ahead.' The very successful establishment and development of this new museum we owe to Colin Ford. Both Colin in Bradford and John Coiley in York had been careful to make their museums part of the community and Jack said that he very much liked the relationship the museum had with its visitors, and looked forward to a subject new to him. He continued to serve on this committee until 1990, four years after I retired, maintaining splendid relationships with Colin and John. The three of us became his friends for life. The reading room at the NRM was named after Jack, and he was one of the first Honorary Fellows of the Science Museum.

This account of Jack's very considerable involvement and contribution gives not a great deal about the man himself, only hinting at his ability to inspire great affection. I could go on to mention his quality of mind, that he was extraordinarily knowledgeable, that we were always consulting him and went on doing so, that he was both a very private individual and very good company, but I still felt there was something I was missing. Then I thought I had it. I rang Colin. Had Jack himself written something which would complete the picture? 'I have lots of letters,' said Colin, seeming perfectly to understand. 'I could look at them.' A few hours later

Colin read me a letter from Jack in answer to a request from him for thoughts on creativity in museums. Here are two short extracts from Jack's letter:

> I detect creativity when some object that I am shown, or the way in which it is displayed, hits me, suggesting that someone with an eye and a mind had had a hand in it, so that what might have been a very ordinary statement of fact starts ideas and speculations...

and then:

> Somebody has thought about an object, or a type of object, or a theme or an idea. He has spotted something, treated it with intense interest, and communicated the process to the visitor.

In these short pieces I, for my part, joyfully detect the man I knew.

Chapter 1

Jack Simmons:
the Making of an Historian

Michael Robbins

The first time I met Jack Simmons was one evening in September 1929 when we both walked in as new boys – King's Scholars – at Westminster School. There were 40 King's Scholars altogether. The eight of us elected that year were thrown together in a common room, and we tried to find out about the others. Jack and I had never met before. He was a south Londoner from Carshalton in Surrey; I was a north Londoner from the Hampstead Garden Suburb. We soon discovered that we were both interested, in a boyish kind of way, in looking at railways; but we came from two different worlds of railways. The Southern was very much a developing entity of its own kind at that time. That first evening, he said, 'I bet you can't tell me the headcode of the South London line from Victoria to London Bridge.' I said, 'Well, I can. It's 2.' It was actually easy to remember – all the others were letters. After that we found we had a good deal in common. Neither of us cared for sport. Jack already had a problem with his twisted spine (scoliosis), though that did not prevent him from doing a lot of things. We were expected to participate in sports – cricket or football, or rowing. If there was a football match against Charterhouse or another school, we were all turned out to watch and shout for our side. Jack and I became adept at dodging the column, and on our way to and from the playing field we sat together at the back of the bus and talked about many things, including politics.

Jack had already been in the school for a year in one of the boarding houses (Grant's). He counted as a year older than I was, even though in fact he was only ten days older, because my birthday on 7 September and his at the end of August fell on either side of the crucial 1 September date line. As he had been in the school for a year before becoming a King's Scholar, he moved up a year after two years in College, which meant that we lost touch a little. Both of us started in the Classical Fifth form, but at the end of the first year (after what was then called School Certificate) we had to make a choice between classical and modern (science), or joining a small bunch of historians. Jack went for history. They had not even got a form room – the group met in a corner of the library. There were about eight of them at a time. There were two masters. Lawrence Tanner, who was closely associated with the Abbey establishment, was one of them and a good historian.[1] The other was John Bowle, who was later a professor and wrote a history of the British Empire.[2] I stayed on the classical side.

Jack went on to get a Westminster scholarship to Christ Church, Oxford, in 1933, and he went up that year. I was a year behind and got one in 1934. I found Jack and some other friends there when I arrived. He was doing straightforward Modern History. I ought to have followed a four-year course, but I managed to persuade the Christ Church tutors to let me follow the most unusual course of doing Greats without Mods[3] – something they regarded as very irregular if not immoral. So both of us were on three-year courses, with Jack a year ahead. However, he then fell sick and had to spend a good deal of time in the Acland Nursing Home, so he got leave to drop back a year, and he took his degree in 1937, as I did.

We had already done a certain amount of scribbling about railways together. The Oxford University Railway Society had begun in John Betjeman's day as a sort of on-train dining club on the Great Western Railway, but behaviour had become so uproarious that the proctors closed it down. It therefore had to be refounded in a quite different form, and when I came along it had been running for several years with a programme of lectures, visits, and so on. J.G. Griffith[4] was secretary before me, and I was followed by John MacInnes,[5] who died rather early. In 1936 I got Jack to give a lecture about the proposed railway station for Oxford which was not built in 1863. In preparation for this talk he went into the city archives and examined the drawings and records. I hope that a copy survives somewhere.

I had taken to running a little quarterly periodical called *Locomotion*. Jack was on the committee of foundation members. We were rather schoolboy number-grubbers. Various other friends were involved, including Roger Kidner. He and I founded the Oakwood Press together, and Roger went on with a long string of publications.[6] At

---

1   Lawrence Tanner, Master of the History Form, Westminster School, 1919–32, was Keeper of the Muniments, Westminster Abbey, 1926–66, and Librarian of the Abbey 1956–72. He wrote a history of Westminster School (1934, 3rd edn 1951) and a number of books about the history of the Abbey and its treasures.

2   John Bowle was Senior History Master at Westminster School 1932–40. He held a number of Visiting Professorships (in 1949, 1961, 1965, and 1966) and was Professor of Political Theory at Bruges 1950–67. His many publications include *The Imperial Achievement*, 1974.

3   The course for the BA degree in Classics (*Literae Humaniores*) at Oxford is known as 'Greats', a term applied to its Final Examination, normally taken at the end of the fourth year. Honour Moderations ('Mods') is the first public examination, taken at the end of the fifth term. The name 'Mods' is also applied to the course (mainly linguistic and literary) leading to the examination.

4   J.G. Griffith, MA (Oxon.) 1939 (New College), became Fellow and Tutor in Classics, Jesus College, Oxford; University Lecturer in Classics; and Vice-Principal of Jesus College. In 1980 he was Public Orator.

5   John MacInnes (1915–57) was in naval intelligence in the Second World War. Subsequently he became Assistant Chief Education Officer in North Riding, Yorkshire.

6   Two recent articles in the NRM *Review* (95, Spring 2001, p. 29 by B. Knowlman, and 96, Summer 2001, p. 31 by M. Robbins) celebrate Roger Kidner's 'remarkable record' as a railway historian and publisher. The Oakwood Press had its origins in 1931 with the appearance of the first issue of *Locomotion*, then typewritten. Apart from a six-year

the time in the middle 1930s when Jack and I were at Oxford the Oakwood Library – short histories of individual railways – was begun. I did the North London as the first of the series; D.S.M. Barrie wrote on the Taff Vale as the second; and Jack was on the point of doing the Maryport & Carlisle. But the war stopped all that.

In 1937 Jack had gone to Paris for the winter to brush up his French and I went to Vienna to learn German. I was fairly busy in the two years 1938 and 1939, and I did not see much of him then. We all knew there was going to be a war, but of course we could not know exactly when it would be. It must have been at that time that Jack became very much interested in imperial history, with Reginald Coupland. He became a sort of personal assistant to Coupland, though not called that. He moved his home to Boar's Hill outside Oxford to be near Coupland. It was obvious that he could not be called up – he was not fit enough for that. So he spent the war years reading and ghosting for Coupland, who had all sorts of interests. He also met Margery Perham and did a joint work with her.[7] He did a lot of reading in the British Museum for one of Coupland's books, *Livingstone's Last Journey* (1945) and he involved himself very much with the whole of the 1880s and 1890s imperial period. He also, of course, kept his private interest in railways going, and he finished the Maryport & Carlisle book (Oakwood Library no. 4) in 1947.

Leicester came into the picture in the summer of 1946, when the University College set out to look for a professor of history. Jack must have been in a particularly favourable position – there were very few people at that time who were ready for such a post. They were mainly recovering from the war. Jack was the perfectly right person; but he was also lucky to be in the corridor at the moment the door opened. He was offered the post. He liked the principal, F.L. Attenborough, and his sons, and got to know them quite well. W.G. Hoskins was already on the staff (as lecturer-in-charge of the Department of Economics and Commerce), but he did not apply for the chair.

The first time I visited him at Leicester, Jack was living in a hotel in the New Walk and his mother was still back at Boar's Hill – this was before they had a house at Leicester in Stoneygate off the London Road. I met Hoskins and I got the impression that there were no hard feelings between them at all about the chair. Hoskins did not want it – he wanted to get on with writing about Midland peasantry and Wigston and so on, and he also resented every minute of the time he was away from Devon. These two were strong characters, but they hit it off quite well. This was in a way surprising, and it led to Jack starting to think about the range of history that he wanted to master.

---

interruption during the Second World War, the Press and Kidner have been active in producing books on transport history ever since. The Press's first book-length publication, in 1936, resulted from Kidner's particular interest in light railways, as did its series of Light Railway Handbooks (the first were hand-printed at Kidner's home at Sidcup: he moved house from the Oaks to Southwood, hence Oakwood). It was Kidner who wrote and published the very first pocket 'trainspotter's guide' (*How to recognise Southern Railway Locomotives*, 1938), the prototype of that popular type of publication. Since the War, Kidner has had an unbroken connexion with the Press. From 1953 to 1984 he was its sole proprietor; he is the author of 29 of its titles and co-author of three more. He remains in regular touch with the present owner to whom its high standards of research and presentation have been passed on.

7   *African Discovery: an Anthology of Exploration*, Faber & Faber, 1942.

His inaugural lecture, which contains the clues to much of all this, was called *Local, National, and Imperial History.*[8] That was given in 1948, and in it he mapped out his programme, though not in detail. He showed that he had a concept of the three areas having a relation with one another. This is a key document in understanding Jack; at that time he was looking pretty broadly at history (though not outside the empire). He was never especially interested in European or United States or world history.

Soon after his arrival at Leicester, Jack was impressed with the need for a countrywide series of books on England, county by county, with a few separately on cities, providing both a history and a gazetteer. He persuaded the firm of Collins to take the project on, somewhat on the lines of their successful *New Naturalist* series. I did one on Middlesex (1953) – not my line at all, really – and W.G. Hoskins (who had become Reader and Head of Leicester's new Department of English Local History from 1948 until 1951 when he was appointed Reader at Oxford) did a very good one on Devon (1954, since reprinted). But the intended authors were slow to deliver, and Collins lost patience and threw up the series. Jack was a firm editor who provided clear guidance about what he wanted. He himself was to have written Berkshire, but that never came off. It was a great loss that no more volumes appeared.

Things began to change a good deal for Jack in the early 1950s, because he thought that transport history – not railway history – was neglected by his academic colleagues. It had no journal of its own and there were few contributions on the subject in the economic history journals. So Jack proposed to the University College of Leicester that it should publish a journal devoted to the history of transport in the broadest sense, and he asked me to join him as joint editor. The first issue came out in 1953. We did a sort of balancing act. I tried to bring in people from the world of operating transport who had broad interests, like Peter Masefield (air) and John Elliot (rail). At the same time we took a more strictly academic path, with contributed articles seriously referenced, book reviews, current publications, and accessions to documentary collections. The *Journal of Transport History* came out twice a year. We got along like that until about 1963/4. I managed to find just enough time to play my part when I was Secretary of the London Transport Executive. But in 1965 I got a job there which made me sure that I must pull out because of professional pressures: there was not enough time to do my part of the Journal job properly. I told him that I must stand down, which he understood, and I notified Leicester University, which took absolutely no notice. I did not even receive a letter of acknowledgement, let alone thanks. Soon the Journal stopped appearing, and for a time there was a gap.[9] In 1975 there was a commemorative number, with a list of Jack's published writings down to that date, when Jack retired.[10]

It had become evident that Jack had extraordinary strength in the writing of the railway history of Britain (not Ireland, and sometimes also leaving Scotland to other

---

8   Published as a pamphlet by Leicester University College in 1950; reprinted in J. Simmons, *Parish and Empire*, Collins, 1952.

9   The last volume of the first series was vol. VII for 1965–6, of which part four was published in 1970. The first volume of the new series was 1971–2.

10 *Journal of Transport History*, n.s., 3, 1975–6, pp. 145–58. This volume includes a valuable appreciation of Jack Simmons by H.J. Dyos (pp. 133–44).

hands). He knew a great deal about the subject; he knew what railways were, how they operated, and so on. On the whole he liked them – mainly, but not exclusively, the old steam railways. He was willing to put in a great deal of time to his research and writing. For example, the little Maryport & Carlisle book was a work of real craftsmanship, with nothing skimped at all. Perhaps most importantly, he had an enormous depth of reading. The list of references at the back of his books shows what a staggering variety of sources he could call upon. There are the expected ones, of course – the official materials that became available only slowly in this period with the opening up of the railway archive in the Scottish Record Office and the historical records of the British Transport commission, transferred later to the Public Record Office. But there are also literary sources, like Henry James, R.L. Stevenson, and plenty of others. All this was worked together very skilfully into a narrative.

His year as Acting Vice-Chancellor of the university[11] in 1961–2 had made it difficult for him to do the research and writing that he wanted to do, but after his retirement he came back to them again. He had in mind a four-volume series on the railway in England and Wales. The first volume of this (*The Railway in England and Wales 1830–1914*) appeared with Leicester University Press in 1978, and the further volumes were to be *Town and Country*, *Mind and Eye*, and a volume treating the community at large and the part played by the railway in its social, economic, and political life. However, that arrangement collapsed, and Jack had to find different publishers for the remaining volumes as free-standing projects. The second volume, *The Railway in Town and Country 1830–1914*, came out with David & Charles in 1986. The third volume, *Mind and Eye*, appeared from Thames & Hudson in 1991 as *The Victorian Railway*.

The series is marked by a particular personality and feeling and approach – a civilized approach to the place of the railway in British society that had not been demonstrated before. There had been technical histories, mostly of locomotives and rolling-stock (the 'hardware' approach), and there had been the gossipy history as written by Hamilton Ellis. But there had been nothing similar to the comprehensive railway history that Jack produced. He turned to all kinds of sources that had not previously been used. His achievement was to make it clear that railway history was an area of the social history of England and Wales that deserved looking at much more closely, and he showed how to do that. Then in 1997 there appeared the indispensable *Oxford Companion to British Railway History*, of which Jack was co-editor with Gordon Biddle and to which he made by far the most contributions (more than 300; about one-third of the volume). It was a triumph.

Most boys who get interested in railways probably begin with a love of locomotives. Jack began to write in *Locomotion* in the 1930s with an article about the Ivatt locomotives of the Great Northern Railway; in 1936–7 he wrote about the naming of locomotives, and he went on to biographical notes about Dionysius Lardner and George Bradshaw. Signs of interest in museums followed, with pieces on

---

11   Jack was Pro-Vice-Chancellor of the university from 1960 to 1963, and Acting Vice-Chancellor between the departure of Charles Wilson (VC 1957–61) and the arrival of Fraser Noble (VC 1962–76).

the Great Western collection at Paddington (he always had a special feeling for the GWR) and the Rastrick collection in the University of London (this one unsigned).[12]

Jack Simmons also showed a very strong interest in the museum side of things, in collections. There is a reading room named after him at the National Railway Museum at York. He was also involved with transport museums in western Europe. Throughout, his strength was to be broad, liberal, and civilized in his approach to what could easily become a humdrum, inward-looking subject. He was no great technician, and he did not want to be. But he spotted and lamented that there was no decent history of British railway signalling (though the Friends of the National Railway Museum are now doing something about that). Jack was always looking over the fence at the impact of the railway on the country through which it passed, and he was concerned too with the indirect effects the railway had on people's lives. For example, he and I argued about how far the spread of fish and chips could be put down to the railway. There was fish and chips in Oldham in the 1860s; where did the fish come from, and how?

These occupations did not prevent him from doing things that he thought were part of the duty of a good citizen within Leicester itself. He wrote a guidebook to the city, which went into several editions, and a history (*Leicester Past and Present*, 2 volumes, 1974), and he served on the committee of the Leicestershire Archaeological Society – he edited its *Transactions* from 1948 to 1961 and was its president from 1966 to 1977. He was chairman of the Leicester Local Broadcasting Council from 1967 to 1970.

So Jack was not solely a railway historian; but the subject was never very far from his mind. The aspects of the railway in history were aspects that had on the whole been ignored by other people. Railway historians tend to grind out information from statutory and other sources about the preliminaries, the birth, and perhaps the early years of a railway, and then they pay no attention to developments until closure many years later. This is the kind of thing you get from official records. But there is the tedious reading of the daily and weekly press to get more and fill in the middle. Jack always paid attention to the official record; but after the Maryport & Carlisle book (which was inclusive enough to mention Wilkie Collins and *The Woman in White*) he went on making more of the subject. He saw the railway as a part of the wider society rather than as an entity in itself. The impact of the railway on the landscape and townscape was coupled with that. He drew attention to those aspects more sharply than anyone had done before. *Mind and Eye* as a proposed title is a clear indication.

Jack (and I) got into the railway via locomotives, as small boys used to do. We went together to Clapham Junction, Euston, Paddington, and so on, writing down the engine names and numbers and noting the trains being hauled. We used to love getting a 'cop' – seeing an engine we had not seen before. One can still see fathers and sons at places like York doing much the same. It was engines that started it. Jack's comprehensive approach made it clear what the railway contributed in its time to the way that this country developed. There was a great risk that this would never be done; the societies and their periodicals concentrate on bits of specific information. But

---

12   For a list of Jack's contributions to *Locomotion*, see the bibliography of his published works at the end of this volume, p. 284.

Jack's work was quite different. He brought to bear his gifts to synthesize, to link together facts of a widely disparate character from very different sources, and he deployed a huge range of reference to produce a very convincing whole. This allowed him to set his railway interest in a much broader context in a way that other writers could not, or did not, do. Comprehensiveness was his hallmark. Indeed, the strength of Jack's work as a railway historian came precisely from the fact that he was not just a railway historian.

SECTION I
THE RAILWAY:
ORIGINS AND WORKING

Chapter 2

Pre-locomotive Railways of Leicestershire and South Derbyshire

Marilyn Palmer and Peter Neaverson

Jack Simmons is one of the most eminent of British railway historians. His main interests, like those of most railway historians, lay in locomotive-hauled railways. In the opening chapter of *The Railways of Britain: an historical introduction*, he writes: 'The history of railways, as we know them today, begins in the year 1830'.[1] But, as he admits, transport using wheeled vehicles on rails has a much longer history, tracing its origins back to the growth of mining activity on the continent in the early modern period.[2] There is clear evidence in Germany for miners pushing trucks on rails from the fifteenth century onwards, but this paper is concerned with trucks on rails drawn by animal power, usually horses. Such pre-locomotive railways are often known as tramways, tramroads, or waggonways: it is the intention of the authors to retain the term 'railway' in this chapter since it causes the least confusion.[3]

These early railways consisted of parallel wooden rails along which four-wheeled waggons were drawn by horses. In Britain, the earliest for which there is documentary evidence was built c.1600 by Huntingdon Beaumont for bringing coal for sale in Nottingham from the pits at Strelley, Bilborough, and Wollaton.[4] Beaumont was, of course, from the Leicestershire coal-owning family based at Coleorton, and a considerable innovator in mining techniques both in the midlands and later in the north-east coalfield. It was in this area that the real development of these wooden railways took place from the second half of the seventeenth century onwards and in Shropshire during the eighteenth century. These lines were built to convey coal to the Rivers Tyne and Wear for shipment in the first case and coal and ironstone to ironworks and to the River Severn in the second. Within the last decade, archaeological evidence has been found for at least three such systems, at Bedlam in

1   Simmons, J., *The Railways of Britain: an historical introduction*, 2nd edition, Macmillan, 1968.
2   Lewis, M.J.T., *Early Wooden Railways*, Routledge & Kegan Paul, 1970.
3   For a discussion of nomenclature, see Baxter, B., *Stone Blocks and Iron Rails*, Newton Abbot: David & Charles, 1966, pp. 18–21.
4   Smith, R., 'England's first rails, a reconsideration', *Renaissance and Modern Studies*, iv, 1960, pp. 119–34.

the Ironbridge Gorge,[5] Bersham in North Wales,[6] and Lambton Colliery on Wearside.[7] These wooden rails wore very quickly, despite the often-used addition of an extra strip of wood and later iron on the bearing surface of the rail, and cast-iron rails, pegged to heavy stone blocks set in a foundation of broken stone, were first introduced in Coalbrookdale in 1767. The only documentary evidence for a wooden railway in the area under discussion was at Measham, most likely at a colliery belonging to Joseph Wilkes.[8] This had probably been replaced with iron rails by 1799, when Wilkes proudly showed a party from the Grand Junction Canal around his colliery and demonstrated the performance of a single horse in pulling five tons on a gradient of 1 in 115.[9] This railway appears to have had edge-rails and waggons with flanged wheels, similar to those used in Shropshire and South Wales.[10] It is unlikely that the Measham collieries were the only ones with railed transport in this period, and there is some evidence from aerial photographs and field walking for possible railway systems in the Swannington and Coleorton areas which connected to the turnpike road network.[11]

The canal mania of the later eighteenth century boosted the construction of new pre-locomotive railways. Canals which sought to penetrate hilly ground and involved building large numbers of locks proved too expensive, and it became common for Acts of Parliament authorising the construction of canals to include provision for the construction of railways to the more remote mines and quarries. Extensive systems of railways were constructed, for example, in conjunction with the canals of South Wales.[12] Similarly, the railways which are the subject of this chapter were built to link the collieries and limestone quarries of north-west Leicestershire and south Derbyshire to the burgeoning canal system in the final decade of the eighteenth century and the first decade of the nineteenth. Farey wrote in 1817: 'Within the county of Derby, or near it, I believe, there is not any Public Rail-way under Act of Parliament, separate and distinct from the Canals'.[13]

These railways have been the subject of historical study for the past hundred years, some authors having the benefit of seeing some of the lines still in existence in the last years of their operation. The earliest of these railway historians was Clement E.

---

5   Jones, N.W., 'A Wooden Waggon Way at Bedlam Furnace, Ironbridge', *Post-Medieval Archaeology*, 21, 1987, pp. 259–62.

6   Grenter, S., 'A Wooden Waggonway Complex at Bersham Ironworks, Wrexham', *Industrial Archaeology Review*, XV/1, 1993, pp. 195–207.

7   Ayris, I., Nolan, J. and A. Durkin, 'The Archaeological Excavation of Wooden Waggonway Remains at Lambton D Pit, Sunderland', *Industrial Archaeology Review*, XX, 1998, pp. 5–22.

8   Farey, John, *General View of the Agriculture and Minerals of Derbyshire*, Vol. 3, Board of Agriculture, 1817, p. 288.

9   Marshall, C.F.D., *The History of British Railways down to the Year 1830*, Oxford, 1938, pp. 41–2.

10   Farey, ref. 8, p. 288.

11   Baker, Denis W., personal communication.

12   Rattenbury, G., *Tramroads of the Brecknock and Abergavenny Canal*, Oakham: Railway and Canal Historical Society, 1980.

13   Farey, ref. 8, p. 283.

Stretton, who first drew attention to the historical importance of the Leicester & Swannington Railway in 1891[14] and followed this by a more comprehensive article on the county's early railways in 1900.[15] There followed a series of articles in *Railway Magazine*, the first, in 1900, by E. Brand on the railways of the Ashby Canal,[16] followed by two in 1938 and 1939 by Charles E. Lee on the Belvoir Castle[17] and the Swannington and Ashby Canal railways.[18] In the late 1930s two major books on the evolution of railways included for the first time substantial information on horse-drawn railways, including the East Midland lines.[19] Research on these continued after the war, with Temple Patterson's comprehensive study of the Leicestershire canals,[20] making use of newspaper sources and locally available canal minute books, whilst Robert Abbott used similar sources to write an account of the railways of the Leicester Navigation Company.[21] At the same time, other authors began to make use of archive material held by the British Transport Commission, together with fieldwork along the railways themselves.[22] More recently, with the growth of industrial archaeology, more comprehensive fieldwork has been carried out, notably by Brian Williams on the Leicester Navigation and its railways,[23] Leicestershire Industrial History Society on the Leicester & Swannington Railway at the time of its sesquicentenary,[24] and by the authors on the Ashby Canal railways.[25] Following the acquisition of the Calke Abbey estate by the National Trust, local authors have produced a series of guides to the Ticknall tramway and limeyards.[26]

A surprising amount of research has, therefore, been carried out on the pre-locomotive railways of Leicestershire in the twentieth century. Much of the work, however, particularly that before 1960, was undertaken from the local perspective

---

14 Stretton, C.E., *Notes on the Leicester and Swannington Railway*, lecture printed as a pamphlet, 1891.

15 Stretton, C.E., *Early Tramroads and Railways in Leicestershire*, reprinted from the *Burton Chronicle* as a pamphlet, 1900.

16 Brand, E., 'Some forgotten tramroads', *Railway Magazine*, 6, 1900, pp. 211–15.

17 Lee, C.E., 'The Belvoir Castle Railway', *Railway Magazine*, 82/492, 1938, pp. 391–4.

18 Lee, C.E., 'Swannington: One-Time Railway Centre', *Railway Magazine*, 85/505, 1939, pp. 1–9.

19 Lee, C.E., *The Evolution of Railways*, 1938; Marshall, ref. 9.

20 Patterson, A.T., 'The Making of the Leicestershire Canals, 1766–1814', *TLAS*, 27, 1951, pp. 67–99.

21 Abbott, R., 'The Railways of the Leicester Navigation Company', *TLAHS*, 31, 1955, pp. 51–61.

22 Clinker, C.R. and G. Biddle, 'Swannington and Ticknall Today', *Railway Magazine*, 98/612, 1952, pp. 263–6; Clinker, C.R. and C. Hadfield, 'The Ashby-de-la-Zouch Canal and its Railways', *TLAHS*, 34, 1958, pp. 53–76; Clinker, C.R., 'The Leicester & Swannington Railway', *TLAS*, 30, 1954, pp. 59–114.

23 Williams, B.C.J., *The Forest Line*, Loughborough: WEA, 1974, and *An exploration of the Leicester Navigation*, Sileby: Leicester Navigation 200 Group, 1994.

24 Leicestershire Industrial History Society, *An Early Railway*, Leics. Museums, 1982.

25 Palmer, M. and P. Neaverson, 'The Ticknall lime industry and its transport system', *Bulletin of the Leicestershire Industrial History Society*, 10, 1987, pp. 5–21.

26 Usher, H., *The Ticknall Limeyards*, 1995, and Holt, G., *The Ticknall Tramway*, 2nd edition, 1996, both published by the Ticknall Preservation & Historical Society.

without reference to the national context. Two seminal publications by Bertram Baxter[27] and Michael Lewis[28] in 1966 and 1970 respectively gave an overview of both the history and the extent of pre-locomotive railways in Britain. This present chapter is able to take cognisance of these two publications, together with the recent archaeological work which has taken place, and attempt to place the local pre-locomotive railways in a wider perspective.

The Railways of the Charnwood Forest Canal

The construction of a canal across Charnwood Forest was prompted by rivalry between the coal owners of Derbyshire and west Leicestershire for the sale of coal in the growing industrial towns of Loughborough and Leicester. The opening of the Soar Navigation from the Trent to Loughborough in 1779 enabled coal from the Erewash valley to be sold much more cheaply in these towns than that brought by packhorse from west Leicestershire. The proposed extension of the Soar Navigation to Leicester was therefore bitterly opposed by the Leicestershire coal owners, amongst them several aristocratic landowners as well as mine lessees in Coleorton and Swannington. Consequently, the Leicester Navigation Company had to include in its plans a branch canal from Loughborough westwards. Because of the terrain, it was soon recognized that a canal was impractical and Christopher Staveley in 1791 produced plans for a realigned composite canal and railway system across Charnwood Forest (Figure 2.1).[29] Engineered by William Jessop, the canal was opened in 1794 but was never successful in enabling the Leicestershire coal owners to gain additional markets and many soon turned their attention to the potential of the Ashby Canal which gave them access to the growing Midland canal system. Those in the Coleorton and Swannington areas continued serving their local markets using a pre-existent transport network eventually linking to the Leicester & Swannington Railway which opened in the 1830s.

At its western end, the Forest Canal divided into two branches at a junction-house, the ruins of which still exist (SK 427187).[30] The northern branch terminated in a basin from which railways would connect to the limestone quarries at Barrow and Cloud Hills. There is no documentary or physical evidence that the rail branches projected on Christopher Staveley's map were ever built. The short distance from the small Barrow Hill quarry to the basin could easily have been worked by horse and cart. At the same date, proposals for the Ashby-de-la-Zouch canal further west included a branch to the more important Cloud Hill, which was eventually constructed as a railway. Hence Farey commented in 1817 that 'had this last extension [to Cloud Hill]

27   Baxter, ref. 3.
28   Lewis, ref. 2.
29   LLRRO QS.72/1, A Plan of the intended Navigation from Loughborough to Leicester with the proposed Water level and Railways from Swannington, Cole Orton and Thringston Commons, Barrow Hill and Cloud Hill to the Loughborough Canal basin, surveyed in 1790 by Chr. Staveley Jn^r, certified 13 June 1791.
30   Davis, P.M. and R.E. Holmes, 'Toll House (Canal Junction House), Osgathorpe', *Leicestershire Industrial History Society Bulletin*, 4, 1980, pp. 14–28.

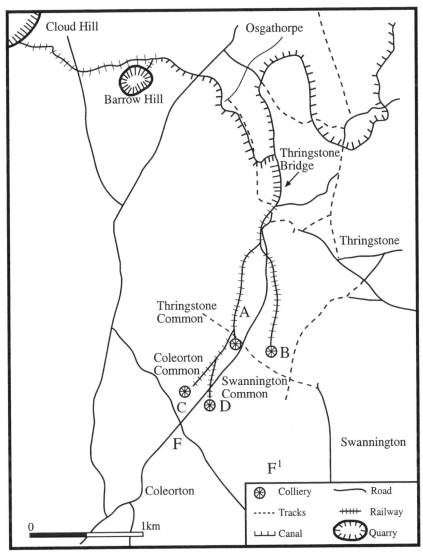

A = Mr Boultbee's proposed new Foundation; B = Messrs. Raper & Fenton's proposed new Foundation; C = Mr Burlem's new Foundation; D = Mr Boultbee's new Foundation; F and F¹ = existing fire [steam] engines.

Figure 2.1 **Sketch map showing the western end of the Charnwood Forest Canal built by the Leicester Navigation Company. Two canal branches may be seen, together with the proposed railways to the limestone quarries and to collieries in the Coleorton and Swannington areas** (redrawn from Christopher Staveley's plan, 1791).

been executed, junctions would very nearly have been effected with the Ashby-de-la-Zouch Rail-way and also the proposed Breedon Rail-way'.[31]

Staveley's map shows that the shorter southern branch of the canal terminated at Thringstone Bridge, whence a railway ran south with several branches proposed to Coleorton and Swannington Commons. Parts of these branches are also shown on the Thringstone and Peggs Green Enclosure map of 1807[32] and there is clear evidence on the ground for their construction, particularly the extant double hedgerows and embankments, and the deep cutting on the Swannington branch (SK 42121775). The evidence from aerial photographs and field remains indicates that the transport network was much more extensive than that shown on either map, and it is likely that the new railways were intended to link existing railways or stone roads to the Forest Canal. It is possible that the 'new' and 'proposed new' colliery foundations shown on Staveley's map were part of a promotional exercise to convince potential investors in the Navigation, whereas numerous pits already existed, some of them long-established with at least two steam engines and an extensive drainage sough. For most of the 1790s, the canal company was writing to the mine operators notifying them of the imminence of the canal's completion and urging them to open their 'new foundations'.[33]

John Farey described the edge-rail that was used on the new railway branches:

> The Rail-ways belonging to this Company are single, have bars [rails] flat at the top, and the wheels are cast with flanches, inside, for keeping the Trams upon them. The bodies of the Trams were made to lift off, or to be placed on their wheels, by means of Cranes erected in Forest-lane and Thringstone Bridge wharfs, so that the Bodies of the Trams only, stowed close together could be carried in the Boats on the Water Level.[34]

His description of these containers resembles those used on the Little Eaton Gangway from the Derby Canal, built in 1795, with which he must have been familiar. The contract specified that the rails were to be of cast iron, 3 ft long and 28 lbs in weight, and laid on 6 ft long oak sleepers, with a 'pad' at each end to which the rail was fastened by oak pins.[35] Unlike the railways from the Ashby Canal, which are considered below, there are therefore no surviving stone blocks marking the trackway.

At the eastern end of the canal the longest section of railway in the system was constructed, over two miles down to the Soar Navigation basin in Loughborough: this replaced the expensive canal locks originally proposed. Part of the line is still a public footpath from Nanpantan towards Loughborough. Many writers have suggested that William Jessop used fish-bellied edge-rail for the first time on this line, which would have strengthened the rail in between the sleepers, but this cannot be substantiated.

---

31   Farey, ref. 8, p. 377.
32   LLRRO, 13D/40/3/12.
33   LLRRO, 3D42/4/2, Leicester Navigation Minute Book, 6 December 1793, 9 April 1799 and so on.
34   Farey, ref. 8, p. 379.
35   Stevens, P.A., *The Leicester and Melton Mowbray Navigations*, Stroud: Sutton, 1992, p. 30.

Leicester Museums have in store such a rail reputed to come from the Nanpantan railway, but regrettably the documentation is inadequate to prove this.

It was presumably by these means that, in October 1794, the first boatload of coal passed along the Forest Canal from Burslem's mine at Coleorton and reached Loughborough and thence Leicester.[36] Under the terms of the Act, this enabled shipments of Derbyshire coal to move along the Leicester Navigation. There is little documentary evidence for further shipments along the Forest Canal, although another boatload of Burslem's coal was recorded six months later. Subsequently, 400 tons of his coal were stockpiled at Thringstone Wharf, implying some failure in the canal section unless it was priced out of the Leicester market, and the Navigation Company was obliged to purchase the coal at cost.[37] Documentary evidence suggests that the canal leaked substantially, and Jessop recommended the construction of the reservoir at Blackbrook.[38] Although this had been part of the original scheme of 1791, it had not been built, which may help to account for the inadequacies of the canal. The new reservoir burst in February 1799 and, although it was rebuilt within two years, the coal owners of West Leicestershire appear to have given up their endeavours to use the Forest Canal. A dispute between Sir George Beaumont and his tenant Joseph Boultbee led to the closure of the latter's mining activities and consequently he had no coal to send along the canal. As Farey said:

> all these Coal works were suspended, and the trade on these expensive Rail-ways, and the Water level was entirely discontinued, and yet remains so, I believe. When I viewed the Forest in August 1807, the canal was without any water in it.[39]

Although Farey is incorrect in implying that mining ceased in the area, the field evidence certainly supports his contention that they had been expensive railways to build. The system remained in this derelict condition; some of the rails and sleepers were lifted in 1820 and subsequently offered for sale. Pressure from Lord Stamford for the re-opening of the system and the provision of the promised connection to Breedon and, more importantly, the proposals for the Leicester & Swannington Railway, led the Navigation Company in the early 1830s to investigate the possibility of regenerating it. Proposals considered included making a horse-drawn railway throughout, using the rails still in their possession, or conversion to steam traction. None of these was implemented and the line was finally abandoned in 1836.[40] Nevertheless, mining continued in the Coleorton and Swannington areas using their earlier transport systems, and there is even some evidence that new railways were built, such as the system at Peggs Green which connected to the Hinckley–Melbourne Turnpike.[41] Field evidence shows that this was eventually connected to the Coleorton extension of the Leicester & Swannington Railway, which finally provided an adequate transport outlet for the expanding coalfield in the 1830s.

---

36   LLRRO, 3D42/4/2, Leicester Navigation Minute Book, 24 October 1794.
37   Abbott, ref. 21, p. 53.
38   LLRRO, 3D42/4/2, Leicester Navigation Minute Book, 4 June 1795.
39   Farey, ref. 8, p. 380.
40   Abbott, ref. 21, pp. 54–9.
41   LLRRO, Map DE1760/47/26, undated.

The Railways of the Ashby-de-la-Zouch Canal

Cargoes of limestone from quarries in the ownership of Lords Stamford and Ferrers and the Burdett and Harpur-Crewe estates were one of the main potential sources of revenue for the Ashby Canal Company. In 1787 William Jessop had reported on a canal and railway system linking Lord Stamford's Breedon Hill quarry to the River Trent.[42] This was never executed, and it would appear that this important quarry was the only one in the area which never achieved a link to a canal or railway system. In 1792 the newly-formed Ashby Canal Company held its first meeting at the Queen's Head in Ashby and discussed 'the utility of a canal from the Limeworks at Ticknall to Ashby Woulds, and from thence to unite with the Coventry Canal at or near Griff':[43] ensuing discussions included connections to other limeworks at Cloud Hill and Staunton Harold and even the collieries at Coleorton. These branches would have involved considerable engineering works, with locks up to a summit level supplied by a pumping engine and reservoir and then a tunnel, followed by another series of locks down to three level branches. The Act for them was obtained in 1794 and the level section of canal was opened through to Ashby Woulds in 1798. It was realised by then that the available funds were inadequate for the completion of the various branches as canals. Their replacement by railways was proposed and Benjamin Outram of the Butterley Ironworks in Derbyshire was consulted. Sir Henry Harpur, impatient to market his Ticknall limestone, had already contacted Outram, who advised that

> I am of the opinion that the most advantageous Mode of conveying Lime and Limestone from Ticknall to the Canal now executed, and Coals and Slack in return from the Woulds to Ticknall, will be by a Cast Iron Railway of the best construction.[44]

Outram was engaged by the canal company and constructed the railways shown in Figure 2.2. The replacement of the canal branches by railways not only reduced the length of the line but also saved costs, although the proposed connection to the Coleorton collieries was not made immediately. By October 1802 a double track had been built from Willesley Basin to Old Parks Tunnel and single-track branches to Cloud Hill and to Ticknall, with sidings into the various limestone quarries and collieries en route. The total length of track was over 12 miles.[45] Farey confirms that the track, unlike that of Jessop's forest line railways, was a plateway:

> The Rail-way bars are flanched, three feet long, weight 38lb, and are spiked down into blocks of stone of about 1½ cubic feet each, by means of an oak plug inserted into a hole driven into the stone.[46]

---

42 PRO, RAIL 803/12/1, 10 September 1787.
43 PRO, RAIL 803/1 Minute Book, 30 August 1792.
44 Derbyshire County Record Office, D.2375M 113/43, Observations on the Practicability and Expediency of a Railway from Ashby Woulds to Ticknall, received from Mr Outram 19 August 1798.
45 PRO, RAIL 803/3 Minute Book, 5 July 1803.
46 Farey, ref. 8, p. 299.

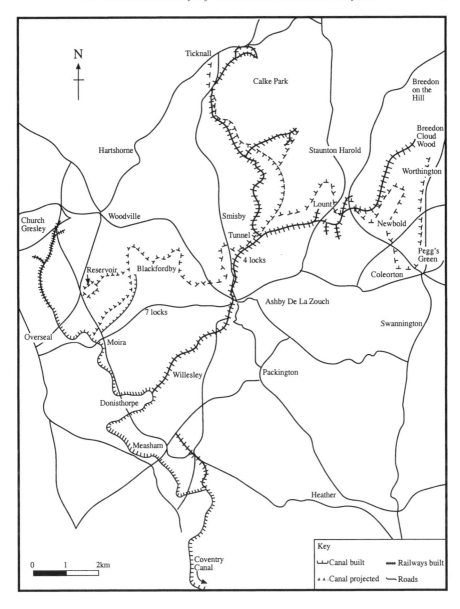

Figure 2.2   Sketch map showing the proposed branches of the Ashby Canal to the limestone quarries and collieries at Breedon Cloud Hill, Staunton Harold, Ticknall, and Lount. The system of railways eventually constructed is also shown (redrawn from Robert Whitworth's survey plan of 1792, LLRRO DE421/4/22)

**Plate 2.1   A section of the Ashby Canal railway, Ticknall branch, showing the
plate rails with points and crossover sections** (Science Museum/
Science & Society Picture Library)

In fact, lighter rails of 30 lb were used on the single-track lines north of Old Parks
tunnel (Plate 2.1). The gauge was 4 ft 2 in, and a ballast of small stones was laid
between the double row of blocks to provide a footing for horses. Large numbers of
the stone blocks still survive, some still containing the oak plug and spike described
by Farey.

The Limestone Railways

North of the tunnel at Old Parks, the railway divided into two main branches, to
Ticknall and to Cloud Hill. On the Ticknall branch there were two other tunnels,
both now in Calke Park. One is about 50 yards long and the other, 138 yards long, is a
cut-and-cover tunnel under the carriage drive from Ticknall village to Calke Abbey,
which now forms part of a footpath. At the east end of the tunnel, the line again
branches into two (Figure 2.2). The southern portion served the Harpur-Crewe limeyards with sidings which were laid and removed as different quarries were exploited.
Once the railway was opened, batches of lime-kilns were constructed in a worked-out
quarry area; these were built in an elongated horseshoe pattern with the railway laid
at two levels for both charging and emptying the kilns. Such a track layout is shown

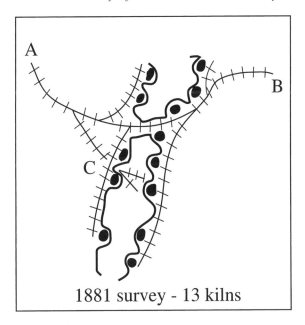

1881 survey - 13 kilns

Figure 2.3   **A typical layout of 13 lime-kilns in the Ticknall limeyards served by railway from the Ashby Canal, based on the 1881 OS 25-inch map. The railway from the canal enters at A on a high level from which sidings connect to the tops of the kilns for charging purposes, the line continues to the east at B. A short spur passes between two kilns at a lower level at C to facilitate unloading the kilns. At least two other similar arrangements of kilns have been located at Ticknall.**

in Figure 2.3, and at least two such systems remain in these yards along with embankments, cuttings, and overbridges.

The northern portion was built to serve Sir Francis Burdett's limeyards in Paddock Wood. This crosses the street in Ticknall by means of a brick bridge, which, along with the tunnel under the drive, is now a Scheduled Ancient Monument (Plate 2.2). The construction of the bridge was investigated by Leicestershire Industrial History Society in 1990 at the request of the National Trust when it was repaired. It was found that tie-bars inserted in the bridge subsequent to its building had broken the clay puddle which lined the bridge, similar to that of a canal. A new membrane and stainless steel tie-bars were inserted and the blocks carrying the single-track railway replaced. A horseshoe of lime-kilns within Paddock Wood was doubtless also served by the railway, although little trace remains. This railway north of the road was eventually extended back over it by a level crossing to serve other limeyards.[47] A

47   Lichfield Joint Record Office, Ticknall Tithe Apportionment map, 1843.

**Plate 2.2   The bridge c.1802 carrying the Ashby Canal railway over the main
street in Ticknall** (authors)

substantial embankment exists, alongside which were placed four kilns worked by
one of the few freeholders, together with substantial chambers underneath it for lime
storage.

The survey of the limeyards at Ticknall by the authors identified 11 separate
limeyards, with at least 37 kilns, not all of which were served by the railway. One of
these, in Margaret's Close, south of the main limeyard, was found to have a short
railway linking the quarry to a bank of four lime-kilns. Archaeological investigation
indicated this was also a plateway, but of 21 inch gauge (Plate 2.3). The blocks were
set in a solid matrix of limestone chippings, which would have minimised sideways
slippage, always a problem with this type of track.[48] It is conceivable that within the
limeyards similar lines were laid down and were then lifted and moved as required.

The Act had allowed for the construction of a line from the Ticknall branch to the
Dimminsdale limeworks and lead-mines on the Staunton Harold estate whenever
Lord Ferrers requested it. This option was not taken up until 1828, when Lord Ferrers'

---

48   Palmer and Neaverson, ref. 25; Palmer, M. and P. Neaverson, 'The Ticknall lime industry:
     excavation report and further historical evidence', *Bulletin of the Leicestershire Industrial
     History Society*, 12, 1989, pp. 5–16; Marshall, G., M. Palmer and P. Neaverson, 'The
     History and Archaeology of the Calke Abbey Lime-yards', *Industrial Archaeology
     Review*, XIV/2, 1992, pp. 145–76.

Plate 2.3   An excavated stretch of the stone blocks supporting the 21-inch gauge railway in Margaret's Close limeyards at Ticknall (authors)

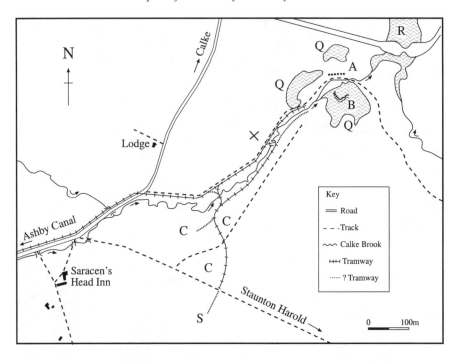

**Figure 2.4   The railway branches at Dimminsdale and Heath End, from a site
survey in 1994. The railway linked to the Ticknall branch and
divided at a bridge (X) to serve limeyards (Q) belonging to the
Calke (A) and Staunton Harold (B) estates as well as coal pits (C)
and a pit shaft (S).**

agent met the canal company engineer and a survey followed rapidly.[49] The line ran
across Calke estate land beside the Heath End road, and this may have prompted Sir
George Crewe to seek a connection to his own limeworks on the northern side of the
Calke Brook in Dimminsdale. The line terminated at a bridge over the Calke Brook
and the final links to the limeyards were the responsibility of the landowners.
Archaeological evidence indicates that a line was laid on each side of the brook, and
that additional lines were laid to small coal pits beside the back drive to Staunton
Harold mansion, known as Heath End colliery. The probable extent of the system is
shown in Figure 2.4. Documentary evidence in the Calke and Staunton Harold estate
papers indicates that the limeyards in Dimminsdale were disused and the private lines
were removed in the mid-1870s. The 1881 Ordnance Survey 25 inch to 1 mile map
shows the rails still in place as far as the Calke Brook, and the colliery site cleared.

---

49   PRO, RAIL 803/4 Minute Book, 1 April 1828; 2 September 1828.

There seems to be no foundation for suggestions previously made that the line remained open until 1891.[50]

The other branch railway from Old Parks tunnel ran for just over four miles to Cloud Hill limestone quarry, finally achieving the rail link that the Charnwood Forest Canal had failed to provide. It also served Lord Ferrers' collieries at Lount and short branches were built to other collieries in the area. The real interest of this line is the conversion of the track on part of the line following the construction of the Coleorton Railway from the bottom of the Swannington incline on the Leicester & Swannington Railway. The Coleorton Railway was opened in 1833, primarily to serve Beaumont's collieries in the Coleorton and Peggs Green area, but also connected with the Cloud Hill branch of the Ashby Canal railway at Worthington Rough.[51] The Coleorton Railway was 4 feet 8½ inch gauge and used edge-rail, in contrast to the 4 feet 2 inch gauge plateway of the Ashby Canal Company's railway. So, limestone from Cloud Hill or Ticknall would have to be transhipped into different wagons for onward transport on the Leicester & Swannington Railway. This was anticipated by the canal company, which in February 1833 considered Lord Stamford's proposals to lay an edge-railway from Cloud Hill to connect to the Coleorton at Worthington Rough. This was not immediately carried out, and in 1837 the Ashby Canal Company considered a proposal from the railway company 'to lay down rib [edge] rails parallel with the tramroad of this company along the branch to Cloud Hill from the point of the junction with the Coleorton railway with that branch'. The canal company went one stage further by proposing

that the company's engineer prepare and estimate of the expense of relaying the whole of the present line from Willesley Basin to Ticknall and Cloud Hill with rib [edge] rails instead of the present tram rails and to estimate the increase of traffic if they do replace the rails.[52]

Presumably the canal company could not afford this expense and the final solution was to lay a dual-purpose edge and plate rail, the cost of which was probably borne by Bostock, Lord Stamford's tenant at Cloud Hill, but part funded by the Leicester & Swannington Railway, which stood to gain considerable additional traffic. The connection was completed by August 1840.[53]

The Ashby Canal Company and its railways were sold to the Midland Railway Company in 1846, and the Act giving effect to the sale stated that the canal was to be kept in good order. The Midland also bought the Leicester & Swannington Railway and extended it from Coalville to Burton via Ashby in 1849. The following year Charles Hastings of Willesley Hall gave notice that he required possession of the land through which the canal railway ran from Willesley basin to Ashby. This section was closed in 1850, which meant that the former canal company's railways were isolated

---

50   Brand, ref. 16, p. 214.
51   LLRRO, QS.73/5, Plan and section of an intended railway or tramroad commencing at or near the termination of the Leicester and Swannington Railway near the northwardly end of the village of Swannington and terminating at the Ashby de la Zouch Railway at or near a place called Worthington Rough. 30 November 1832.
52   PRO, RAIL 803/5, Minute Book, 5 September 1837.
53   Clinker, C.R., 'The Leicester and Swannington Railway', *TLAS*, XXX, 1954, pp. 74–5.

from the Ashby Canal. Goods still carried on them had to be transhipped on to the standard-gauge line either in Ashby or at Worthington Rough on to the Coleorton Railway. This situation prevailed until 1874, when the railway to Cloud Hill was replaced as part of the Midland's new connection between Ashby and Derby. This entailed some realignment and the re-boring of Old Parks tunnel. The old railway to Ticknall was not converted and remained as a plateway until its final closure in 1915: goods had to be transhipped at a wharf at the north end of Old Parks tunnel.[54] The amount of traffic involved was probably quite small, since the Dimminsdale traffic had ceased in the 1870s and the railway system had been lifted in the Ticknall limeyards by 1899.[55] The line had probably been retained as a convenient system on the Calke estate and in the final years token journeys were made to preserve the right of way.

The Colliery Railways

Beyond Willesley Basin, the Ashby Canal passed through Measham, Donisthorpe, and Ashby Woulds to terminate north of the newly-established village of Moira. Ashby Woulds was enclosed in 1800 and developed by the Earl of Moira for both coal and ironstone production. He took advantage of the canal to open a large iron furnace and a series of coal-shafts between 1804 and 1806.[56] A complex system of railways was opened to link the pits, furnace, and lime-kilns, the first documentary evidence for which is a sketch made by Lord Moira c. 1811.[57] Traces of this system are still visible as low embankments in the furnace area, now a heritage site maintained by the local council. As further pits were sunk north of the furnace, the railway system was extended to them: Bath and Marquis Pits in 1812–13, followed by Rawdon Pit in 1821, and Newfield Pit in the early 1830s (Figure 2.5). There is no evidence for the type of track used in the Moira area, but in 1818 the foundry attached to Moira furnace supplied 300 rails for the Ashby Canal Company's railways which would have been plateway, and it is possible therefore that the Moira system was also a plateway.[58] Although the Midland Railway's Ashby to Burton line opened through Moira in 1849, Rawdon, Marquis, and Newfield collieries continued to make use of the canal and their old railway systems for some years.

There were two other railway connections from the Ashby Canal, one at Ilott's Wharf south-east of Measham serving coalmines at Measham Fields, and a series from the canal head at Moira to collieries and potworks in the Gresley area. The former, referred to earlier,[59] used edge-rails instead of plateway and was probably the first railway to be linked to the canal in 1799. A later standard-gauge railway eliminated it, but a section of the route survives as a double-hedged track to the Ashby to Measham road. The other system from the canal head was not built until

54   Clinker and Hadfield, ref. 22, pp. 21–2.
55   Palmer and Neaverson, ref. 25, pp. 16–17.
56   Cranstone, D. ed., *The Moira Furnace*, North-West Leicestershire District Council, 1985, pp. 7–20.
57   Hastings Papers, Dumfries House, Box 35/6.
58   PRO, RAIL 803/4, Minute Book, 3 March 1818.
59   See Farey, ref. 8, p. 288.

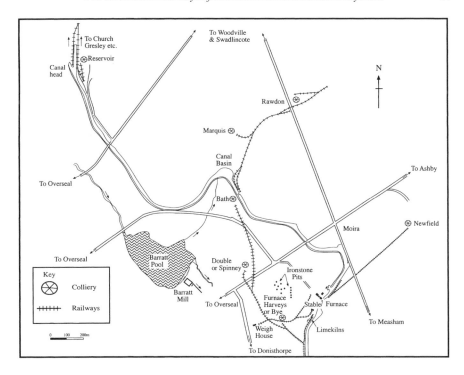

Figure 2.5 **The railways connecting ironworks and collieries on Ashby Woulds to the northern end of the Ashby Canal, showing their possible full extent between 1800 and 1840. The earliest railways are those close to the iron furnace and canalside pits, and later railways were constructed to other coal pits as they opened. A further railway system serving collieries and potworks in the Church Gresley area ran from the head of the canal.**

the late 1820s and connected to a pre-existent railway constructed by Dewes at his Swadlincote collieries. This extension from the Ashby Canal terminus proved very profitable but was eventually superseded by a large network of standard-gauge mineral lines.

The railways built by the Ashby Canal Company therefore achieved its objective of moving both coal and limestone to the canal, enabling it to survive as a viable transport system well into the steam railway age. The upper end of the canal was a victim of its own success, its closure being caused by mining subsidence. The longevity of the horse-drawn railway system has resulted in considerable field evidence of its former existence, adding considerably to the knowledge that can be derived from documentary sources.

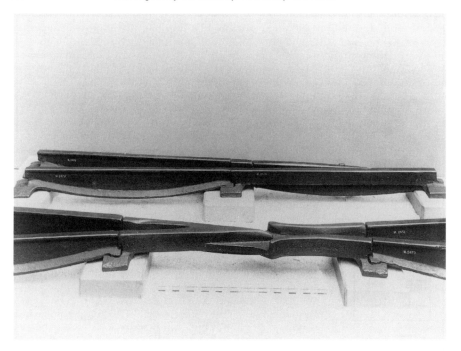

**Plate 2.4   Sections of the Belvoir Castle railway track of c.1815 as
reconstructed by the National Railway Museum. The upper fish-
bellied edge-rails include a point system on the right, while the
lower rails show a crossover** (Science Museum/Science & Society
Picture Library)

The Belvoir Castle Railway

An intriguing early railway in east Leicestershire is the private line constructed for the
Duke of Rutland from the Grantham Canal to Belvoir Castle. This was originally
thought by Stretton to have been opened in 1793, but the canal itself was not finished
for another four years. Later research has shown that the railway was opened in 1815.
Nor is there any real evidence for Stretton's description of the duke's assessment of
sections of both edge-rail and plateway and choosing the former.[60] Certainly the
Belvoir line used cast-iron edge-rails from the Butterley Ironworks,[61] similar to those
on the Nanpantan to Loughborough connection of the Leicester Navigation. Sections
of the railway and wagons are preserved in the National Railway Museum at York

---

60   Stretton, C.E., *Early Tramroads and Railways in Leicestershire*, reprinted from the
      Burton Chronicle, 1900.
61   Hadfield, C. and A.W. Skempton, *William Jessop, Engineer*, David & Charles, 1979,
      p. 172.

Plate 2.5   A load of peat being transported on the Belvoir Castle railway: the castle may be seen in the background (Science Museum/Science & Society Picture Library)

(Plate 2.4) and also in the cellars at the underground terminus in the castle itself. The railway was in use until 1918, and the photograph shown in Plate 2.5 may represent one of the final journeys.

The Leicester & Swannington Railway

At the same time as some of the horse-drawn railways discussed in this chapter were being built in the early nineteenth century, pioneering colliery owners in north-east England were using the new steam locomotives on their railways. Various Leicestershire mineral owners are known to have travelled to northern England to see these new railways, and they regarded them as the way forward. The first locomotive-hauled public railway was opened between Stockton and Darlington in 1825, but the early locomotives were used only on level sections of track, with stationary steam engines and even horse traction working on inclines. The Leicester & Swannington Railway is a prime example of these early hybrid railways, since the Act of 1830 permitted the use of 'stationary and locomotive steam engines, horses or other adequate power':[62] the Cromford & High Peak Railway in Derbyshire had similar provisions and made considerable use of stationary steam engines on the inclines and horses on the level sections.

George and Robert Stephenson, responsible for the construction of colliery railways in the north-east and the Liverpool & Manchester Railway of 1830, were engaged to oversee construction of the Leicestershire railway. In common with the latter, wrought-iron fish-bellied edge-rail was used, here resting in chairs on diagonally-placed stone blocks in cuttings and transverse wooden sleepers on open or embanked sections where settlement was likely to occur. The line was standard gauge (4 feet 8½ inches), over 15 miles long, with two incline sections at Bagworth and Swannington, and a tunnel at Glenfield just over a mile long. Horse traction was not a regular feature of the line, its use being restricted to connections to quarries and collieries such as Groby, Bagworth, Ibstock, and the pits worked by the Stephensons at Snibston. The Leicester & Swannington Railway was bought by the Midland Railway in 1845 and subsequent realignment as part of the Leicester to Burton line made it part of the national mainline system.[63]

Conclusion

The pre-locomotive railways of Leicestershire and South Derbyshire can therefore be seen as a microcosm of the national development of this kind of railway. There is some documentary evidence for the use of wooden track, although this is nowhere near as extensive as in other areas of Britain. Most of the local system developed in conjunction with the construction of canals in the late eighteenth and early nineteenth centuries, the railways greatly extending the economic value of the canal system by enabling goods to be transported in hilly areas where canals were not a viable proposition. Horse-drawn railways also survived the advent of locomotive railways, some lines continuing to operate until the end of the nineteenth century to serve local needs. The Leicester & Swannington Railway is also an early example of a hybrid railway, making use of gravity and powered inclines and horse traction as well as locomotives.

This chapter has also indicated the value of field evidence in conjunction with documentary evidence, since many of these early railways were not recorded on maps or referred to in documents. It is clear that lines were often laid as needed and then moved to another position as quarries or mines were exhausted. The maps drawn up by surveyors to meet the demands of parliamentary acts for canal construction indicate intent rather than actuality, and the field evidence derived from aerial photographs or field walking can indicate both lines which pre-dated the canal system, as in the Swannington area, or proposed lines which were never built, like those to Barrow and Cloud Hill limestone quarries from the Charnwood Forest Canal. A surprising amount of field evidence survives for Leicestershire's pre-locomotive railways, ranging from double hedgerows, footpaths following the original lines,

---

62   Clinker, ref. 53, p. 62.
63   Leicestershire Industrial History Society, ref. 24, p. 23.

embankments, tunnels, bridges, and even stretches of stone blocks which mark the original track. Such archaeological evidence must be used in conjunction with maps and documents if the history of the county's pre-locomotive railways is to be understood.

Acknowledgements

We are grateful to Denis Baker for sharing his wide knowledge of the Swannington area and Fred Hartley for commenting on a first draft of this article. Debbie Miles-Williams of the School of Archaeology and Ancient History at Leicester University kindly redrew the maps. The Science Museum Picture Library was helpful in tracing early photographs.

Chapter 3

The Transport Geography of the Wigan Coalfield: the Canal and Railway Contributions

David Turnock

Introduction

As a distinct sub-area within the wider south-west Lancashire field, the Wigan coalfield has a complicated geological structure because the main anticline is broken up by a series of NW–SE faults. Many seams come to the surface in a relatively small area extending from Ince to Standish, providing ideal conditions for early mining while hindering the modern industry (Holcroft n.d. p. 8). Fault lines have also acted as sluices for damming back water accumulating in old mines. The Lower Coal Measures, found on the margins of the anticline, were rarely thick enough for commercial mining and have left only a modest legacy of 'day eyes' near the main outcrops. However, the Middle Measures were much more extensive: especially the Arley seam, which was the first to be exhausted after early mining development in the east at Haigh and Worthington and in the west at Orrell and Winstanley (Figure 3.1). Higher up comes the cannel which was mined from the sixteenth century: John Leland mentions Mr Bradshaigh's activities at Haigh in 1538. Cannel was widely used in the late sixteenth century and seventeenth century as fuel and ornament (of remarkable hardness) when Wigan had more outcrop workings than other parts of South Lancashire. In 1595 the Rector complained that the burgesses were 'digginge coal pittes and taking coals out of the same to great value' (Powell 1986 p. 10). Indeed, cannel was much sought after until the incandescent mantle killed demand by the middle of the nineteenth century and by this time fortunes had been made around Haigh and Winstanley (Clegg 1957–8; Hawkes 1945). However, major investments like the Haigh Sough were needed, and miners had to wear a gauze mask for protection against splinters which left blue marks on the face and body (Worswick 1971).

It is important to stress that well into the eighteenth century the Wigan coalfield was quite isolated and intermittent working was dominated by the landowning gentry, most notably the Protestant Bankes and Bradshaigh families. No outside capital was drawn in, and the collieries generally had a short working life, rarely producing more than 1 000 tons annually (Worswick 1971). The local brass, iron, and pewter

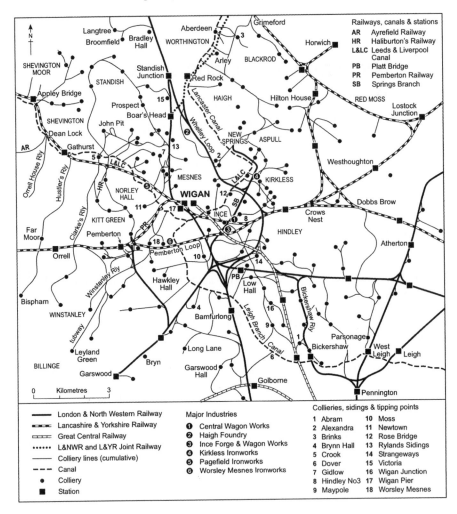

Figure 3.1   The Wigan Coalfield: cumulative map of mines, canals, and railways

industries were not dynamic and Wigan was far removed from the Liverpool transport system comprising the Mersey/Weaver navigation and the Prescot turnpike (Barrow 1935; Walker 1939). Attempts to sell coal along the Warrington–Wigan–Preston turnpike (1720), with broad-wheeled wagons replacing packhorses, were not wildly successful, although they did attract Arthur Young's memorable comment on 'this infernal road' which travellers were cautioned to avoid 'as they would the devil'. It was only in the middle of the eighteenth century that the situation began to change to the point where outside capital was attracted into the business. The local entrepreneurs were much involved in the struggle to enlarge their market area, with Liverpool the main pole of attraction (Parkinson 1952), a situation which helped to ensure that improvements were cumulative and wasteful duplication of facilities was generally avoided; for the railways were broadly complementary with one another and with the canals that preceded them. This chapter traces the process of development through the eighteenth and nineteenth centuries, paying particular attention to the relations between the canals and railways. Such was the transformation that output increased from some 12 000 tons for the Middle Douglas area in 1680, rising slowly to 18 500 tons in 1740 and then more rapidly to 400 000 tons in 1800. Clegg (1957–8 p. 796) has produced an inventory indicating an output of 4 000 000 tons in 1863 at the approach to the peak years which lasted until the First World War when this study terminates.

Large collieries were developed at the start of the canal age when the first steam engines were installed (Langton 1979). But still in 1851 there were many small producers. A list of collieries 'around Haigh' (actually larger than the area covered by this chapter) referred to 103 separate workings with an average annual production of 30 000 tons (around 100 tons per day) but 58 per cent were producing below this average (Taylor 1954 p. 120) However, the larger pits became progressively more dominant and the big growth in output in the late nineteenth century came largely from a new generation of pits of up to 900 yards depth sunk between the 1860s and 1890s at Alexandra, Garswood Hall, Golborne, Ince Moss, Maypole, Pemberton, and Wigan Junction. There were also changes in business structure, notably through the creation of the Wigan Coal & Iron Company (WC&IC) by amalgamating the Kirkless Hall Coal & Iron Company with the collieries of the Earl of Crawford & Balcarres (at Haigh), the Standish & Shevington Coal & Cannel Company, and Alfred Hewlett (Peden 1967; Turner 1976). Investment in new pits, in preference to the growth of steel production, was encouraged by the 'golden age' of early 1870s (the period of the Franco-Prussian War) when work was plentiful (the price of steam-coal in Manchester rose from 5s 11d per ton in 1869 to 17s 2d in 1873). But subsequent overproduction – with short time and 'starvation level' wages by 1880 – precipitated a growth of union activity and increased strike action, as in 1881 and 1893.

The geography was also changing, because the most productive measures turned out to be those occurring above the cannel: the Upper Coal Measures of Ince and Pemberton, which comprised a good depth of general-purpose bituminous coal worked from the early nineteenth century with the Watt engine for pumping saturated strata reached through impermeable boulder clay. This coal was worked progressively southwards from the outcrop with deeper mines sunk in the Abram and Golborne areas where a particularly intricate network of railways emerged (Ridyard 1972). But it was 'common coal of very inferior description' and the dirt had to be removed first by the pit-brow lassies working on the 'picking belts' and later by washeries installed from the 1860s. This had consequences for the landscapes in terms of tipping. This was strictly controlled through leasing conditions at Orrell, where landowners valued their amenity (Anderson 1975 p. 197), but Ince became a 'sordid mess' when longer-term collieries induced owners to accept lucrative royalties and move to homes elsewhere (ibid. p. 199).

The Canal System

The Douglas Navigation has its origin in a survey carried out in 1712 by Thomas Steers. However, landowners concerned with its effects on their meadow-land successfully opposed the bill in 1713 and delayed its passage until 1720. Financial problems then slowed progress, and the navigation opened in stages between 1738 and 1742. Supported by the Bankes, Bispham, and Bradshaigh families, the navigation gave Wigan coal a wider reputation, since it could be sent by 'bulk carriers' of 60 tons to Tarleton for trans-shipment and coastal delivery to Barrow and Liverpool. But only cannel could compete against Whitehaven coal, given the modest size of the vessels and seasonal hold-ups. The project was undercapitalized and might not have been completed but for the enthusiasm of Alexander Leigh, who thought London could be supplied from the Ribble via Liverpool (Ferber 1983). The extra capacity was belatedly provided by the Leeds & Liverpool Canal (L&LC) which offered a direct route to Liverpool and made it possible to develop the productive measures in general without bias to cannel.

The greatest impact was evident in Orrell and Pemberton, on the west side of Wigan, because Blundell and Hustler (the dominant coalmasters of the area) formed a cartel which prospered primarily through shares in the canal company rather than profit on coal. John Hustler was actually a Quaker wool merchant from Bradford who, as one of the canal promoters, became involved in coal mining in the Douglas Valley (Langton 1979). For the rest of the century, the two families reinforced their locational advantage on the canal with control of Liverpool retail prices by selling low enough to ensure that competition could not be waged by the Bridgwater Canal (Harris 1956; 1969). But at the same time the price had to be high enough to capitalize the expansion of collieries with use of the longwall mining system and the adoption of steam engines. The first steam engine was installed in 1769 at Orrell Hall (Fire Engine Colliery), with four more in the 1770s, two in the 1780s, and seven listed in the 1790s (Anderson 1975 p. 201). In a situation where miners were in short supply (piece-rates did not fall with increases in productivity), such investment would have been complicated had the canal been used by a large number of small producers in the Wigan area. Meanwhile, the Haigh estate was effectively cut out of the market because higher costs were incurred through a location on the east side of Wigan (beyond the initial canal terminus), with additional problems through low water and insecure premises. Attempts were made to develop local markets, notably by setting up the Haigh Foundry in 1790 (Birch 1953).

The original L&LC plan was to cross the Lancashire Plain and follow the Ribble valley towards the Pennines, a route favoured by Yorkshire interests. But this 'Parliamentary Line' of 1770 did not appreciate the significance of Douglas Valley coal and gave too much emphasis to the limestone traffic. A direct Hindley–Liverpool canal project surfaced in 1771 to tap coalfields south of Wigan and reduce rising coal prices which the inefficient Sankey Navigation was unable to control. The L&LC Lancashire lobby therefore secured the Wigan route through acquisition and development of the upper section of the Douglas Navigation. The Liverpool–Wigan canal opened in 1774. East of Wigan there was a dilemma between an independent route across Red Moss (difficult in terms of lockage and water supply) in order to make a connection with the Manchester Bolton & Bury Canal and a joint effort with the Lancaster Canal whereby a longer route would rule out a link with Manchester via Red Moss but reduce the amount of new construction and facilitate a coal supply to the towns of north-east Lancashire. In 1793 the L&LC asked parliament for powers to complete the canal through Blackburn, Accrington, and Burnley – a longer route but involving less new construction (Farrington 1970). This was approved although it was 1816 before the company built a flight of locks through Ince to meet the Lancaster Canal, which built a short extension from Whelley to Kirkless in 1816. Meanwhile, the L&LC found an alternative way of linking with Manchester through the Leigh Branch, which was authorized in 1819 and opened in 1820, long delayed beyond the initial negotiations in 1803 because Bridgwater interests feared the end of their Liverpool–Manchester monopoly and tried to secure tolls of 1d per ton-mile for merchandise (Clarke 1990 p. 129).

The foundry now became something of an irrelevance for the Haigh estate and smelting was discontinued by 1815, whereupon the works became purely a foundry and engineering business. With a substantial interest in the Lancaster Canal, which had opened to Walton in 1798 and Preston in 1803, thanks to the Walton–Preston tramroad (Biddle 1963; Hadfield & Biddle 1970), Balcarres (now the owner at Haigh) could adopt the Blundell–Hustler strategy and supply Preston with coal below the prices required by the competition. By keeping the market well-supplied with moderately-priced coal the Haigh estate minimized the risk of undercutting by speculators 'founded in the extreme of wildness and folly' (Earl of Crawford & Balcarres 1933–4 p. 17) and made it possible to gain control of minerals over a wide area by buying land and reselling the surface rights, for 'our part of Lancashire was divided into any number of small freeholds unworkable as mining units and the amalgamation of these parcels was necessary for working the shallow seams economically' (ibid. p. 16).

The canal attracted mineral lines immediately. The first used narrow gauge (that is, narrower than the later standard of 4 ft 8½ in) and wooden rails. But heavy wear of rails led to the use of a double system – employing a replaceable second rail, with ballast then raised to cover the sleepers and protect them from the horses' hooves. Wrought or cast iron strips could also be laid to reduce wear and friction, though the nails were always working loose and so plates were usually restricted to curves to lessen the load on horses. Before the end of the eighteenth century wrought or cast iron edge-rails and cast iron plate-rails were in use, with stone sleepers (Anderson 1975 p. 111). Once the standard-gauge mainline railways were in place, the canal links used the same technology, and, where distances were convenient, collieries liked to retain the option of despatch by both rail and canal. So canal transport of coal persisted right through the subsequent history of mining in the Wigan area.

There were numerous tipping points along the canals and they were sometimes referred to as piers. The most famous was Wigan Pier, previously known as 'Bankes' Pier' although it was actually built in 1822 for Thomas Claughton and acquired by Meyrick Bankes of Winstanley, who was thereby able to secure access to his collieries at Worsley Mesnes and Leyland Green (c. 1845) after being (predictably) refused access to the Pemberton Railway owned by Blundell (Hannavy & Winstanley 1985). Bankes went on to deepen his shafts in 1848, the year in which the L&LC forged a connection with Mersey and opened coalyards at Seacombe and Tranmere. However, the earliest canal railway was Hustler's wagonroad from Far Moor Pits

Orrell to Gathurst in 1776. 'After it had been extended in 1840 to their new Brownlow Colliery at Billinge Hill [it] was four miles long and had an overall fall of 470 ft to the canal at Gathurst' (Anderson 1975 p. 120). However, Dean Lock a little further downstream was the destination of lines from Ayrefield and Orrell House collieries in the mid-1770s.

A further railway was built to Crook at the beginning of the nineteenth century to bring coal from John Pit in Orrell. This line became famous for experiments with steam locomotives in a bid to reduce fodder costs during the Napoleonic Wars. Robert Daglish became manager of Clarke's Collieries c. 1810, and in 1812 he converted the wagonway to iron rails and stone sleepers in order to run a Blenkinsop rack locomotive. 'Yorkshire Horse' was duly built at Haigh Foundry to the patentee's specification, with a toothed driving wheel engaging with specially-cast cogged rails (Anderson 1975 p. 111) – the first locomotive to run in Lancashire (Bankes 1939 pp. 57–8). Eventually three 'travelling steam engines' were retained for well over 30 years, saving £500 per annum over the use of horses. Coal was also delivered to Crook from Standish Hall Colliery and legged through a 1 000-yard tunnel (in use c. 1810–40) on the last stage of the journey. Coal was also delivered to the canal at Crook from Norley Hall on the south side from 1845 and by the 1860s there were lines from John Pit and Standish (Robin Hill) on the northern side. Finally, it was 1827 before Blundell's Pemberton Collieries began their railway to Seven Stars Bridge; previously all Pemberton coal was disposed of by landsale or carted to the canal. Since Pemberton was one of three coke producers at the turn of the century (along with Bankes and Hustler), the railway and sidings were ballasted with hundreds of tons of unsaleable coke breeze (Anderson 1964–7).

It is not possible to detail all the arrangements. Suffice it to say that by the 1840s there were tips at six other places between Crook and the centre of Wigan, also at eight points on the section thence to Top Lock (including Whalley's Basin, where coal arrived from Ince Hall Colliery). In going round the south east of Wigan instead of directly through the town, the canal secured land more cheaply and also provided excellent opportunities to develop coal traffic as mining expanded at Ince and Kirkless. There were also nine tips on the section of Lancaster Canal as far as Aberdeen Basin at Adlington, including the main facilities for Haigh Estate coal (much of it from Aspull Moor) at New Springs. On the Leigh Branch only the Hawkley Hall Colliery was producing coal in the 1840s and sending it to the canal at Moss. However, as mining developed around Platt Bridge, Abram, Bickershaw, and West Leigh, the collieries sought canal tipping points as well as access to the mainline railways. In sequence from Wigan the collieries were Moss, Park Lane, Brynn Hall, Garswood, Strangeways, Low Hall, Maypole, Abram/Bickershaw, West Leigh, and Parsonage. It was fortunate that this canal should prove to be so conveniently aligned for the collieries of the later nineteenth century. The only problem was that the scale of activity caused immense subsidence problems. After 90 feet of coal had been removed the canal was left standing like an embankment as the surrounding land fell away and great flashes formed, Pearson's to the east and Scotsman's to the west. The trackbed of a railway that once crossed over the canal now lies 15 feet beneath it (Clarke 1990 p. 201). Thus while the L&LC generated a corridor of growth across the Pennines, the same can be said of the local scale of activity in the Wigan area.

The Railway System

Steam railway building in the area started with the Wigan Branch Railway from the Liverpool & Manchester (L&MR) at Parkside in 1832 (Greville 1953). Coal featured prominently in the calculations because it was intended that Springs Branch (to serve collieries in Ince) would open at the same time. As it happened the project was delayed until 1838, the same year that the company, the North Union Railway (NUR), reached Preston. Springs Branch connected with the canal at Top Lock, although in a rare case of blatant obstructionism the Lancaster Canal sought to frustrate this by extending its canal by a few yards to force the railway company into building a bridge over it. The 1834 Act for the NUR was significant in giving colliery proprietors and landowners powers to make new sidings and connections – and to run their own trains. These rights were effective until the London & North Western Railway Act 1888, although the right seems to have been exercised only occasionally after c.1870 (Townley et al. 1991–2 p. 19). However, Springs Branch was an exception and the WC&IC was also allowed to use the main line between Rylands Sidings and Broomfield Colliery.

The L&MR meantime had a link with Bolton by virtue of the Kenyon & Leigh Junction Railway (1831) which connected with the Bolton & Leigh Railway, conceived in 1825 to link the collieries of Atherton & Leigh with the Bridgwater Canal (Patmore 1964). The latter included inclines worked by stationary engines which were not replaced by deviations for adhesion working until 1885. So it was understandable that Bolton should seek a direct Bolton & Preston Railway (secured in 1844), although to begin with fares had to be the same as for the much longer journey via the L&MR. The original idea of a route incorporating the Lancaster Canal's tramway across the Ribble Valley was changed to a junction with the NUR at Euxton. These various railways were taken over by the Grand Junction in 1845 which in turn amalgamated with the London & Birmingham and Manchester & Birmingham to form the London & North Western (L&NWR) the following year. In 1864 the L&NWR opened the direct line between Warrington and Wigan, avoiding the detour over the L&MR between Earlestown and Lowton (Nock 1968); and in the same year the line from Eccles to Springs Branch Junction opened simultaneously with a connecting line from Tyldesley through Leigh to the Bolton–Kenyon line at Pennington.

The mania years of the 1840s brought a clutch of plans for east–west lines through Wigan to connect the Liverpool and Manchester areas, but the only concrete results were the appearance of the Liverpool & Bury Railway (L&BR) and the Manchester & Southport Railway (M&SR), authorised in 1845 and 1847 respectively (Marshall 1969–70). However the L&BR was taken over in 1846 by the Manchester & Leeds, which changed its name to the Lancashire & Yorkshire Railway (L&YR) the following year. It then took over responsibility for both the Liverpool and Southport lines and proceeded immediately with L&BR, which was opened in 1848, though it used the M&SR alignment into Wigan instead of the authorised route to the south of the town. The Wigan–Southport line followed in 1855, but only after Southport interests obtained a writ to spur the company into action following the economies of the post-mania years. The direct line from Pendleton to Hindley was delayed until 1888 and was then immediately quadrupled the following year when passenger services started. The Pemberton Loop, recalling the original L&BR concept (though not the exact alignment) was also opened in 1889 and a link between Dobbs Brow and the Hindley & Blackrod line (referred to below) was ready in the same year. The net result was a major growth of traffic where the Blackrod, Bolton, and Pendleton lines converged at Crows Nest Junction east of Hindley and the Pemberton Loop diverged from the Wigan line at Hindley No 3. This section was widened to four tracks in 1887.

With rising output in the 1860s, coal owners looked for improved outlets to Blackburn and, especially, Liverpool (via St Helens). There was much support from local coalmasters (including the Earl of Crawford & Balcarres, John Lancaster, and Alfred Hewlett) and also from the L&NWR for the Lancashire Union Railways Company (LURC) proposal for a line from Blackburn to Parr near St Helens passing east and south of Wigan, connecting with the main line at Boar's Head on the northern side of Wigan and at Bamfurlong to the south (with a crossing at the latter place as well). There would also be connections with Springs Branch and the L&YR which made its own proposals for a Blackburn–Chorley–Hindley connection and also a branch to Horwich (Marshall 1970). Parliament however sanctioned a simplified set of projects which restricted the LURC to the section south of Adlington, making it necessary to use the existing L&YR line to Chorley. The Chorley–Blackburn and Adlington–Boar's Head line would be built under joint L&YR and LURC ownership. The rest of the LURC programme was authorised, as was the L&YR Horwich Branch, with a direct approach from Wigan via Hindley (Turton 1962). The L&YR line from Hindley to Red Moss and the Horwich branch opened in 1868 while joint lines between Boar's Head and Blackburn opened the following year, along with the LUR from Haigh Junction to Ince Moss and St Helens and the Pemberton branch. A further connection between the LUR and the L&NWR at Standish was completed in 1882. The LURC was then absorbed by the L&NWR in 1883. The new layout handled large amounts of coal between Bamfurlong (Moss Sidings) and Garston, with additional traffic originating from the Pemberton Branch which gave access to Norley Hall and Winstanley collieries as well as Pemberton itself (Patmore 1961). The branch actually crossed the Winstanley system and a spur was easy enough to install, although additional contact came through the Pemberton Collieries' Tan Pit railway which Meyrick Bankes found he had the right to purchase. However, subsidence steepened the gradient across the main line from the original 1 in 86 to 1 in 42 which meant that empties had to be banked where trains exceeded 35 wagons. Meanwhile, the so-called 'Whelley Loop' was used as a freight bypass after adequate access was provided at the southern end in 1886 and Bamfurlong Sorting Sidings were subsequently laid out at this point in 1895.

Meanwhile, a new challenge was mounted in the 1870s when the Wigan Junction Railway (WJR) was conceived as a means of providing a stake in Wigan for the Cheshire Lines Committee (in the joint ownership of the Great Northern, Manchester Sheffield & Lincolnshire, and Midland Railways (CLC)). The idea was to connect Glazebrook on the CLC's Manchester–Liverpool line with the centre of Wigan and also with the WC&IC collieries at Rylands Sidings, already served by the L&NWR main line. The extension beyond Wigan was refused, but the line opened for freight to Strangeways Colliery in 1879 (with spurs to the LURC and Springs Branch by the end of 1882) and passenger trains reached Wigan in 1884, although the new Central

Station did not open until 1892. Further attempts were made to extend northwards and powers were obtained in 1883 to gain access to Preston via the West Lancashire Railway, but they were not exercised, apart from the short extension to Wigan Central just referred to, and the idea was abandoned in 1896. However, a final abortive bid was made by the Wigan & Heysham Railway in 1907 to run from the Central Station to Preston and Lancaster, with a branch to Blackpool.

In the early 1880s the WJR may have been driven in part by the expansionist climate created by the Lancashire Plateway project: a bid, which failed in parliament in 1882 and 1883, by Liverpool interests to block the Manchester Ship Canal and retain a monopoly on cotton through an improved inland distribution system under their control. However, by the 1890s competition was forcing company profits down and excessive duplication was therefore avoided. But the WJR stimulated the L&NWR to improve its arrangements in the Bickershaw area, first by a connection from Pennington (between Bolton and Kenyon Junction) with the Wigan–Tyldesley line (1885), incorporating the formerly private Bickershaw Railway, and then by upgrading it to passenger standards, with a flyover at Pennington providing a continuous loop from Tyldesley to Hindley which passenger trains began using in 1903. However, it should be added that railway politics were complicated in the early 1890s by the Manchester Ship Canal's inspiration of a line to Ince, including feeders from the L&NWR and WJR at Bickershaw, to the Partington Coaling Basin. This was turned down by parliament in 1895. It is also significant the WJC started a stampede to improve railway facilities in the centre of Wigan. The extension from Darlington Street to the new Central Station (1892) was followed by the L&NWR quadrupling of track from the south to Wigan in 1894, with an enlarged North Western station opened in the same year (with double track on the Whelley Loop, quadrupling resumed at Standish in 1895) and the L&YR provided a new Wallgate Station in 1896.

The railway network was in part a response to the existing mining industry, and pits that were operating at the start of the railway age quickly forged connections with the main lines. Springs Branch drew coal from collieries at Ince and Kirkless while the Wigan–Preston railway served some collieries on the Mesnes and drew traffic from John Pit and Standish with Rylands Sidings as the main exchange point. The coal from Haigh meanwhile was fed on the L&YR Bolton–Euxton line at Grimeford (as well as Springs Branch), while the Wigan–Liverpool line was convenient for Norley Hall, Pemberton, and Winstanley. Opening dates are notoriously difficult to obtain, but it would appear that the collieries took the earliest opportunity to connect with the public railways. In the case of Winstanley the 'shunt near the Arches in Pemberton' was operating by 1851 and the cost of a new tramway is mentioned in the account for 1848 (Bankes 1962 pp. 183–4). However, the Winstanley system continued to be narrow gauge and it was only in 1867 that a third rail (providing for narrow and standard-gauge) allowed standard gauge wagons to be loaded at the collieries.

But the main lines were also part of the context within which decisions were made about sinking later pits, and the L&NWR system between Tyldesley and Wigan was particularly influential. Arrangements were somewhat simplified by the formation of larger companies, especially the WC&IC which combined the Haigh and Kirkless Hall interests and led to a continuous private railway from Kirkless to Aberdeen, linking with the L&YR, L&NWR, and LURC at the south end and also with the

L&YR and LURC through Grimeford at the north end. The Company's Standish interests were rationalized to produce a single exchange point at Rylands Sidings on the L&WR West Coast Main Line as already noted.

Canal and Railway in the Late Nineteenth Century

The canal made good profits in the early years. 'The whole area around Wigan was truly opened up by the canal with coal output increasing tremendously' (Clarke 1990 p. 196). The company (L&LC) was not directly involved in carrying until after the Canal Carriers Act of 1847, but its toll revenue from merchandise and minerals rose steadily. In the case of coal receipts the annual amounts were £20–40 000 during 1790–1810 and £40–60 000 during 1820–40 before exceeding £70 000 in the early 1840s. Meanwhile, John Hargreaves was a prominent local carrier of merchandise in the 1830s and 1840s, negotiating both canal and railway leases to gain maximum efficiency, including his own warehouse in Wigan at the point where the railway and canal crossed. His empire was somewhat undermined in 1845 when the railway leases (in respect of the Bolton & Leigh and Kenyon & Leigh Junction) were stopped by the amalgamations which created the Grand Junction. However, the canal faced growing railway competition which forced toll reductions until negotiations in 1848–9 led to agreement whereby the railway companies leased the canal merchandise traffic. In 1850 wharfs and cranes were taken over, while the boats were purchased outright – a fleet which the canal company had taken over from bye-traders in 1848 when they were in financial difficulty. The new arrangement was a good deal for the canal since the substantial annual payment helped stabilize company finances and while some traffic was switched there was still considerable short-distance traffic when the arrangement was brought to an end in 1874.

Meanwhile, the L&LC was able to retain a significant interest in coal transport, with the added advantage of unified control over its entire route after the lease of the southern section of the Lancaster Canal was negotiated in 1864. Indeed, the railways were unable to compete because the canal served factories built along its banks. Rather, the railways had to develop new traffic with customers establishing premises along their lines. 'It also took some time for railways to organise the use of their track sufficiently to undertake the transport of bulk goods for the slow speed of this type of train interfered with the running of passenger traffic which was then producing most of their income' (Clarke 1990 p. 159). The Liverpool traffic was particularly strong since virtually all the town's coal arrived by canal for many years, rising from 200 000 tons in 1800 to 270 000 in 1832, and increasing, despite railway competition, to a maximum of 1.2 million tons in 1858 and then holding at around one million tons through the 1860s (Clarke 1990 pp. 148–50).

The Wigan area maintained a significant share of the market in north-east Lancashire despite the supplies from local pits around Accrington and Burnley. Much of Blackburn's coal originated from Haigh (boatmen on this length were known as 'Haigh cutters'), though there were smaller amounts from canalside collieries at Aberdeen, Arley, and Adlington. Finally, there was competition between L&LC and Lancaster interests over the Fylde and Irish trades which the Douglas Navigation had initially served. Coal continued to be sent down the Douglas after 1774, but there was competition from the Lancaster, especially after the Preston route was fully open and the Glasson Dock connection was in place in 1826.

Clearly, the balance tipped in favour of the railways from the late 1840s, and L&LC coal traffic peaked at 1 900 000 tons in 1866 (two thirds of this generated in Lancashire); though this was more than double the 900 000 tons carried in 1850 (again with virtually two-thirds from the Lancashire side). However, the railway had difficulty competing in north-east Lancashire until the Great Harwood loop eased congestion at Accrington while the canal continued to improve efficiency, for example with some boats fitting seven-ton boxes for quick unloading. Some colliery companies gave up their canal boats (for example, Ince Hall Coal & Cannel Company in 1872), whereupon the capacity was acquired by the L&LC ready for the ending of the railway lease in 1874 (Clarke 1990 p. 199). But most were happy to keep their options open and Maypole colliery was able to sell coal to the new Whitebirk power station in north-east Lancashire in the 1920s, conveyed by the Blackburn canal carrier Dean Waddington & Co. There was a sharp fall in coal revenue in the late 1840s, but it fluctuated within the £30–40 000 band during 1850–70 and yielded £20–30 000 in 1870–1910 (Clarke 1990 p. 139). Meanwhile the canal company position was strengthened by the fact that debt arising from capital raised to build the canal was paid off by 1847.

Ince Hall Coal & Cannel Company suggested steam tugs for the Liverpool traffic in 1852 and one was tested in 1853. The tug 'Conqueror' (a converted passenger boat) started taking trains of four 40-ton barges from Appley Bridge to Liverpool, but, because of problems with wash, speed was restricted to two miles per hour. When Burch's patent screw propeller was tested in 1857, the L&LC Steam Tug Company was set up in 1858 by coal owners and merchants involved in the Liverpool trade and four tugs were built, each capable of towing six boats at two miles per hour from Appley Bridge to Liverpool. Twenty tugs were planned, but the business was not profitable and the company failed. However, no action could be taken over a request from the Coal Association of South Lancashire in 1864 that the canal from Wigan to Liverpool should be improved at the time the Lancashire Union Railway was being promoted. But the L&LC continued to innovate although the benefits were probably more evident in respect of merchandise than the mineral traffic. Steam power was successfully introduced in 1878 when W. Wilkinson of Wigan converted a barge into a tug by installing a steam-driven propeller. His company produced a simple robust engine using a boiler produced by another Wigan company. Eventually the L&LC owned 46 steam barges. They found that one steam engine would replace ten horses and reduce costs considerably (Roberts 1965). However, steam was not entirely satisfactory. Tugs continued to work regularly to Liverpool, but it was not economic to run them frequently on the Wigan–Blackburn section. Hence steam tugs were withdrawn in 1905 and replaced by horses and 'marines' on call 24 hours a day at Blackburn and Wigan. 'The merchandise traffic, especially between Liverpool and North East Lancashire built up rapidly after 1874 as the canal successfully took trade away from the railway', for customers felt the canal provided a faster service and had better warehousing (Clarke 1990 p. 211). And efficient operation of steam-powered cargo flyboats in Liverpool in the 1880s secured much cotton and grain for Manchester with cloth as a return cargo, at the expense of the East Lancashire Railway.

Conclusion

Clearly, there could be no hopes for Wigan until the canal age provided the technology necessary for competition in a regional market. It was fortunate that although there was no potential for a continuous canal along the west coast, given the proximity of the coastal shipping route, a regional network developed which afforded outlets for Wigan coal in Liverpool, Preston, Blackburn, and the Manchester direction. Local coal-owning interests were much involved and helped to ensure that the original route for the L&LC was changed for their benefit. This crucial decision was also significant in terms of the route taken through the Wigan area, which proved convenient for the existing mines and also for subsequent developments. Meanwhile, in the railway age, it was inevitable that Wigan would benefit from being at the intersection of north–south and east–west routes. Such a network was beneficial for the coal industry – the Wigan–Preston railway opened up coal reserves untouched for want of cheap conveyance – but was also justified in the interest of long-distance travel. Two separate companies emerged around the two axes, but their interests in Wigan were largely complementary and relations appear to have amicable for the most part.

Local coal-owning interests became involved when capacity increases were deemed necessary, and the LURC programme was clearly geared to the needs of the industry. It was sensibly modified by parliament to harmonize with the L&YR as well as the L&NR interest. The amenity considerations of landowners did not pose significant problems and a cut-and-cover tunnel built for the Whelley Loop in the Haigh Plantations of the Earl of Crawford & Balcarres was opened out in 1883 when subsidence occurred. The new capacity also proved to have a wider significance. While some services did not survive long (Whelley–Liverpool passenger trains introduced in 1872 lasted for only two months), the Liverpool–Wigan line was useful for L&NWR Liverpool–Scotland trains until the L&YR Liverpool–Preston connection became available after the First World War when Grouping produced the London Midland & Scottish Railway. And Scottish banana trains from Garston Docks, which started in 1913, continued to use the Wigan route for half a century. With the addition of the link with Standish, the Whelley Loop became a useful bypass route, especially in view of the burrowing junction layout which made it ideal (in preference to Euxton) for feeding both L&NWR and L&YR trains from the Manchester area on to the main line once the necessary junctions were in put in place during the 1880s. And after Grouping, heavy summer-weekend Blackpool excursion traffic from the Pennine textile towns used this route as well. Finally, the exotic influence provided by the WJR was suitably restricted to additional capacity in the main area of coalfield expansion and a welcome stimulus for better stations in the centre of Wigan. The result was a cohesive transport system which survived largely intact until the 1960s.

Bibliography

(*THSLC* = *Transactions of the Historic Society of Lancashire & Cheshire*)

D. Anderson 1964–7, 'Blundell's collieries', *THSLC* 116, pp. 69–116; 117, pp. 109–43; 119, pp. 113–79.

D. Anderson 1975, *The Orrell coalfield 1740–1850* (Leek: Moorland Publishing Company).

J.H.M. Bankes 1939, 'Records of mining in Winstanley and Orrell', *Transactions of the Lancashire & Cheshire Antiquarian Society* 54, pp. 31–64.

J.H.M. Bankes 1962. 'A nineteenth century colliery railway', *THSLC* 114, pp. 155–88.

W. Barrow 1935, 'The industrial development of Wigan' (University of Liverpool dissertation).

G. Biddle 1963, 'The Lancaster Canal tramroad', *Journal of the Railway & Canal Historical Society* 9, pp. 88–97.

A. Birch 1953, 'The Haigh Ironworks 1789–1856: a nobleman's enterprise during the industrial revolution', *Bulletin of John Rylands Library* 35, pp. 316–34.

M. Clarke 1990, *The Leeds & Liverpool Canal* (Preston: Carnegie Press).

H. Clegg 1957–8, 'Some historical notes on the Wigan coalfield', *Transactions of the Institution of Mining Engineers* 117, pp. 784–99.

J.T. Clossick 1970, 'Economic evolution of Wigan' (University of Salford BSc dissertation).

Earl of Crawford & Balcarres 1933–4, 'Haigh cannel' (Manchester: *Transactions of the Manchester Statistical Society*).

J.H. Farrington 1970, 'The Leeds–Liverpool canal: a study in route selection', *Transport History* 3, pp. 52–79.

T. Ferber 1983, 'Robert Holt Leigh: landed gentleman – benefactor and new industrialist' (Manchester Polytechnic thesis).

M.D. Greville 1953, 'Chronological list of the railways of Lancashire 1828–1939', *THSLC* 105, pp. 187–201.

C. Hadfield & G. Biddle 1970, 'The Lancaster Canal', in C. Hadfield & G. Biddle, *The canals of North West England* (Newton Abbot: David & Charles), pp. 182–211.

H. Hannavy & J. Winstanley 1985, *Wigan Pier: an illustrated history* (Wigan: Smiths Books).

J.R. Harris 1956, 'Liverpool canal controversies 1769–1772', *Journal of Transport History* 2, pp. 158–74.

J.R. Harris 1969, 'Early Liverpool canal controversies' in J.R. Harris (ed.), *Liverpool Merseyside: essays on the economic and social history of the port and its hinterland* (London: Frank Cass) pp. 78–97.

A.J. Hawkes 1945, *Sir Roger Bradshaigh of Haigh, Knight & Baronet 1628–1684* (Manchester: Chetham Society Publication 109).

F. Holcroft n.d., 'South West Lancashire coalfield' (University of Manchester Geography Department dissertation).

J. Langton 1979, *Geographical change and industrial revolution: coalmining in South West Lancashire 1590–1799* (Cambridge University Press).

J. Marshall 1969–70, *The Lancashire & Yorkshire Railway* (Newton Abbot: David & Charles) 2 volumes.

J. Marshall 1970, 'Lancashire Union Railways', *Railway Magazine* 116, pp. 190–3; 254–7; 316–8.

O.S. Nock 1968, *North Western: the saga of the premier line of Great Britain 1846–1922* (London: Ian Allan).

C.N. Parkinson 1952, *The rise of the port of Liverpool* (Liverpool University Press).

J.A. Patmore 1961, 'The railway network of Merseyside', *Transactions of the Institute of British Geographers* 29, pp. 231–44.

J.A. Patmore 1964, 'The railway network in the Manchester conurbation', *Transactions of the Institute of British Geographers* 34, pp. 159–73.

J.A. Peden 1967, 'History of Kirkless ironworks' (University of Liverpool dissertation).

P. Powell 1986, *Wigan town centre trail* (Wigan Metropolitan Borough Council).

R. Ridyard 1972, *Mining days in Abram* (Leigh Local History Society).

G. Roberts 1965, 'Steam haulage on the Leeds & Liverpool', *Journal of the Railway & Canal Historical Society* 11(3), pp. 8–11.

A.J. Taylor 1954, 'The Wigan coalfield in 1851', *THSLC* 106, pp. 117–26.

C.H.A. Townley, F.D. Smith & J.A. Peden 1991–2, *The industrial railways of the Wigan coalfield* (Cheltenham: Runpast) 2 volumes.

D.H. Turner 1976, 'The Wigan Coal & Iron Company' (Glasgow: University of Strathclyde dissertation).

B.J. Turton 1962, 'Horwich: the historical geography of a Lancashire industrial town', *Transactions of the Lancashire & Cheshire Antiquarian Society* 72, pp. 141–50.

F. Walker 1939, *Historical geography of South West Lancashire before the Industrial Revolution* (Manchester: Chetham Society Publication 103).

M. Worswick 1971, 'The Wigan coalfield 1750–1850' (Liverpool: Edge Hill College of Education dissertation).

Chapter 4

Rolling Stock, the Railway User, and Competition

Michael Harris

Railways were promoted during the nineteenth century as commercial undertakings and necessarily conducted their business in response to the demands of the market, but for as long as they were dominant in the provision of land transport, how responsive were they to their customers' requirements? And how well did they match the provision of passenger and goods vehicles to their users' needs?

The basic motivation for ordering rolling stock was essentially reactive. Railway chief mechanical engineers – otherwise titled locomotive, carriage, and wagon superintendents – usually submitted annual statements detailing those locomotives and other rolling stock being proposed for condemnation in the succeeding year. In response to this programme, traffic officers made their recommendations to the board for replacement vehicles, and usually these were charged to the railway's revenue account.

Only at times of expansion, or in response to competition from other transport modes, was the cost of new stock charged to the capital account. The Great Western Railway's (GWR) construction of gangwayed coaches for express services in the late 1890s and the London & North Eastern Railway's (LNER) so-called High Speed Trains of the late 1930s respectively represent the two circumstances.[1] During the mid/late 1930s expenditure on some new carriages and modifications to existing stock benefited from government grants made under financial assistance programmes to relieve unemployment.[2]

Passenger Vehicles

Examples where the provision of new rolling stock was justified in terms of the need to upgrade services to meet road competition include 30 gangwayed coaches for the

---

1   GWR Carriage Lots identifying those vehicles whose construction was charged to revenue or capital account: archive held by the National Railway Museum. Financial provision for LNER carriage construction features in PRO RAIL 390/871.
2   Grants were made available under Government Assistance Works schemes such as the Railways (Agreement) Act, 1935.

LNER's London–Cromer service whose construction was approved in 1928 ahead of the normal carriage-building programme.[3] Also, the LNER Board approved the building of five new excursion trains at the depth of the Depression in 1933, with the aim of safeguarding its share of traffic to major sporting events by providing stock that was measurably superior to road coaches. In this case, the cost of the new trains was justified as being in part-replacement of aged suburban stock due for withdrawal.[4]

Letters from passengers adversely comparing a company's stock with its competitor's often provided useful ammunition for improving the fittings and facilities of new stock, or retrospective modifications to existing vehicles. Railway managers claimed in justification that their railway needed to remain competitive. In 1933, a complaint from a passenger who compared LNER carriages unfavourably with those of the London Midland & Scottish Railway (LMS) influenced the LNER's decision to fit reading lights to its new gangwayed stock.[5]

These are some of the ways in which railways contemplated the provision of new rolling stock or changes to specification. In addition, there was progressive introduction from the earliest days of a range of facilities accorded or sometimes denied to passengers.

As early as 1839 Lieutenant Peter Lecount farsightedly advocated in *A Practical Treatise on Railways* that some railway carriages should be reserved entirely for ladies and that saloon coaches should be provided for invalids. If railways were unable to allow a reasonable time for passengers to take refreshments at stations in the course of their journeys then, said Lecount, a refreshment coach should be attached to trains, and accordingly through-communication would need to be provided from one end of the train to the other.[6] It took 60 years for such facilities to become commonplace in Britain. Long-distance trains had continued to make their stops for passengers to detrain at principal stations, there to bolt down meals. As an alternative, there were luncheon baskets, which from March 1875 could be obtained from Midland Railway (Midland) stations.

Not until the late 1890s was general-service rolling stock equipped with lavatories, which during the previous decade had begun to be included in new stock. Railways were loath to provide separate ladies and gentlemen's lavatories, but in 1921–22 they were provided in new London & North Western Railway (LNWR) gangwayed stock.[7] The newly-formed LMS soon discontinued the practice, which had not been adopted by the other Big Four companies – though the earliest Pullman cars supplied for use in Britain in the 1870s had been so distinguished.[8]

During the final years of the nineteenth century, train heating using steam supplied at low pressure from the engine had become a feature of rail travel. From the 1870s

---

3   Minutes of meeting of LNER Joint Locomotive & Traffic Committee, 16 February 1928, PRO RAIL 390/14.
4   Ibid., 5 January 1933.
5   Minutes of meeting of LNER Superintendents and Passenger Managers, 21 February 1933, PRO RAIL 390/241.
6   *A Practical Treatise on Railways* features in PRO RAIL 1007/103.
7   LNWR papers on passenger facilities, PRO RAIL 1007/103.
8   B. Haresnape, *Pullman – Travelling in Style* (1987), p. 40.

foot-warmers had been obtainable from stations, and they remained in use into the first decade of the twentieth century. With regard to lighting, the introduction of oil lamps in compartments dated from 1842.[9] The Lancashire & Yorkshire Railway (LYR) had used Newall's patented coal-gas lighting from 1862 and found it to be 50 per cent cheaper to use than oil, and to provide superior illumination. The LNWR adopted coal-gas lighting for some Liverpool-area local trains in 1870 but discontinued the experiment after a gas cylinder exploded on a train.[10] Pintsch's compressed oil-gas system proved much more successful and was used from the late 1870s. Although electric lighting first made an appearance in Pullman cars in 1881, it was not generally a feature of British trains until the 1890s, one reason being that pressure oil-gas lighting gave a better light than the early electric filament-bulbs.

Savings in operating costs were the main motivation for switching to the wholesale use of an improved passenger amenity. In 1904 an LNWR report put the savings from the replacement of gas by electric lighting on 1724 vehicles as £13 468, say £800 000 at today's prices.[11] However, the LNWR was unable to quantify the savings made by introducing train heating, although it was noted that at main stations stokers were needed to tend boilers that supplied hot water to fill the company's 20 000 foot-warmers, which alone cost £10 250, say £615 000 at today's prices.[12] With the abandonment of foot-warmers in favour of train heating the railway companies must have made appreciable savings.

As to those facilities demanded by passengers, smoking compartments were high on the list. From 1 October 1868, the *Regulation of Railways Act* required that all railway companies except the Metropolitan 'shall in every Passenger Train where there are more carriages than one of each class provide smoking compartments for each class of passenger unless exempted by the Board of Trade'.[13] This legislation may have been provoked by a case of 1861 when Huddersfield magistrates had refused to convict an offending smoker on the grounds that the railway company concerned – the LNWR – had not provided 'smoking carriages' and that railway directors and officials were 'in the constant habit of smoking on trains'.[14] In later days the number of smoking compartments in the booked formation of a train was in the order of two-thirds smoking to one-third non-smoking.[15]

Evidence of the shortcomings of early British railway carriages is scant, so we must be grateful for William Buddicom when he wrote to Thomas Brassey in May 1868 and compared British and French practice.[16]

A very great improvement may be made in the form of stuffing of the English carriages. At present or at all events until lately the backs had no shape, no attempt being made to make them fit the backs of the traveller. The seats have or had a bad form, they ought to incline

9   PRO RAIL 1007/103.
10   Ibid.
11   Ibid.
12   Ibid.
13   Ibid.
14   LNWR correspondence dated 13 May 1861, PRO RAIL 410/1566.
15   SR operating instruction dated 8 March 1924, PRO RAIL 414/646.
16   Correspondence: Carriage Improvements and Comparisons, PRO RAIL 410/1570.

from the front towards the back so as to throw the weight of the body on the thighs or rather off the knee joint. The French carriages are far superior in this respect, as also in many others.

When it came to appearances, the interiors of most mid-nineteenth-century British carriages were spartan, made less appealing by their comparatively low interior height. An 1851 specification for a London Brighton & South Coast Railway (LBSCR) second-class vehicle gives the height from bottomside (the bottom rail of the bodyside framing) to the top of the roof boards as just six feet.[17] The interior partitions were sometimes disfigured by advertisements, as George Smith of the LBSCR, at New Cross, pointed out to the chairman and directors of his company: 'Regret to call your attention to the disgraceful state of the interior of some second and third-class carriages, Mr Edwards having acted contrary to his agreement by posting placards on the ends and sides. I hope some steps will be taken to prevent him affixing placards to the new stock.'[18] From the mid-nineteenth century the lot of the passenger markedly improved. In his review of North Eastern Railway (NER) carriage stock dated 1877, Chief Engineer T.E. Harrison noted that 'more carrying capacity has generally been provided in the new carriages, and more room for each passenger; more expensive fittings have also been introduced'.[19]

British companies reacted slowly to developments in other countries: usually it was the United States that set trends, not least with Pullman cars. Their use had stimulated the provision of Britain's first dining-car in 1879 by the Great Northern Railway (GNR). Other railways reacted cautiously. It was not until 1891 that the Great Western Railway reviewed the experience to date with restaurant cars on both the LNWR and GNR. The GWR's Traffic Department report of that year concluded that dining-cars could not be worked at a profit; they would not be made use of 'to a large extent', and would add to the weight of trains. 'On these grounds the officers do not recommend the adoption of dining-cars on this company's lines. If however the directors decide to provide such cars they recommend they should be run as follows.' A list of services followed.[20] There seemed no question of eliciting the views of the passengers as to whether they might use restaurant cars were they to be provided.

Sleeping cars were introduced in central Europe during the early 1870s, and in 1873 one was put into traffic by the North British Railway between Glasgow and King's Cross. The Pullman Car Company's agreement with the Midland Railway stung the GNR to experiment with sleeping cars in the mid-1870s. In January 1873 the GNR's Locomotive Committee heard from the general manager that

there is reason to believe that the Midland was contemplating the use of Pullman sleeping and drawing room cars. A number would be ready to run by the end of March. In view of these facts the general manager submitted for consideration whether the GNR should not

17　Correspondence from Locomotive Department, New Cross, LBSCR, PRO, RAIL 414/ 622.

18　Ibid.

19　Memoranda to NER directors, PRO, RAIL 527/229.

20　M. Harris, *Great Western Coaches from 1890* (1985), p. 50.

be prepared to place some sleeping cars on their system for travellers from Scotland and York.[21]

The railways combining to provide Anglo-Scottish services on either side of the country respectively formed the West Coast Conference and the East Coast Conference, which concentrated on their own matters and were usually fiercely competitive. Yet the rivals joined forces in a common aim to resist any moves to bring in third-class sleeping accommodation. The Conferences met to discuss the matter in 1906, and again in October 1909 when they 'unanimously agreed that the undesirability of running third-class sleepers is still maintained'.[22] The reason was obvious: first-class sleeping cars were very profitable and the availability of third-class berths would apparently result in what the railways termed an 'abstraction' of business. First-class sleeper travel was so valued that as early as 1909 the East Coast companies were able to justify the additional cost of equipping the latest sleeping car with pressure ventilation and heating.[23]

Railways were responsive to demands for the provision of club carriages which were run on limited services on the LYR, LNWR, LBSCR, and NER. A club carriage first made its appearance in 1895 on the LYR's Blackpool–Manchester services following an approach from first-class season-ticket holders. In return for a guarantee that they would purchase a specified number of annual tickets, the regular travellers asked the railway to provide a carriage for their exclusive use, and users were charged a supplementary fee. Election by the club committee was mandatory and there were strict rules.[24]

In 1910 a number of businessmen living on the North Wales coast travelling to and from Manchester joined together to form a club. For their accommodation the LNWR attached two saloon coaches to a morning train from Llandudno, and back in the evening from Manchester. An attendant served tea from a pantry in one of the cars and members had keys to personal lockers.[25] The heyday of the club saloons was before the First World War, and they declined in use during the 1930s, although as late as 1935 a new club saloon was built for the Manchester–Blackpool working. It seems surprising that the use of club carriages was not more widespread on British railways, given the volume of first-class commuting, particularly to and from London.

The late 1890s saw the almost simultaneous introduction on British railways of a range of innovations which significantly increased passenger comfort and convenience. This transformation in just two decades was described graphically by the Great Eastern Railway's Locomotive Carriage & Wagon Superintendent, A.J. Hill.[26] He reviewed increases in the average expenses of the principal British railways

21   Minutes of meeting of GNR Locomotive Committee, 16 January 1873, PRO, RAIL 1007/103.
22   Minutes of meeting of West and East Coast Conferences, 8 October 1909, PRO, RAIL 172/10.
23   Minutes of meeting of East Coast Traffic Superintendents, 1 September 1909, PRO, RAIL 172/10.
24   J. Simmons and G. Biddle (eds), *Oxf. Comp.*, p. 91.
25   *North Western News*, summer 1910 issue, PRO, RAIL 410/2014.
26   Presidential Address to Institution of Locomotive Engineers, 12 January 1914.

since 1892. They were of the order of 37 to 40 per cent, and Hill listed the main causes as 'higher speeds, greater loads, and larger and more complicated engines'. He pointed out that the cost of repairs and renewals, both per passenger-mile and per vehicle, had increased by at least a quarter between 1892 and 1912. In 1892, the mainline vehicles on the Great Eastern Railway (GER) were mostly six-wheeled; twenty years later they were bogie stock and, in the case of a third-class carriage, the tare weight of carriage per passenger had nearly doubled. Whereas in 1892 the GER had operated just one dining-car train, in 1912 thirteen such trains departed from Liverpool Street daily. Said Hill, 'The expenses incidental to trains of this sort, which of course must be corridored throughout, are out of all comparison with... twenty years ago.' He continued, 'the internal fittings are much more luxurious than they used to be and the provision of lavatory accommodation on all trains travelling any considerable distance is another additional expense while the improved lighting of our trains must not be forgotten.'[27] When comparing the various pre-Grouping companies, Hill noted that the expenses of those railways with a large proportion of mainline carriages were greater than recorded by the GER, LBSCR, and London & South Western (LSWR), none of which spent nearly so much on their fleets of predominantly suburban and local vehicles.

Despite the improvements incorporated in carriages, when in 1905 the Midland Railway analysed the costs of new carriages built in its own workshops and by contractors over the five preceding years, it found that their building costs had declined as a result of the availability of cheaper materials.[28] No doubt this situation also encouraged other railways to restock their carriage fleets.

Carriages with 'through corridors' were introduced from the 1890s. More accurately they were described as gangwayed, for there were concertina gangways at each end of the vehicle to allow passage throughout the train. The majority of these carriages were of the traditional British layout with an internal corridor running alongside individual compartments. Exceptions were dining-cars with and without kitchens, in which there was a centre passageway, on either side of which seats faced tables, and saloon vehicles of various kinds, usually for party travel.

Slightly later came the so-called 'open' coaches for normal service. Probably the earliest dated from 1900–01, when the GWR built two sets of clerestory-roofed open coaches for its Paddington–New Milford boat trains. They were also notable for being the first public-service trains on that railway with electric lighting. Tables could be fitted to each bay of seats and the idea was that light refreshments could be served to all passengers. The intention seems to have been to develop excursion traffic to Ireland.[29] The New Milford stock proved unpopular and indeed on its debut had attracted unfavourable comment from the railway trade press. In 1920 a GWR employee who was involved with train operating and the marshalling of trains commented that the travel experience offered by the stock was 'so uncomfortable, particularly during winter months, that it has been known for passengers presenting themselves at Paddington for Ireland (who had previously paid for a reserved

---

27   Ibid.
28   Proof of Evidence from D. Bain for rating appeal, 1905, PRO, RAIL 491/875.
29   M. Harris, *Great Western Coaches from 1890* (1985), pp. 54–5.

first-class compartment) refusing to travel in the centre-corridor train, and deciding to go by the LNWR route'.[30]

The GWR took another gamble with its 69-70 foot-long side-corridor 'Dreadnought' stock, introduced from 1904 and put into service the following year on the company's premier express, the *Cornish Riviera Limited*. The wide, elliptically-roofed bodies were modern in appearance. In another departure from practice, the vehicles' layout meant that entry and exit was solely by means of doors in end and central vestibules, the carriages being devoid of exterior doors to the compartments. Apart from the stock for the premier GWR expresses, three sets were built for excursion work.[31] The GWR observer of 1920 earlier referred to reported that 'from the passengers' point of view they (that is, the 'Dreadnoughts') were most unpopular coaches... difficulties and complaints... arose in consequence of passengers scrambling to get in before detraining passengers had alighted'. As to the 'Dreadnoughts' built for excursion work, he alleged that a train of this stock had been provided once for a football special conveying about 1 000 supporters from London to Birmingham and 'on the return journey the scrambling and difficulty of the passengers to entrain by the limited entrances resulted in free fights, one or two doors being broken off'.[32]

Innovations tended to be applied first to special stock for premier trains. In 1928 the LMS built ten semi-open firsts and brake firsts for use on the *Royal Scot* and other prestigious trains. The brake vehicles featured deep leather armchairs and settees, all accompanied by occasional tables. Luxurious though they were, if overblown in style, these carriages proved unpopular with passengers and saw limited service.[33]

During the 1920s and 1930s the Anglo-Scottish expresses of the LMS and LNER were the recipients of a number of high-specification vehicles. There was good reason: until 1932, the schedules of day expresses on both companies' routes were subject to a minimum 8½-hour timing between London and Edinburgh and London and Glasgow. Competition was accordingly conducted in terms of special facilities on-board trains, self-evidently features that would attract press coverage. Some of the facilities proved problematic.

Known as 'super-firsts', new first-class carriages were built in 1930 for the LNER's *Flying Scotsman* with armchair seating for six passengers in each attractively furnished compartment. The operators reported 'considerable complaints' from passengers, and after no more than three months' service the carriages were altered with fixed seating.[34] The next year, another pair of 'super-firsts' equipped with an early form of air-conditioning entered service on the *Flying Scotsman*. No separate under-seat heating had been fitted and passengers complained of being cold. Additional steam-heated radiators were swiftly fitted.[35]

---

30 GWR (London) Lecture & Debating Society, meeting 30 December 1920, PRO ZPER 38/ 13.
31 M. Harris, *Great Western Coaches*, pp. 59 et seq.
32 GWR (London) Lecture & Debating Society, meeting 30 December 1920, PRO ZPER 38/ 13.
33 R. Essery, D. Jenkinson, *The LMS Coach* (1969), p. 74.
34 Minutes of meeting of LNER Superintendents and Passenger Managers, 21 October 1930, PRO RAIL 390/239.
35 Ibid., 15 December 1931 and 19 January 1932, PRO, RAIL 390/240.

In close cooperation with suppliers, from 1930–39 the LNER persevered with developing a successful system of, first, pressure ventilation, and then the more advanced air-conditioning, for use in its prestige passenger stock. The reason was that steam heating was frequently ineffective in the rear vehicles of lengthy trains, and the steam supply could not be increased in pressure without risking injury to passengers.

Other innovations at various times on the LNER's Anglo-Scottish expresses included coaches wired for wireless reception and for relaying music from gramophone records played on board (1930–35); ladies' retiring-rooms (from 1928 and perpetuated until 1963); hairdressing salons (1928–39); cocktail bars (1932–39); cinema cars (1935–39), and shower cubicles in sleeping cars (from 1930). The cocktail bars proved popular and attracted new business. The on-board wireless and cinema cars were initially well used, but patronage soon declined. The two hair-dressing salons each lost the LNER £500 annually. Nor, other than their employment on the summertime *Northern Belle* luxury cruising train, was much use made of the shower cubicle fitted in eight or so first-class sleeping cars, an ingeniously designed facility that was expensive to install.[36]

Complaints from passengers sometimes sparked off a review of standard carriage fittings and furnishings. In 1926, a Mr Lupton complained to the LNER about the type and nature of third-class lighting and seating. This provoked the company to turn out a newly-built standard vehicle with a changed upholstery design and floor covering, less sombre paintwork, and improved light fittings and ventilation. What became known as 'Mr Lupton's Complaint' (with its overtones of a recent American novel) occupied the attentions of LNER managers for the next year or so but the improvements it had provoked were adopted, coincidentally with the provision of more adequate interior heating in all new vehicles, and enhanced lighting in first-class stock.[37]

The Royal Commission on Transport reporting in 1928 had little to say about the needs of passengers and the way they were accommodated in railway rolling stock. An exception was its recommendation that British railways should be under a statutory obligation to provide a seat for each passenger joining a mainline train at its starting-point. The implication was that as road coach stage services guaranteed a seat for passengers the railways should do likewise. The Railway Companies Association rejected the idea, commenting that the railways 'did their utmost to provide adequate accommodation'. The Ministry of Transport sided with the railways and the recommendation was dismissed.[38]

One of the few general initiatives to provide a better standard of third-class seating on mainline trains came with the introduction from 1932 of three-a-side seating in LNER and LMS side-corridor mainline stock. The GWR and Southern Railway (SR) retained four-a-side seating for mainline third class.

The Railway Clearing House (RCH) was the forum for decisions reached by the railways on more general issues affecting passenger travel, such as the provision

---

36   M. Harris, *LNER Carriages* (1994), pp. 60–1, 92, 113.
37   Minutes of meetings of LNER Superintendents and Passenger Managers, 28 September 1926 et seq., PRO, RAIL 390/237.
38   Royal Commission on Transport, PRO MT6/3332.

of smoking accommodation. During 1930, the GWR called the RCH's attention to the volume of complaints received regarding passengers smoking in non-smoking compartments. The railway superintendents agreed that non-smoking compartments should in future be labelled 'Smoking Prohibited', either on the windows or inside compartments, or both. This led to the use of the once-familiar red triangular window-label.[39] The use of 'Smoking' labels continued, and they were insidiously sponsored under contracts between railway companies and cigarette manufacturers.

In general, changes to carriage design were not a subject for discussion within RCH committees. An exception was the contentious subject of all-steel carriages. Until the early 1950s, the majority of vehicles on British railways had steel underframes, while their bodywork was constructed of largely hardwood framing to which was screwed light-gauge steel exterior panelling. The interiors were almost entirely made of wood. The publication during 1929 of the Ministry of Transport reports into the Charfield and Darlington accidents stimulated press concern at dangers associated with these collisions, these being fire occurring as a result of the involvement of largely wooden, gas-lit vehicles, and at Darlington the incidence of telescoping.

Late in 1929 a high-powered RCH meeting considered whether British railways might make greater use of steel in passenger and freight stock. Among those attending this meeting were representatives from the steel industry which was simultaneously making representations to the Ministry of Transport for the railways to substitute steel for wood in carriage construction.[40] The railway managers stated that steel was already used for about 75 per cent of the total weight of British carriages (a specious argument because the underframe and running gear clearly constituted the bulk of a vehicle's tare weight) and raised objections to the increased use of steel in mainline carriages on the grounds that it would increase both the cost of their construction and vehicle weight, that the interiors of carriages would heat up more slowly in winter and too quickly in summer, that steel construction was not as suitable for the traditional British design of side-corridor carriage, and that corrosion would prove a problem.[41]

There was public concern during the late 1930s at the risks presented by carriages of composite steel and wood construction, sufficient to instigate Parliamentary questions, in answering which government spokesmen not only upheld the railways' arguments but 'refined' them to include the questionable statements that 'dynamics cannot be eliminated by the use of all-steel construction which is not necessarily stronger than composite construction', and that 'what we call wooden coaches in England are 75 per cent steel'.[42] Thus was public safety and interest ill-served during the 1930s by railways, the MoT, and government. Although London Underground and some suburban electric stock was all steel, it was not until 1951 that new mainline stock in Britain had steel-framed bodies.

---

39 Minutes of meeting of Carriage & Wagon (C&W) Superintendents held at the Railway Clearing House (RCH), 29 April 1930, PRO, RAIL 1080/761.

40 All-Steel Coaches, PRO MT6/3301.

41 Minutes of meeting of Joint Technical Committee Railway C&W representatives held at RCH, 6 December 1929, PRO, RAIL 1080/761.

42 All-Steel Coaches, PRO MT6/3301; refers to House of Lords debate, 20 April 1937, and House of Commons debate, 30 July 1937.

Improvements to basic carriage design made by the Big Four companies – principally the LMS – involved the adoption of flow-line production methods allied to the use of modern machinery to supersede traditional coach-builders' methods. There was some use of new technology: in place of riveted construction, electric-arc welding was adopted for fabricating underframes that were both stronger and lighter, and also for bogies. Trials resulted in the evolution of strengthened timber body-framing.[43] Projections and recesses were progressively eliminated from exterior body-panelling, and also from interior woodwork and fittings, not only to reduce building costs but make cleaning and routine maintenance easier.[44] The drive to cut production costs during and after the Depression sometimes led to the use of cheaper materials and reductions to specification so that the required number of vehicles could be produced within a reduced budget for the annual building programme.[45] In time, the LNER, for one, discovered that higher quality stock, for which passengers were charged a supplementary fare, had the effect of stimulating an increase in passenger traffic. The introduction in 1935 of the first of these expresses, the *Silver Jubilee*, was said to have contributed to a 12 per cent increase in all Newcastle–London ticket sales in just ten months. During a four-week period in 1938, all four LNER supplementary-fare expresses – the *Silver Jubilee, Coronation, West Riding Limited*, and *East Anglian* – turned in a combined profit of £16 700: £900 000 at today's prices.[46]

During the Second World War the Big Four companies set up committees to recommend what might be required for post-war services. This led to a number of initiatives in rolling stock design. The LMS considered improvements to the layout of its restaurant cars, to give more room for the movement of passengers and staff and to make the interior appear more spacious, as well as introducing better lighting, heating, and ventilation, and improved internal fixtures and furnishings. Two vehicles were altered and refurnished in 1946.[47] The LNER set up a committee to report on the design of rolling stock. Dated 1944, its conclusions were that (side-corridor) compartment stock was 'substantially more popular', that carriages with entry and exit through end-vestibules only were preferable, that interiors should be refitted at intervals in order to keep up to date with improvements in road vehicles, that carriage windows should be made lower and wider, and that preferably lighting should be diffused, supplemented by reading lights. The committee 'understood' that the company's future carriages would be all-steel.[48]

---

43   Dr P.L. Henderson, 'Development in Carriage Design' paper, LMS Chief Mechanical Engineer's Conference, 1934, PRO, RAIL 418/181.

44   T.R. Howard, 'Design and maintenance of carriages and wagons' paper, LMS CME's Conference, 1935, PRO, RAIL 418/182.

45   Minutes of meeting of LNER Joint Locomotive & Traffic Committee, 4 January 1934, PRO, RAIL 390/14.

46   Memoranda from Sir R. Wedgwood, Chief General Manager, LNER, PRO, RAIL 390/987.

47   LMS memorandum 'Experimental Restaurant Cars', dated December 1946, PRO 1007/103.

48   Report on design of rolling stock, Postwar Development of Passenger Traffic, National Archive of Scotland LNE 4/302.

For possibly the first time the public's opinions were deliberately canvassed for the future passenger-stock of both the SR and the LNER.

Completed in September 1945, the first of the SR's all-new mainline gangwayed carriages was put on public display shortly afterwards at London's Waterloo and Victoria stations. Those entering the carriage were handed a questionnaire headed 'Southern Railway – Getting into its stride again – A NEW CARRIAGE – tell us what you think of it!' Visitors were invited to say what they liked and did not like.[49] Fewer than 10 per cent of an estimated 25 000 visitors completed the questionnaire, the findings being that they preferred compartments to saloons, wanted reading lights in compartments, and heating provided in corridors and the lavatories. Such features duly appeared in what became known as Bulleid stock which was built between 1946 and 1951.[50]

The LNER's prototype post-war mainline carriage entered service at the start of 1945. A questionnaire was distributed to passengers travelling in this first-class vehicle. With the assistance of illustrations comparing the interior of a pre-war carriage with the 1945 prototype, passengers were asked for their preferences regarding the interiors of carriages – compartment or open; seating design; window type; lighting; luggage racks and interior furnishings. At least 17,000 completed questionnaires were received.[51]

After 1945 the Railway Companies Association commission for post-war standardization of coaching stock arrived at standard dimensions for mainline carriages, as well as the best arrangement of interiors and the layout and location of doors. With nationalization, this work was re-examined and altered to take into account the use of all-steel construction. The objectives to be attained by the new standard stock were, said the Railway Executive, 'to provide stronger... safer carriages with better passenger amenities, without increase in tare weight or initial cost and a reduction of maintenance costs'. For the first time in the history of British passenger-stock history, overriding aims were set out.[52]

From the mid-1950s BR was concerned with making improvements to the interior layout and design and to the heating and ventilation of future standard mainline carriages. During 1957 experimental versions of the existing standard designs that featured some of these principles were put on public display in London.

In the early 1960s a radical rethinking of carriage interiors had been considered essential in the face of increasing competition from cars and domestic airlines on long-distance routes. In 1963 outside design consultants were engaged to produce a mock-up of a carriage section which went on public display as the Carriage of Tomorrow, many of the features of which were incorporated in the experimental XP64 train which went into service the following year. Public surveys were conducted on board the train, as a result of which, and also from service experience,

49   Leaflet in PRO, RAIL 1017/1.
50   D. Gould, *Bulleid's SR Steam Passenger Stock* (1980), p. 10.
51   *Design for Comfort*, LNER, 1945; see also M. Harris, *LNER Carriages* (1994), p. 102. Minutes of meetings of LNER Superintendents and Passenger Managers, 12 December 1945, PRO, RAIL 390/247.
52   S. Smith, Institution of Locomotive Engineers Paper No. 499 'Standardisation of Coaching Stock', 17 January 1951.

modified designs of seating, lighting, lavatories, doors, and other internal fittings from the XP64 carriages were adopted for the BR Mark 2 carriages built in quantity between 1964 and 1975.[53]

The 1960s marked a period when greater care was taken to sample public opinion when embarking on the design of passenger rolling stock. Opinion surveys conducted during the late 1960s revealed that passengers considered full air-conditioning was a priority on long-distance rail journeys, and all new mainline InterCity stock was so equipped after 1971.[54]

Freight Wagons

Wagon design in Britain was largely determined by private-owner wagons on account of their sheer volume – there were 700 000 in 1925. 'Private owner' was defined by the RCH as 'a party for whose purpose the wagons are used upon the railway'.

Of these private-owner wagons, coal wagons were the most numerous, their design being determined by the standard specifications drawn up by the RCH. That for the 12-ton standard mineral wagon dated 1923 sets out the basic conditions:

> the body of coal wagons to be built strictly in accordance with the standard drawing, the number and type of doors fitted to be at the option of the Owner. The capacity of wagons for conveying commodities such as iron ore, limestone &c., will be varied to suit such commodities and bodies may be of timber or metal as desired, but the approval of the Wagon Superintendent of the registering company must be obtained.

Underframes, their fittings, and running gear had to be built in accordance with the specification.[55]

Before 1948 it was the railway authorities rather than owners or builders who closely controlled the design of wagons used upon the system. The designs for wagons for specialized traffics had to be submitted to the RCH for approval. Typically, these were tank wagons of various types, coke, and salt wagons. The designs were put forward either by the railway company intending to register them or by a wagon builder. If the private owner's wagon was accepted by a railway operator, a register-plate was affixed to each side of the wagon. Standardization had the effect, however, of imposing the lowest common denominator in terms of technological progress. Obsolescent and none-too-well maintained grease-lubricated wagons which were capable only of running at slow speeds survived well into the motor age.

Reference was made earlier to the GER's Locomotive Carriage and Wagon Superintendent, A.J. Hill, and his review of increases in railway expenses. Speaking in 1914, he made the telling point that although the cost of wagons had hardly increased over the previous 20 years the cost per goods train-mile had risen by over 50 per cent. It indicated, he said, that 'there has been comparatively little development in

53   M. Harris, *British Rail Mark 2 Coaches* (1999), pp. 15 et seq.
54   Ibid., pp. 67–8.
55   Railway Clearing House (RCH), Specification for 12-ton mineral wagon, PRO, RAIL 1085/58.

the size and weight of wagons used'. He remarked that 'it had been found impractical to introduce to any great extent the large wagons... used in America'.[56]

In the previous decade Vincent Raven had reported to the North Eastern Railway on his study visit to North American railroads. He said that they achieved faster loading of wagons and that this was not possible in Britain 'owing to the smaller type of wagon. Even were we to obtain larger wagons it would be some years before we could be in the same position as the Americans'. To improve productivity he advocated the use both of standard British-type box vans permanently coupled in pairs and higher-sided wagons.[57] Later, the NER was to experiment with bogie covered vans, as did the GWR.

After the First World War there was little progress with the basic design of common wagon types. From the railways' point of view, even with wagons of ten-ton capacity it was difficult to obtain full payloads, and the companies contended that for most traffics there was little point in introducing higher capacity wagons. Full loads were certainly obtainable for coal and minerals, yet few collieries were equipped to handle 20-ton wagons, despite the fact that these were not that much larger than ten-tonners.[58] The GWR was successful however in introducing twenty-ton wagons for coal and coke traffic in South Wales and offered a rate rebate of 5 per cent on all coal carried in such wagons.

In 1927 the government appointed a Standing Committee on Mineral Transport under the chairmanship of Sir Arthur Duckham. In its first report, the Committee was unanimously of the opinion that the principal obstacle to the use of higher capacity wagons was that ports and collieries and iron and steelworks were 'not equipped to take 20-ton wagons on a scale that makes their general use practicable and economic'.[59] The Committee recommended that there should be a deadline after which the manufacture of wagons smaller than 20-ton capacity should be discontinued. The government was urged to make available money towards the cost of improving wagon-handling terminals at ports, collieries, and works. The 1929–31 Labour government backed the Committee's recommendations and eventually offered financial assistance in the form of grants towards the provision of improved handling facilities. There was qualified support from the railways, but industrial users provided the stumbling block, declaring that without a guarantee of lower rates for traffic conveyed in larger wagons they were not prepared to invest in improved facilities. Exemplifying a frequently adversarial relationship between the railways and their users, this potentially worthwhile scheme died in Britain's political and economic crisis of 1931.[60]

Some less ambitious initiatives involving users and providers of more specialized wagons were more successful. From 1927, in place of milk churns conveyed in vans, producers began to introduce considerably more hygienic tank wagons with glass-lined barrels. For their part, the railways welcomed an opportunity to stem the

---

56   Presidential Address to Institution of Locomotive Engineers, 12 January 1914.
57   Vincent Raven, report on visit to America, PRO, RAIL 527/432.
58   GWR (London) Lecture & Debating Society, meeting 24 February 1921, PRO ZPER 38/13.
59   Standing Committee on Mineral Transport, PRO MT6/3319.
60   Ibid.

loss of milk traffic to road transport. Not without difficulties, a satisfactory design of underframe was eventually evolved through the agency of an RCH committee. Investment was shared: the underframes of the tank wagons were provided and maintained by the railway companies, the special tanks being the responsibility of the milk producers.[61]

Sometimes, a radical redesign was instigated by users. In 1929 the Federation of British Wholesale Fishmongers and the RCH began discussions on the design of a standard covered fish-van that would be capable of carrying the maximum possible payload on a given underframe length. An initial specification was modified progressively in response to suggestions from the fish trade and railway companies, the final design being approved by all parties during the spring of 1932.[62]

Nationalization brought a thorough review of wagon design and with it new standard types for quantity production. The more consensual deliberations of the RCH were replaced by the businesslike decisions of the Railway Executive. In 1950 the Ideal Stocks Committee set out the preferred design, capacity, and types of freight rolling stock for the unified system, arriving at best practice from a range of Big Four designs and sometimes adopting new dimensions and features.[63]

In the late 1950s BR and banana importers jointly evolved a design for an improved banana van. Increased insulation, combined with a change to loading the fruit at an increased temperature, meant that the need for the steam heating was eliminated, something that had been necessary previously to ensure that bananas in transit were kept within defined temperature limits.[64]

More far-reaching cooperation between the railways and their freight users drew inspiration from a train operation of the early 1950s based on the use of a new type of specialized hopper wagon to transport imported iron ore between the docks and Consett steelworks. From this came discussions between BR and the coal and electricity-production industries for an improved method of supplying the forthcoming generation of power stations with coal, and involving the automated loading and discharge of wagons while a train was moving slowly. This led to the introduction from 1965 of the so-called merry-go-round operation which featured a completely new design of high-payload hopper wagon.[65]

Standardization of passenger and freight rolling stock types had been a consequence of nationalization, but from the 1970s and 1980s an increasing responsiveness to users meant that once more the range of rolling stock types began to widen, a process that has accelerated following the privatization of British Rail.

61   Minutes of meetings of Carriage & Wagon Superintendents held at RCH, 26 July 1927 et seq., PRO, RAIL 1080/761.
62   Ibid., 14 August 1930 et seq.
63   S. Parkhouse, Institute of Transport Paper 'Railway Freight Rolling Stock', 5 February 1951.
64   E. Gooding, 'The Distribution of Bananas by Railway' paper to BR(WR) London Lecture & Debating Society, 22 January 1959.
65   R. Munns, *Milk Churns to Merry-go-Round* (1986), pp. 148 et seq.

Chapter 5

A Note on Midland Railway Operating Documents

John Gough

The railways were among the greatest industries of the nineteenth century and they were accordingly among the largest employers. A railway is, however, not like a factory – the work-site is by definition dispersed over many miles and communication cannot be on the basis of going out of an office door to give instructions to someone on an adjacent shop floor. The dispersed nature of the organization brought with it a need for good and clear written communications. Furthermore, the railway placed a special emphasis on safe operation, which necessarily brought with it a need for a discipline in its working practices more like that to be found in the military than in most other spheres. Right from the start, there were many areas of railway activity in which written communication for both operational and commercial purposes was essential. With this, of course, came the requirement for literacy across a broad spectrum of staff. The written communication covered both the commercial and the operational sides. Copious records were also required for management purposes, many of them in quantities sufficient to justify printing. A full account of the internal literature of the railway remains to be written.

This chapter is concerned with one aspect of the operational literature – the documents that pertain to the everyday safe working of the system. In particular, it deals with those publications used by drivers, signalmen, and staff working on the track. This is the paperwork that shows how the railway was operated as a transport system to carry traffic and conduct its trade. It is not concerned with the provision of rolling stock or staff for services ('diagraming'), with the selling of transport services to the customer, or with the business management of the company.

The whole subject of working documents on the railway is one that has received surprisingly little attention. There appears to be no account of the origin of the working – as opposed to the public – timetable, there appears to be no account of the origin of the graph timetable and the development of its use in the railway industry, and there is certainly no full study of other operating documentation such as the rules and regulations, the appendices to the working timetable, and the operating notices issued at regular intervals to keep staff informed of special conditions along the line.

In an essay on working timetables,[1] Jack Simmons examined the significance of these documents for the study of the railway, commented on the little attention they have generally aroused, and showed how much information might be derived from their study. It can safely be asserted that for anyone wanting to discover the detail of how the railway operated on a day-to-day basis, how it delivered transport services to its customers, how intensively its assets were exploited, there is no substitute for a study of the working documentation. Only there can the intertwining of passenger and goods services, of fast and slow services, the use of single and multiple lines, the management of junctions, the detail of the infrastructure facilities provided and the maintenance of those facilities, and many other aspects of the way in which the decisions taken at board and committee level or by the officers were daily implemented on the ground be clearly followed.

My intention here is to consider the working documents of a single company, the Midland Railway. This was one of the largest railway companies: it carried a huge goods and mineral traffic in addition to operating major competitive express passenger services and a full range of local trains; it extended widely throughout England, and it reached into Scotland, Wales, and Ireland too. This was also a company that looked after all its records, including the operational materials. It maintained and retained bound reference sets of its working documents over many years. Timetables of all sorts, and documents such as engineering notices, have always been essentially ephemeral. Early timetables had a life of one month only, later ones several months (though normally still with monthly supplements appearing). An expired timetable is good only for rubbish: no one runs a railway or plans journeys on the basis of out-of-date information. However, at the Midland's headquarters in Derby a number of sets of the working timetables and at least one set of the public timetables, in each case with all supplements and many single-sheet notices included, were bound up for retention. The appendices to the working timetables were bound in with the timetable volumes for the year in which they were issued, and separate copies were also provided with a hardback binding for retention. At least one set of the weekly operating notices (the notices to all those working on the line itself) was also bound up. Unfortunately, the bound-up documents are printed on the same low-quality, short-life paper used for the routine production of documentation designed to be jettisoned on expiry – but nevertheless, at least they *were* preserved. The entire run of what seems to have been the master-set of working timetables is still available, along with a substantial proportion of the public set.[2] For the engineering notices a

---

1   'Working Timetables', in *The Express Train and Other Railway Studies*, Thomas and Lochar, Nairn, 1994, pp. 194–212.

2   The volumes of working timetables for 1855 to 1922 (with the exception of that for 1861) are held in the Public Record Office (PRO) under RAIL 963. The tables for the first half of 1915 are missing from the volume that claims to contain them, and the issue for December 1869 is defective, with a section missing, along with any supplements. The volumes all cover both passenger and goods trains until 1917. From 1918 there are separate volumes for passenger and goods trains. The volume for 1854 is in the Roy F. Burrows Midland Collection Trust (Ashton under Hill, Worcestershire), and that for 1861 is held by the Derby Industrial Museum. These two volumes both appear to be strays from the main collection now in the PRO. The run of public timetables is not as complete. The main

lengthy run, from 1886 to 1922 (and, in fact, on to 1927) survives.[3] As far as the writer knows, no similar collection of such notices survives from any other railway company. This routine information about the maintenance and development of the infrastructure is invaluable for what it tells us about the railway on the ground, and it is devoutly to be wished that means of converting it into digital format can be found (despite the huge number of pages involved) to avoid risking loss of evidence through any further deterioration of the already-brittle paper.

At the start of a 1907 article on the working timetables of the Great Western Railway the then editor of the *Railway Magazine* defined the operational material of the railway as comprising the rule-book, the working timetables, the appendices to these timetables, special instructions, weekly notices, and other special notices.[4] Little comment is required on the first of these. Everyone working on the railway (and not just the operational staff) has been bound by his company's rule-book from the beginnings of the industry. Adopted for a company by resolution of the board, distributed to all, and essentially the basis on which all employees work, the rule-book sets out the framework in which the railway is operated.[5]

---

collection is in the PRO under RAIL 962. There is a significant holding in the British Library which extends the evidence of the PRO set back for some years but lacks the additional material bound in to those volumes. And there are public timetables in various private collections, including an important run of very early ones, from 1849 to 1852.

3   This set is held in the library of the National Railway Museum (NRM). It seems possible that the volume for 1886 may, in fact, have been the earliest to be bound.

4   G.A. Sekon, *Railway Magazine*, October 1907, pp. 339–48 ('Working Timetables– 1. Great Western Railway'. This was the first part of an article on the working timetables of that railway, which in turn was the first of a series on the working timetables of a number of the major companies.)

5   The earliest rule-books seen for companies which later became constituents of the Midland Railway are from the North Midland Railway in December 1840, printed by William Bemrose of Derby and entitled *Signals and Regulations to be observed by the Enginemen, Guards, Policemen, and others employed on the North Midland Railway*, and for the Midland Counties Railway in April 1842, printed by R. Allen of Nottingham and entitled *Rules and Regulations to be observed by the Enginemen, Guards, Policemen, and others employed on the Midland Counties Railway*. The earliest Midland volume seen dates from April 1848, is printed by Bemrose, and is entitled *Rules and Regulations for the Guidance of the Officers and Men in the Service of the Company*. This already extends to almost 70 pages. These three items are all in the Roy F. Burrows Midland Collection. The gradual development of a standard rule-book is described by Philip Bagwell in Chapter X 'Standardisation: Railway Operation' of his *The Railway Clearing House in the British Economy 1842–1922*, George Allen & Unwin, London 1968, pp. 221–249. The direct successor to the rule-books of the early years is the *Master Rule Book for Personalised Rule Book Series (GO/RT3001 to GO/RT3013)*, Railtrack PLC (Safety & Standards Directorate), London, August 1999.

The Working Timetable

The earliest internal timetable of a railway company that seems to have survived is a small booklet entitled 'Time Tables. Merchandize Trains. March, 1846' issued by the London & Birmingham Railway.[6] In this we find the timings of the merchandise, cattle, and coal trains running on the company's system. Departure and arrival times are given, although no passing-times for intermediate points are shown, and there are instructions as to how trains are to be worked relative to other traffic on the line. Because the timings for the goods traffic are published separately from those for passenger trains, it is not however possible to form a clear picture from this one document of all the traffic passing along the line. The same is true of several publications from the early 1850s, which is the period in which we really begin to see significant evidence of the development of what we might recognize as a working timetable. There is a volume of Lancashire & Yorkshire Railway 'Time Tables for June, 1852' marked 'For the use of the company's servants' and printed by Bradshaw and Blacklock.[7] Although this is clearly a working timetable in function, the way in which the information is presented is still in many ways closer to that of a contemporary public timetable book than to later working books. Goods and passenger trains appear in separate tables. Station arrival and departure times are not separately shown. Non-station locations are not included. There are no distances. Passing-times for intermediate locations are not supplied. But trains are numbered.

There is interesting evidence from Scotland. From the Caledonian Railway a 'Time Table of Passenger, Cattle, Goods, and Mineral Trains, for the exclusive guidance of Guards, Enginemen, Police, Platelayers, Labourers, &c.', to come into effect from 1 January 1853, survives.[8] Not only is this numbered 'No. 65', but it also carries the instruction 'Time Bills of an earlier Date to be destroyed'. There is the suggestion of a monthly sequence: the first section of the Caledonian Railway was opened for traffic between Carlisle and Beattock in September 1847 (with the routes through to Edinburgh and Glasgow being opened in February 1848), and if the timetable for that month were number 1 and there were monthly issues, then that for January 1853 would indeed have been number 65. From the Edinburgh & Glasgow Railway there are also 'Time Tables of Passenger, Goods, Cattle, &c., Trains' in a volume to come into effect on 1 January 1853 and issued 'For the Special Guidance of the Company's Officers and Servants'.[9] There follows a strict injunction that 'Every Officer and Servant must make himself conversant with this Table each month, in order that he may be thoroughly acquainted with all Alterations that may have been made'. Staff are also warned: 'Last Month's Tables to be destroyed'. Both of these

---

6   NRM, BTC 1196/62. The surviving copy appears to have been issued to one I. Ward, at Camden Town station.
7   This Lancashire & Yorkshire timetable is also in the NRM collection, where it is to be found under the mark BTC 923/58. It is a small 24-page volume in a black oilcloth cover, with a title page but no index.
8   National Archives of Scotland (NAS – formerly the Scottish Record Office), BR/TT(S)54/64.
9   NAS BR/TT(S)59/1.

early Scottish working timetables (although that title is not used) were printed by Hugh Paton, of Edinburgh, who also printed public timetables.

It is not clear when working timetables were first introduced – when it was felt necessary to differentiate between timetables showing the passenger service available for public use or listing goods trains for the benefit of a company's staff on the one hand and on the other hand timetables for operational purposes showing all traffic running on the line in its full timings, including intermediate times. Neither the term 'working timetable' nor the variant 'service timetable' appears in the *Oxford English Dictionary,* and so no evidence for a date of first appearance is to be obtained from that work. The emergence of a timetable with passing-times for intermediate points as well as arrival and departure times for the stations at which trains called implies the development of a discipline far removed from the state of affairs indicated in the 'Regulations of the Grand Junction Railway' in 1837: 'Passengers intending to join the Trains at any of the stopping-places, are desired to be in good time, as the Train will leave each Station as soon as ready, without reference to the time stated in the above table, the main object being to perform the whole journey as expeditiously as possible.'[10]

It is the Midland Railway that appears to have provided us with the earliest modern-type working timetables that survive. These are three volumes entitled *General Time-Tables* and covering the months of February 1852,[11] September 1852,[12] and October 1853.[13] Printed, as we might expect of Midland documents, by Bemrose of Derby, these include the elements that we associate with the working timetables of later years. There is no indication that the February 1852 issue was the first – it seems to follow in an established monthly sequence. The book is 6¼ inches by 4⅜ inches in size, in landscape format, with a soft card cover, and has 68 pages. Features that were to be common in working timetables for decades to come are already present: distances are given; the table of locations includes junctions and goods-only places as well as stations; trains are numbered (goods and passenger stations in one sequence); trains are timed at each station, even when not calling there; footnotes are used to give additional information, such as details of where and why certain trains shunt, a number of what we might now call trip workings, and so on; and conditional working is covered ('This Train runs through to Colne, on Mondays and Tuesdays, should there be any Cattle at Skipton'). A feature not generally perpetuated in British practice (and given up by the Midland at the start of 1854) is the use of two columns for each train, to present the arrival/passing times and the departure times for each location.[14] Only one table in these early Midland books limits itself to the citation of departure times only, and that is for the line between Abbot's Wood Junction and

---

10   *The Grand Junction, and the Liverpool and Manchester Railway Companion*, Birming-ham, J. Cornish 1837, reprinted by Frank Graham, Newcastle upon Tyne, 1969, p. 12.
11   In private ownership.
12   In the Roy F. Burrows Midland Collection.
13   NRM, uncatalogued.
14   Strangely, one of the companies that did retain this form of layout was the London Tilbury & Southend Railway. When in due course this company was taken over by the Midland, its working timetables were soon changed to the standard layout.

𝔐𝔦𝔡𝔩𝔞𝔫𝔡 𝔕𝔞𝔦𝔩𝔴𝔞𝔶.

GENERAL TIME-TABLES
FOR
SEPTEMBER, 1852.

CONTENTS.

		PAGE
Derby to Leeds	(*Pass. and Goods*)	2 3 4 5
Leeds to Derby	(*Pass. and Goods*)	6 7 8 9
Derby to Leeds } Sunday	(*Pass. and Goods*)	10 11
Leeds to Derby }		
Leeds to Bradford, Skipton & Colne	(*Pass. & Goods*)	12 13 14 15
Colne to Skipton, Bradford & Leeds	(*Pass. & Goods*)	16 17 18 19
Leeds to Bradford, Skipton & Colne } Sunday (*Pass.*)		20 21
Colne to Skipton, Bradford & Leeds }		
Skipton to Lancaster, &c.	(*Pass. and Goods*)	22 23 24
Morecambe and Lancaster	(*Pass. and Goods*)	25
Sheffield to Masbro' }	(*Pass. and Goods*)	26
Masbro' to Sheffield }		
Rotherham to Sheffield }	(*Pass.*)	27
Sheffield to Rotherham }		
Sheffield to Doncaster	(*Pass.*)	28
Doncaster to Sheffield	(*Pass.*)	29
Derby and Nottingham to Rugby	(*Pass. and Goods*)	30 31 32 33
Rugby to Derby and Nottingham	(*Pass. and Goods*)	34 35 36 37
Derby to Rugby } Sunday	(*Pass. and Goods*)	38 39
Rugby to Derby }		

		PAGE
Derby to Nottingham and Lincoln	(*Pass. and Goods*)	40 41
Lincoln to Nottingham and Derby	(*Pass. and Goods*)	42 43
Nottingham to Codnor Park & Mansfield	(*Pass. and Goods*)	44 45
Mansfield to Codnor Park & Nottingham	(*Pass. and Goods*)	46 47
Leicester to Peterboro'	(*Pass. and Goods*)	48 49
Peterboro' to Leicester	(*Pass. and Goods*)	50 51
Leicester to Burton	(*Pass. and Goods*)	52 53
Burton to Leicester	(*Pass. and Goods*)	54 55
Derby to Birmingham	(*Pass. and Goods*)	56 57
Birmingham to Derby	(*Pass. and Goods*)	58 59
Derby and Ambergate to Rowsley	(*Pass. and Goods*)	60 61
Rowsley to Ambergate and Derby	(*Pass. and Goods*)	62 63
Birmingham to Gloucester	(*Pass. and Goods*)	64 65
Gloucester to Birmingham	(*Pass. and Goods*)	66 67
Worcester to Abbot's Wood Junction	(*Pass. and Goods*)	68 69
Abbot's Wood Junction to Worcester	(*Pass. and Goods*)	70 71
Gloucester to Bristol	(*Pass. and Goods*)	72 73
Bristol to Gloucester	(*Pass. and Goods*)	74 75
Whitacre to Hampton }	(*Pass. and Goods*)	76
Hampton to Whitacre }		

PRINTED BY W. BEMROSE AND SON, DERBY.

Figure 5.1 A page from an early Midland Railway Working Time-Table

Worcester, where we are told that passenger trains take about 10 minutes and goods trains about 20 minutes for the run.

These *General Time-Tables* are very different from early copies of the public timetables. The earliest of these that appears to have survived is that for December 1848, entitled simply *Time Tables / from and after / December 1st, 1848*. This small eight-page booklet obviously includes the timetables only of the passenger trains. It also offers full information about fares (first, second, third, and fourth or government class, day tickets, half-price travel for children between the ages of three and twelve, travel in or with one's own private carriage, free luggage allowances, 'Parcels of unfinished Worsted Pieces', etc). It also notes, amongst other things, a number of connecting coach services, draws attention to the facility to send telegraph messages, and, curiously, makes special mention of the refreshment rooms at Nottingham and Peterborough stations. The tables show distances, and they also show off-line points for which the trains connect. The basic format used survived until the mid-1860s, when the first major change in presentation took place.[15]

---

15 For a brief discussion of the development of the Midland Railway's public timetables, see *Midland Railway 150*, edited and published by Roy F. Burrows, Ashton under Hill, Worcestershire, 1994, Chapter 1 (pp. 8–14) 'Midland Railway Public Timetables' (written by the editor).

The first issue of a *Working Time-Table* under that title, and thus the first example I have seen of the use of the term itself, appears to have been that produced by the Midland Railway for January 1854 (if we assume that November and December 1853 saw further issues of the company's *General Time-Tables*, with a major change of style reserved for the start of a new year), and the surviving run of Midland working books is then unbroken right through to the end of the company's independent life at Grouping, thus offering an invaluable insight into the growth of the company's traffic and the way in which it was handled. (Other companies for which long runs survive are the North Eastern Railway and the Great North of Scotland Railway, both starting in 1856, the London & South Western Railway, starting in 1860, the Great Western Railway, starting in 1862, and the North British Railway, starting in 1865.[16] There is reported to have been a London & North Western set that was still in existence until quite recent years, but this is now believed to have been destroyed.)

Unsurprisingly, the printers of the new Midland *Working Time-Tables* are still Bemrose (now 'Bemrose & Sons' rather than just 'Bemrose & Son'). In addition to the change of name there is a total change of format. The page size is 6 inches by 9¼ inches (as cropped – it is significantly larger in uncropped issues) and the format is now portrait-style instead of landscape-style. The covers are in blue paper, that at the front carrying the words 'Midland Railway', the company's arms, and the title 'Working Time-Table for January 1854'. The conventions of presentation are significantly changed: each train is now assigned only a single column, but in order to retain the ability to indicate separate arrival and departure times for larger locations, these are allowed two lines of the table (other locations have a single line to show departure or passing-time only); passing-times are distinguished from stopping-times by the use of a smaller size of type (duly explained on the contents page by a note stating that 'Trains do not STOP at the Stations Marked with SMALL Figures'). The information given is, however, generally just the same as in the *General Time-Tables*, although the changed style of presentation and the generally smaller typefaces mean that only 36 pages are required instead of the 80 of the end of the previous year. For the November 1854 issue the period of validity is altered to 'November, 1854, / and until further notice.' However, a December issue appears as usual. It is a number of years before months are regularly missed, with a 'Notice for...' appearing instead, to provide details of alterations to the previous month's issue.

Over the years the Midland's working timetable evolves and grows rather than undergoes any sudden changes. The first volume to include a large number of leaflets and supplementary notices is that for 1868, but the general layout does not change. Indeed, right up to 1917 the issues of the working timetable are still recognizably the direct descendants of that of January 1854, not merely because the size and format remain constant but also because there is no change in the basic style of presentation. The size of the issues grows considerably over the years, of course, both because of the expansion of the company's system and because of the increase in the number of trains required to convey the ever-growing traffic. Such changes in the presentation as there are tend to be more to do with the order in which the tables come and the way in which they are indexed than in the form of the tables themselves. Thus, for

---

16   For the English companies, in the PRO; for the Scottish companies, in the NAS.

example, from 1 November 1864 the index lists lines in alphabetical order instead of in the previous mainlines-then-branches order. In May 1871 the index is divided into geographical sections followed by a heading for general information instead of having a single alphabetical list including lines and other topics, and a year later, in March 1872 (there was no February issue), the tables themselves were divided up into nine lettered sections. This appears to have been done to avoid the need to issue the complete book to a good many staff who needed to use only a part of it. The fact that some staff received sections only is confirmed by the standard formula that appeared at the end of the Notices for those months for which there was no issue of a new timetable. Thus, for example, the Notice for April 1883 ends: 'One of these Notices must be supplied to each Servant of the Company who has been supplied with a Working Time Table or any of the Sections for February, 1882.'

The timetable certainly grew over time. By the early years of the new century it had reached very considerable proportions – the issue for July 1907, for example, ran to 733 pages; after a contents page, divided into 'North and Manchester Districts', 'West District', and 'East and South Districts', come the timetables, divided up into 14 lettered sections; then a last section, Section R, of 59 pages covers traffic working arrangements. And even a volume such as this was not all, for the company also produced – and had done for many years past – a separate publication detailing all the excursion trains it operated, both those put on for other people and those it marketed itself. The excursion timetable was issued at regular intervals throughout the year and built up to a significant number of pages in the course of a year.

During 1917 the decision was taken to separate the passenger working timetable from the goods working timetable, and the first issue under the new régime was the passenger timetable for 1 October 1917 until further notice.[17] In the first change of size since 1854 the format moves to foolscap, with buff-coloured light card covers and red lettering (and lacking the adornment of an engine). The content is divided into five sections, each with branches following the main line: Carlisle–Leeds, Leeds–Trent; Derby–London, Liverpool–Derby, and Derby–Bristol. A note indicates that the timetables for the Worcester–Swansea and London Tilbury & Southend sections are issued separately. The first separate issue of the goods timetables appears to be that for 1 January 1918 until further notice, and for this the old format is retained. There are 14 sections, and again the tables for Worcester–Swansea and the London Tilbury & Southend line are printed separately.

There was no change to the new form during the remainder of the independent life of the Midland company. It is not clear why such a significant change should have been introduced in 1917, shortly after the major cuts in passenger services made by all the British railways at the beginning of that year, for saving paper does not seem to be an entirely adequate explanation. Rather, it seems more likely that the change was a step in the Midland's reappraisal of its operating practices, even if it is surprising that anything of the sort should have occurred in the midst of the war.

---

17   The copy of the 1 October 1917 passenger working timetable in the Roy F. Burrows Midland Collection was once the property of R.E. Charlewood, and it is inscribed by him: 'First issue of pass[r] WTT in new form, apart from Freight WTT.'

As Jack Simmons has explained, the working timetables show us how the railway functioned as a system in a way that is not possible from the public books. As a broad picture, we can see how the passenger service and the goods service were fitted together on the same lines – both mainlines and branches – how faster and slower passenger services were timetabled over the mainlines so as to allow an attractive express passenger service to be offered, and how intensively the whole system was being used. A particularly interesting feature of the Midland books is that we can see how the company's massive coal trade was worked south to London, from the colliery branches all the way south through Toton and Wellingborough to the London-area yards and beyond. Much finer detail is also available. Thus, for example, because we have the passing-times at various crucial points along the line we can work out what was being demanded of the company's motive power. Aspects of day-to-day operation that could never be picked up from the public timetables can be traced, such as the practice of working round curves and then backing down into stations that was followed at both Derby and Leeds at different times, or the way in which through coaches between London and Birmingham were worked over the Wigston South Curve near Leicester. Details of through-coach workings are unfortunately not shown, and so the working books are in this respect unable to help us answer questions to which the public books too (prior to July 1898) afford no answer. One of the small details that has particularly intrigued the writer is a footnote to the 'Hendon, Hammersmith, and Kew' table that first appears in the April 1874 issue: 'On Sundays a Train leaves Brent for Old Kew Junction at 11.15 a.m., and Old Kew Junction for Brent, at 12.15 p.m. with French Butter from Southampton.' The footnote remains for many years to come. Could there really have been only one train a week of French butter?

The 'Supplementary Working Time-Tables' and the special notices to be found in the bound volumes of working timetables do not merely cover changes to the timetable, the provision of additional or special train services (for holidays, special events, and so on). They also include the notices of the opening of new lines with details of the initial train service on those lines (generally where such openings have not taken place on the first day of a month). For example, 'No. 4, Supplementary Working Time-Table, January, 1870' records 'Additional and alteration of Goods, Coal, & Empty Wagon Trains, between Wellingboro' & Nottingham, commencing on Monday, January 17th, 1870'. In fact, this is the bringing into use for traffic of the Lenton Curve in Nottingham, and this is how it is recorded. The 'Supplementary Working Time-Tables' also record the arrangements made for keeping traffic running under emergency conditions of various sorts. Thus we see the arrangements made for Midland trains to use the London & North Western line between Buxton and Manchester whilst the Midland line was blocked by a landslip at Chapel-en-le-Frith. (A London & North Western Railway notice records that: 'On Monday, June 24th, and until further Notice, the following Midland Passenger Trains will be worked over the L. & N. W. Line between Manchester and Buxton.' Details are given of the way in which the Midland trains were worked.) When there was a fire in the arches under the Leeds Joint Station in 1892, the Midland took London & North Western and North Eastern trains into its own station, which necessitated a number of alterations to its own traffic. When the Settle & Carlisle line was blocked by snow, Midland trains were accommodated on the rival London & North Western line over Shap. The way in which competing companies would actually come to one another's aid in times of difficulty is clearly shown in the working books, which allow us to flesh out the bare reports in contemporary newspapers. Another instance was the Worcester tunnel fall in 1892.

Appendices

The first working timetables showed just train-times and information about a limited number of the multitude of shunting moves required to allow the traffic to flow. However, very soon additional information needed by drivers and signalmen began to be included, and it is this additional information that is the origin of the later distinct publication, the Appendix to the Working Time-Table. The first Midland Railway working timetable to include significant extra information is that for May 1856, which introduces not only a one-page list of up and down lines but also a five-page table of whistle-signals to be given by drivers, both dated 26 April 1856. The list of lines is introduced with the note:

> As it is convenient in the issue of instructions respecting Signal and Working Arrangements, to distinguish the different Lines of Rails by the terms UP and DOWN, and very essential that no ambiguity or misunderstanding should exist on the subject, the following information is given for the guidance of the Company's Servants. / These Regulations to come into operation on the 1st May, 1856.

There is a longer introductory note for the list of signal-whistles:

> Some difficulty having arisen in carrying out Rule No. 64, in the Company's Book of Rules and Regulations, in consequence of misapprehension between Drivers and Pointsmen as to which Line should, in some cases, be considered the Main Line – and which the Branch, the following Regulations have been drawn up for the guidance of Drivers and Pointsmen in the matter, and the particular attention of Drivers is called to the great importance of a strict observance of them. / In order that the Pointsmen may be at all times able to distinguish on which Line a Train is coming, Drivers are instructed, on approaching a Junction, by any of the Lines branching therefrom, to give the same number of whistles they are instructed to do, when approaching the Junction in a contrary direction, and wishing to be turned on to that Line. / The Junction Pointsmen have instructions not to lower their Semaphores for approaching Trains, until the necessary signal has been given, by a Driver, and to report every Train that is stopped, in accordance with such instructions. / These Regulations to come into operation on the 1st May, 1856.

This list continues, with an almost-identical heading, growing in size, through to July 1871. August 1871 saw the issue of a Notice rather than a new working book, and then in September the table of whistle-signals was re-set on the page in portrait format and the first paragraph of the introduction was omitted. The mid-1860s saw the start of the addition of various other sections. First came a 'List of Tranship Vans in January 1865', followed in September of the same year by a 'Table of Gauges' – the structure-gauges not only of the Midland but of other lines as well. A year later, in September 1866, came a table of 'Loads for Engines'.

All this additional information was gathered together into a separate publication in the autumn of 1872. The September working timetable that year was a normal issue in the established pattern, but at the end of the October issue, immediately after the last timetable, we read:

The General Information and Special Instructions for the guidance of Station Masters, Drivers, Guards, and others, hitherto printed in the Working Timetable Book, will in future be printed separately as an Appendix to the Working Book, and will be issued from time to time as circumstances may render necessary. / Each Station Master, Driver, Guard, and other Servant of the Company supplied with Working Books, must take care always to have a copy of the latest issue of the Appendix in his possession.

The first appendix is dated 1 October 1872 – and the first supplement to an appendix followed only one month later, with a date of 1 November 1872.

The list of contents of the first issue forms an interesting record of the sort of information that was thought necessary for all operational staff. It is (with page-numbers) as follows:

Attaching and Detaching of Vehicles where Line is not level	18
Bedford and Northampton Line Special Instructions	22
Block Telegraph System of Train Signalling	24 and 25
Brake Down Vans	30
Distant Signals	20
Engine Head Lights	19
Explosive Signals, Belsize Tunnel	25
Guards to accompany Passenger Trains–Instructions to Station Masters and others as to the Number of	18
Loads for Engines	26 to 29
Single Lines–Train Staff Regulations for Working	23
Table of Gauges	34 and 35
Tariff Van Service	31 to 33
Up and Down Lines	21 and 22
Whistling Instructions, &c.	2 to 18

The Appendix grew rapidly: by No. 7 of 1 June 1875 it had already attained 56 pages and acquired sections dealing with

Breaks on Goods Trains between Rowsley and Woodley;
Caution Signal – Use of, by Platelayers;
Cheshire Lines;
Conveyance of Vehicles behind Rear Vans of Passenger Trains;
Marshalling of Goods Trains;
Sig. Posts – Opening & Closing of;
Signals – Fixing & repairing of, &c.,
Slip Carriages;
Speed of Trains – Maximum rate of;
Speed of Trains – Table to calculate;

Starting Passenger Trains from Stations;
Tail Lamps – Carrying of, on Main Line.

By April 1905 and No. 22 Appendix we find a volume of 584 pages.[18]

The Midland may well have been the earliest English company to move to the issue of a separate appendix – the North Eastern followed suit on 1 May 1873, but it was five years before the London & North Western began, in April 1878 – but at least one of the Scottish companies made the change earlier, for the third issue of the Appendix for the North British Railway is dated 21 June 1872. The Caledonian Railway reached the tenth issue of its Appendix in May 1877, but as we do not know the dates of the nine earlier issues we cannot tell whether its first appendix antedated that of the Midland or not.[19] However, certainly in the case of the Midland, and perhaps for other companies too, the significance of the first issue of an appendix should not be exaggerated, as all the information presented in the separate document had been printed month by month in the working timetables for many years past. The change was merely the production of a separate volume. However, once it had appeared, this volume soon began to expand greatly. Information about signal boxes was brought in, first in the form of those boxes signalling the running lines that were not always open and then as a list of all boxes signalling the running lines with the distances between them; block telegraph signal posts and speaking telegraph posts appear in the No. 10 Appendix, dated 16 August 1877 (and the distances appear the following year). Telephones first make their appearance in the No. 17 Appendix of 17 December 1888.

The company undertook a major change of layout in the appendix in the early years of the twentieth century, at the time when it was introducing a modern control system throughout its operations. The last appendix in the old style was No. 24, of June 1911, although there is one major change here. The issue is divided into lettered sections, from A to N. After ten pages of index and so on, the body of Section A occupies 560 pages. This is the appendix for the Midland system itself. Sections B to N are the 'working over' sections and occupy over 200 pages more. The big change introduced in No. 25 Appendix is the use of a part-graphical layout for signal box and running-line information. The impact on size is considerable: Section A is reduced from 560 pages to just 274 pages (Sections B to N of the previous issue remain in force and are not reprinted). This style of layout was taken up by the London Midland & Scottish Railway for all its constituent parts after Grouping, the first set of LMS Appendices being issued to be effective from 1 January 1931. After nationalization the LMS pattern was extended to the whole of British Railways, the first full set of Appendices for the nationalised railway bearing the date of 1960. Today's Appendices are essentially based on those produced by the Midland Railway in the last years of its independent existence.

---

18   Copies of all the Midland appendices survive, since they have been bound into the volumes of working timetables. One issue, No. 7 of 1 June 1875, was at the back of the January to June 1875 volume in the PRO collection and has been lost, but a copy survives in the Roy F. Burrows Midland Collection.

19   Information on dates of first appearance for companies other than the Midland Railway from the *List of General & Sectional Appendices of the British, Irish, Australian & other Railways*, Signalling Record Society, Research Note No. 8, second edition 1977.

89

Drivers	Must whistle	If going
Approaching Kentish Town Junction on Up Fast Line, on Up Goods Line, and on Up Main Line from Highgate Road	Twice Three times Four times Five short	to St. Pancras Passenger Station „ St. Pancras Goods Yard or Somers Town „ the Metropolitan Line „ Loco. or Carriage Sidings
„ Kentish Town Junc. on Down Fast and Down Goods Lines	Twice Three times Five short	towards Hendon „ Highgate Road into Loco. Sidings

Over **Kentish Town Junction, to or from Tottenham and Hampstead Line—Fifteen miles an hour.**

Through **Slip-points giving communication between Tottenham and Hampstead Line and Up Goods Line, or through Slip-points giving communication between Up Goods Line and Up Passenger Line, at Kentish Town Junction—Ten miles an hour.**

Between **Up and Down Fast and Up and Down Slow Lines at Islip Street Junction—Fifteen miles an hour. From Down Goods to Down Fast Line at Islip Street Junction—Ten miles an hour.**

Approaching Islip Street Junction on Up Slow Line	Three times Four times	on to Up Fast Line forward on Up Slow Line
„ Islip Street Junction on Down Fast Line	Twice Three times Four times	forward on Down Fast Line on to Down Slow Line If requiring to cross to Goods Lines
„ Islip Street Junction on Down Slow Line	Four times	
„ Camden Road Station and St. Paul's Road Passenger Junction on Up Fast Line......	Twice Three times	to St. Pancras Passenger Station on the Metropolitan Line
„ Camden Road Station and St. Paul's Road Passenger Junction on Up Slow Line	Four times	

Drivers approaching St. Paul's Road Goods Junction on the Up Goods Line must sound their engine whistles on emerging from the Camden Tunnel to warn Shunters and others of their approach.

Over **St. Paul's Road Passenger Junction, to or from Metropolitan Line—Twenty miles an hour. Drivers working trains to places south of the Thames composed of 17 to 20 minerals or goods must reduce speed to Six miles an hour at St. Paul's Road Passenger Junction, and enter the King's Cross Tunnel at this speed.   Guards must assist the Drivers with their hand brakes to control the train down the incline.**

Passing St. Paul's Road Goods Junction on Down Goods Line	Twice	on to Down Fast Line at Islip Street
„ St. Pancras Tunnel Box and approaching St. Paul's Road Passenger Jc. on Down Main Line from King's Cross with Light Engines and Trains other than Passenger Trains.....	Three times Four times Five times	into St. Pancras Goods Yard North of Kentish Town, or requiring to enter Loco. Sidings, Kentish Town to Cattle Docks Sidings
Approaching Dock Junction, on Up Goods Line	Once Twice Three times	to Somers Town „ St. Pancras Passenger Station into Cambridge Street Coal Bays
„ Dock Junction on Down Goods Line	Four short	into St. Pancras Goods Yard on Curve alongside North London Incline

Dock Junction—From Down Goods Line to St. Pancras Goods Yard, on Curve alongside North London Incline—Six miles an hour, and no Engine of longer wheel base than 15 feet must work round the Curve.

From West Departure Line to Down Passenger Line at Cambridge Street—Fifteen miles an hour.

When running into or out of Somers Town on any Line and through the Crossover Road between the Up and Down Goods Lines on the North side of St. Pancras Junction Box—Six miles an hour.

Between West Departure Line and East Departure Line, through Crossover Road on North side of St. Pancras Road Bridge—Five miles an hour.

From Nos. 6 and 7 Platform Lines, St. Pancras Passenger Station, on Up Passenger Line to Down Passenger Line, through the Slip Points—Ten miles an hour.

Over Facing Crossover Road and into St. Pancras Passenger Station—Ten miles an hour.

Over Curve near Midland Junction Signal Box in King's Cross Tunnel—Ten miles an hour.

Approaching Highgate Road Junction	Twice Three times	towards Kentish Town „ Gospel Oak

Over Highgate Road Junction, to or from Kentish Town. Also while passing over the curve between Highgate Road Junction and the bridge carrying the Highgate Incline over the main line—Fifteen miles an hour.

Figure 5.2   A page from a Midland Railway Sectional Appendix, giving information to drivers about the approach to London

Operating Notices

Not only is a railway's permanent way very far from being permanent, thus meaning that there is always a requirement for maintenance works, but also through much of the latter half of the nineteenth century and on into the twentieth century the infrastructure was steadily being extended and improved. It has always been essential for drivers, signalmen, track-workers, and others to be kept fully informed about what is happening out on the line – especially where there are safety implications such as a need for extra caution or reduced speeds.

At first, the need seems to have been met by the posting of notices at depôts for drivers to read. Colonel Yolland, reporting for the Board of Trade on an accident at Kettering Junction on 21 November 1874 after an alteration to the junction,[20] explains:

> This alteration in the position of this junction came practically into operation for the working of the traffic on the 1st November last, and the usual notice, herewith enclosed (marked A.), was posted at the corner of the locomotive office at Toten [*sic*], to call the attention of the drivers to the change in the position of the junction, in accordance with the company's rules and regulations.

In finding that the accident was caused by a combination of the train-crew failing to keep a proper look-out while approaching the new junction and 'the great negligence of the driver of the up mineral train, in having failed to make himself acquainted with the alteration which had been made in the position of the junction, as posted up at the locomotive shops at Toten', Yolland comments:

> I understand that the practice followed by the Midland Railway Company in posting up notices similar to A., describing the alteration in the position of the junction, is in general use among railway companies, but, in my opinion, it is not sufficient to provide for the safety of the public. It was stated that there might be as many notices as 200 posted up at the same time, the notices being placed in rotation, and the *latest notices* are put on a particular board; but on the day on which the collision occurred Notice A. was not among the latest notices. It appears to me that every driver when he first has to travel over a portion of line on which such an important alteration has been made, as in this case, in the position of the junction and signals to protect it, should have a copy of such notice given to him, and this I find was actually done, as far as the driver of the Cambridge train was concerned, although as he was constantly employed on this branch it might be assumed that he was fully acquainted with the change which had been made, while the other driver had not passed the spot for rather more than five weeks before the collision took place.

Notice A. itself, reproduced at the end of the report, is entitled: 'Notice to engine-drivers, guards, pointsmen, plate-layers, and others, respecting the opening of the New Junction near Kettering'.

---

20 *Reports of the Inspecting Officers of the Railway Department to the Board of Trade upon Certain Accidents which have occurred on railways during the Month of November 1874*, HMSO London 1875 [C.–1146.] pp. 222–4.

MIDLAND RAILWAY.

For the Information of the Company's Servants only.

OPENING OF
SETTLE AND CARLISLE LINE.

On and from MONDAY, AUGUST 2nd, 1875,

GOODS TRAINS

Will run over the NEW LINE between

SETTLE JUNCTION
AND
CARLISLE,

AT THE TIMES SHEWN HEREIN.

Trains marked "M" do not run on Mondays.

Trains do not stop at the Stations marked with small figures, unless marked "W" where they are liable to stop for Water only.

The following will be the loads for Engines :—

	Goods.	Mineral.	Empties.	
Settle Junction to Carlisle	30	24	40 Inside Frame Goods Engines.	
Carlisle to Settle Junction	25	20	33 Outside	,,	,,

JAMES ALLPORT, General-Manager.

Derby, July, 1875.

Figure 5.3   A Midland Railway notice to staff

It is interesting to observe that 1874 seems to have been a transitional time on the Midland even if it was still general practice not to issue individual notices on other railways, given that we are told that the Toton driver did not receive his own copy whilst the Cambridge man did. We do not know when the Midland began to issue operating notices to individual members of staff on a general basis, but it appears to have been within five years of Yolland's report on Kettering. The title has changed but little. The 'Notice to Enginedrivers, Guards, Signalmen, Platelayers, and others respecting Signal Alterations and other Arrangements' appears as a single sheet in the earliest copies known. The size of the sheet used varied according to the amount of material to be included – paper was not wasted. The last of the single-sheet notices covering a week's activities was that issued on Monday, 11 May 1885, which was 16½ inches by 24¼ inches in size. The notice for the following week, dated 18 May 1885, appeared as a four-page booklet 6⅜ inches by 10 inches in size. Information was presented in the fashion already long established. General items came first, followed by items about specific locations ordered from the north to the south. At the end came a list of the locations where routine re-laying operations were going on.

The weekly notices were at first issued on a Monday, and they covered work to be carried out over the next seven days. With the change to a booklet format came a change to coverage of two weeks' work – so that a job would normally be noted in two successive issues. The last issue to appear on a Monday was that of 20 February 1888. The next issue appeared on Wednesday, 29 February 1888, and from then onwards Wednesday was the normal day of issue. There was a slight hiccough in 1919: the notice for the week ending 18 October of that year records that there had been no printed traffic notice for the week ending 11 October 1919 as the men had been on strike. And the notice for the week ending 4 October is endorsed by hand with the words: 'Strike cancels work'! There was a small change of format in 1920, but no otherwise major changes before Grouping. The issue of these operating notices continued under the LMS, although they had become fortnightly rather than weekly by the time of the Second World War and they had also begun to include traffic information. Weekly issue was resumed under British Railways and has continued ever since.

The significance of the operating notices to the student lies in the detail they provide. Although information on the development of the system can be found in the board and committee minutes, in the Extra Works Books, in the Board of Trade reports, and so on, the precise detail of what was being done and exactly how it affected the layout is to be found spelled out in full and clear detail only in the operating notices. This is not surprising, for the safe working of the system depended on the staff on the ground being clearly informed and responding accordingly. It is in these notices that we can read a full description of exactly what new and changed signalling installations permitted, which fleshes out substantially the bare detail that we find in the descriptions and plans in the Board of Trade reports. It is the operating notices which also provide information about track renewals, bridge reconstructions, subsidence in mining areas, changes at stations (in so far as these affected the working of the railway), and other such routine matters which are vital to the running of a smooth business but too trivial to require mention in the higher-level record.

Chapter 6

Financing the Bagdadbahn:
Barings, the City, and the Foreign Office, 1902–3[1]

P.L. Cottrell

This chapter seeks to celebrate Jack Simmons's outstanding contributions to the development of both imperial and railway history.[2] At the close of the nineteenth century, the railway's strategic importance was being more widely appreciated in terms of its economic and imperial dimensions. As the industrial powers increasingly jostled one another on the world's markets, equations were being broadly drawn up amongst diplomats to discern the interrelations between the wider political influence of individual states and their respective foreign dealings and holdings, including railway shares. It was also being recognized that such interactions could have the greatest importance where local power structures were waning, as in the empires of China, Persia, and Turkey. Furthermore, this increasing identification of the business foundations to political sway overseas spilled out of chancelleries into drawing rooms and public houses.

   In the jingoism induced by the Boer War, Deutsche Bank's agreement in 1899 to construct the Baghdad railway led to a further reappraisal of the impact of the nineteenth century's most powerful technological icon. This line would challenge long prevailing French influence within the eastern Mediterranean, raise strategic political objections comparable to those provoked by the Suez Canal proposal at the

1   I am very grateful to Dr John Orbell and Jane Waller, archivists at ING Barings, for their very considerable help in the underlying researches for this piece, and to a Faculty of Social Sciences, University of Leicester, research grant. I am also indebted to my colleague, Dr B. Attard, for his perceptive comments on an earlier draft.
2   For different facets of Simmons's work as an imperial scholar, see (with M.F. Perham), *African discovery: an anthology of exploration* (London, 1942, 1948, and 1957); 'Rajah Brooke at Burrator', *West Country Magazine*, 1 (1947); 'European expansion in Africa and south east Asia', in M. Beloff (ed.), *History: mankind and his story* (London, 1948); 'The proconsuls', in *Ideas and beliefs of the Victorians* (London, 1948); *From Empire to Commonwealth; principles of British imperial government* (London, 1949); *Parish and Empire: sketches and studies* (London, 1952); and 'Transition in West Africa', in Sir D. Shiels, *The British Commonwealth* (London, 1952).

mid-century and, by partially realizing long-standing English aspirations for an overland route to India, threaten the safety of Britannia's oriental jewel.

The Bagdadbahn's significance within the international tensions that led to the First World War has resulted in many contributions to diplomatic history.[3] Although its role within Deutsche Bank's development has been fully acknowledged by that institution's commissioned historians,[4] the line's economic history has only recently been appraised.[5] Boris Barth's analysis stresses that Deutsche Bank's attempts to realize the project as a pan-European venture were successful through maintaining French participation – principally that of Banque impériale ottomane[6] – until 15 February 1914. This chapter seeks to review the Berlin bank's efforts during 1902–3 through its new head, Arthur von Gwinner,[7] to involve the support of both the British government and the City of London for the construction of the railway's first section. It is an aspect that has only been considered with regard to the feverish adverse press campaign employed by antagonistic cabinet members headed by Joseph Chamberlain that brought about abrupt British withdrawal from the scheme in April 1903.[8]

---

3   See E.M. Earle, *Turkey, the Great Powers and the Bagdad Railway* (New York, 1923); J.B. Wolf, 'The diplomatic history of the Bagdad railroad', *University of Missouri Studies*, 11 (1936); M. Chapman, 'Great Britain and the Baghdad Railway' (Northampton, MA, 1948); and R. Kumar, 'The records of the Government of India on the Berlin–Bagdad railway question', *Historical Journal*, 5 (1962). The most recent is G.G. Schöllgen, *Imperialismus und Gleichgewicht. Deutschland, England und die orientalishce Frage 1871–1914* (Munich, 1984).

4   F. Seidenzahl, *100 Jahre Deutsche Bank 1870–1970* (Frankfurt am Main, 1970); and L. Gall, 'The Deutsche Bank from its founding to the Great War 1870–1914', in L. Gall et al., *The Deutsche Bank 1870–1995* (London, 1995).

5   B. Barth, *Die deutsche Hochfinanz und die Imperialismen. Banken und Außenpolitik vor 1914* (Stuttgart, 1995); and idem, 'The financial history of the Anatolian and Baghdad railways, 1889–1914', *Financial History Review*, 5 (1998). See also H. Mejcher, 'Die Bagdadbahn als Instrument deutschen wirtschaftlichen Einflusses im Osmanischen Reich', *Geschichte und Gesellschaft*, 1 (1975); and W.J. Mommsen, 'Europäischer Finanzimperialismus vor 1914. Ein Beitrag zu einer pluralistischen Theorie des Imperialismus', *Historische Zeitschrift*, 224 (1977).

6   For the development of the Imperial Ottoman Bank (although largely from the perspective of its Paris committee), see A. Autherman, *La Banque Impériale Ottomane* (Paris, 1996); whereas a 'domestic' perspective is given by E. Eldem, *A history of the Ottoman Bank* (Istanbul, 1999). Other treatments include: C. Clay, 'The Imperial Ottoman Bank in the later nineteenth century: a multinational "National Bank"?', in G. Jones (ed.), *Banks as Multinationals* (London, 1990); and idem, 'The origins of modern banking in the Levant: the branch network of the Imperial Ottoman Bank, 1890–1914', *International Journal of Middle East Studies*, 26 (1994).

7   In addition to Seidenzahl, *100 Jahre Deutsche Bank*; and Gall, 'Deutsche Bank from its founding'; see also M. Pohl (ed.), *Arthur von Gwinner. Lebenserinnerung* (Frankfurt am Main, 1975); and J. Lodermann and M. Pohl, *Die Bagdadbahn. Geschichte und Gegenwart einer berühmten Eisenbahn* (Mainz, 1988)

8   R. M. Francis, 'The British withdrawal from the Baghdad railway project in April 1903', *Historical Journal*, 16 (1973).

The present focus is upon the attitudes of the London bankers and financiers involved – Sir Ernest Cassel, Sir Charles Dawkins of J.S. Morgan & Company,[9] and Lord Revelstoke, senior partner in Barings – and their relationships with Lord Lansdowne, the Foreign Secretary, in this ultimately abortive affair. In recently published scholarly business histories of both Barings and Morgans, their respective authors had, understandably, little space to consider in any depth an abortive venture and its surrounding circumstances.[10]

This consideration continues by reviewing the interrelations between the British state and the City. The Baghdad railway project's development as an extension of the Anatolian Railway Company is outlined in the following section. Next is an explanation of how the connection between Lansdowne and Revelstoke – between British foreign policy formation and British financial power – was established in 1902–3 for putative British financing of the Bagdadbahn. The wider background to the embryonic Anglo-German financial alliance of spring 1903 is provided in the next section and points to French diplomacy sharing the aim of internationalizing the Bagdadbahn. The attempt to concert an Anglo-German compact to finance the railway, albeit on the basis of foreign-policy goals, is reviewed and analysed. The various reasons that produced an early rupture are considered, and the retreat of British capital from involvement in the Bagdadbahn is charted. The resulting consequences and implications constitute the final section.

The Interrelations between the British State and the City

At the dawn of the twentieth century, there were prospects for, and reasons why, Britain and Germany might cooperate in the Ottoman Empire. Some City bankers, such as the Rothschilds, continued to fear Russian pressure on the Straits and regarded it a greater menace with the decade-old alliance between Paris and St Petersburg. While freely admitting that he was no Germanophile, 'Natty' Rothschild saw in the protection of British interests in Egypt a sound argument for concert with Berlin.[11] The Boer War, together with the inception of the naval race, had heightened Anglo-German antipathy yet the signing of the Treaty of Vereeniging on 12 May 1902 and Arthur J. Balfour assuming the premiership two months later offered possibilities for a new chapter in Anglo-German relations. If not in terms of shared goals, there

9   Previously a civil servant, Dawkins joined Morgans in 1900 at the instigation of J.P. Morgan, but this proved to be an unsuccesful relationship. Dawkins died in 1909 at the early age of 49. See Y. Cassis, *City bankers, 1890–1914* (Cambridge, 1994), pp. 39–40, 117–18, 294; and V. Carosso, *The Morgans, private international bankers, 1854–1913* (Cambridge, MA, 1987), pp. 443–6.

10   P. Ziegler, *The sixth great power. Barings 1762–1929* (London: 1988), pp. 316–17; and K. Burk, *Morgan Grenfell 1838–1988. The biography of a merchant bank* (Oxford, 1989), p. 124. See also P. Thane, 'Financiers and the British State: The case of Sir Ernest Cassel', *Business History*, 28 (1986), pp. 92–3.

11   N. Ferguson, *The world's banker. The history of the House of Rothschild* (London, 1998), p. 917.

were opportunities, like the Bagdadbahn venture, for confrontation to be modulated by equilibrating cooperation – as through the cosmopolitan world of 'High Finance'.

From the mid-1890s the Foreign Office had increasingly turned to financial institutions for supporting specific British political interests abroad – Hongkong & Shanghai Banking Corporation in China, and Imperial Bank of Persia on the Gulf's northern shore.[12] When Lansdowne was faced with Deutsche Bank's Baghdad railway project, he deployed a somewhat different tactic – to engage the City's mighty resources for obtaining this railway's internationalization to eliminate it as a source of friction in great power relations.

Along with this major objective, the Foreign Secretary was also concerned about the nature of the City's possible involvement since Dawkins's premier banking house, J.S. Morgan & Company, was an Anglo-American enterprise and a participant during 1902–3 in the creation by J. Pierpoint Morgan of International Mercantile Marine Company, New Jersey. This was bringing together American and British trans-Atlantic shipping lines. When it seemed likely that even Cunard might be absorbed within this burgeoning American leviathan, very unusually the British government intervened with funds to safeguard both this leading company's British character and British overseas traders' interests.[13] Lansdowne was equally determined that the Bagdadbahn was not to be another conquest in the 'American' and 'German' invasions of Britannia's realm. Consequently, he wished to ensure that any British financial participation was not managed by either an American or German house within the City, notwithstanding their prestige.[14] He turned to Barings, with which he had close personal relations through being a member of this important merchant bank's inner circle of personal clients.[15]

In some respects fortunate for Lansdowne's approach, J.S. Morgan & Company were associated in Deutsche Bank's Bagdadbahn project with Cassel. He, like Revelstoke, had become a close associate of the British state, being, by 1903, an informal counsellor to the Treasury, with its Joint Permanent Secretary introducing him to a succession of Chancellors. Nonetheless, he was not a typical 'City Man', having no banking house of his own. Rather, he was a financier, working his business through loose relationships – primarily with Rothschilds until the late 1890s. His acceptance within English high society was partly due to his reputation for probity, although Cassel with some of his partners in the many syndicates that he formed sailed very close to the wind, if not to the extent of landing in court.

Cassel's involvement in state policy for the Baghdad railway was not his first connection with the Foreign Office. Following the failures of Hongkong & Shanghai

---

12 See D. Mclean, 'Finance and "informal Empire" before the First World War', *Economic History Review*, 2nd series, 29 (1976).

13 Sir John Clapham, *An economic history of modern Britain*, III, *Machines and national rivalries* (1887–1914), (Cambridge, 1938, rep. 1963), pp. 44–5, 277; and Burk, *Morgan Grenfell*, pp. 59, 74, 104–11, 121, 304. For context, see D. H. Aldcroft, 'The mercantile marine', in idem (ed.), *The development of British industry and foreign competition 1875–1914* (London, 1968).

14 ING Barings, London, The Baring Archive (henceforth TBA); HC 15.1.6, Revelstoke to Cromer ('Uncle Mina'), 4 March 1903.

15 See Ziegler, *The sixth great power*, pp. 294, 299, 305–6, 345, 354.

Banking Corporation and, then, Rothschilds to negotiate with continental European banks an officially inspired loan for China during the immediate aftermath of the Japanese war, Cassel managed to sustain British prestige to a degree by raising a small, £1 million six per cent loan in 1895.[16] Consequently, he was far from unknown to Whitehall during 1902 and 1903. However, as will be seen, Cassel played little further part in the direct negotiations with Lansdowne, his place being almost immediately taken by Revelstoke. Instead, he maintained his peripatetic lifestyle, calling only occasionally on his houses in Newmarket and Park Lane while keeping in touch by letter or telegram, unless his presence was particularly required at a meeting in London or Paris. This social world was also well known to Revelstoke since he was also a member of the 'Smart Set' around Edward VII.[17]

Extending the Anatolian Railway

The Baghdad railway project was the extension of an Anatolian line that the Ottoman authorities had envisaged, primarily for strategic reasons, since the early 1870s. A decade later, they pressed again for an Asiatic railway and sought either British or German financial and diplomatic support. Deutsche Bank's financing of the Anatolian from 1889 came about only after the considerable risks involved had been tempered by, first, Turkish officials agreeing to hypothecate tax revenues for guaranteeing returns on its capital and, second, gaining German diplomatic support.

The Berlin bank's Anatolian railway concession had a basis in the Haida Pascha–Izmid line acquired from a British consortium. Through a separate construction company employing Count Vitali, a French contractor, its first section to Ankara was completed on 31 December 1893. This was ultimately financed through the Bank für orientalische Eisenbahnen, Switzerland, established by Deutsche Bank in conjunction with Crédit suisse and Wiener Bank-Verein. The financial pool was enlarged by Deutsche Bank's informal relationship with Gebrüder Bethmann, which provided business and personal links to the Paris market, above all to Banque impériale ottomane. The resulting approach for Ottoman Asiatic railway development involved three components: a Turkish operating company; a construction company; and a specific financing bank to augment the resources that Deutsche Bank could otherwise obtain. While a German enterprise, the railway's realization involved pan-European connections to secure both funds and, initially, technological 'know how' for its construction.

Even before the Anatolian's line reached Ankara, the Ottoman authorities sought its continuance to Baghdad or the Gulf. However, managements of both the operating company and Deutsche Bank were more concerned with generating traffic. They attempted to gain this through a circuitous branch from Eskisehir to Konya that would also facilitate the substitution of German for English goods on the region's markets,

---

16 Thane, 'Financiers and the British State', pp. 80–2, 83–5. See also Cassis, *City bankers*, pp. 37, 38–9, 42, 72, 177, 181, 257, 293, 296, 304; and D. McLean, 'The Foreign Office and the first Chinese Indemnity Loan', *Historical Journal*, 16 (1973).

17 Cassis, *City bankers*, pp. 256–7.

especially those of the Konya plateau. The 440 km (264 miles) Konya 'branch' was completed during 1896 by Holzmann AG, which had acquired the necessary engineering expertise by previously having partnered Vitali.

Mounting instability in the Balkans led Deutsche Bank to consider liquidating all its existing Ottoman railway interests.[18] However, the bank became entrapped in yet further Asiatic Turkish railway construction by pressures arising from German diplomacy in which Wilhelm II had an increasing hand, the positive reaction of German public opinion, and the Ottoman authorities' continuing attempts to have a line traversing all the Empire's provinces. Increasingly propelled into the Baghdad venture, the beginnings of Deutsche Bank's commitment came at the close of 1899 with its signing of an outline concession.[19] The project might have commenced then through a partnership with French finance but this was disrupted, as an Anglo-German financial alliance was to be in 1903, by a nationalistic press, on this occasion Gallic.[20]

Consequently, one persistent basic problem was amassing the necessary funds for building the highly speculative Bagdadbahn. Deutsche Bank tried to persuade the German government that a clear marker of official approval was required, indicating in autumn 1900 that Seehandlung's involvement would constitute a suitable form. The issue crystallized in January 1902, when Deutsche Bank obtained a more concrete concession for the Bagdadbahn. A number of solutions were open to von Gwinner, who now headed the bank, all involving the pan-European approach that had realized the Anatolian concession. The rehabilitation of Turkish international creditworthiness by the consolidation of Ottoman external debt would free sufficient securities to provide backing for issues of Baghdad railway bonds. This measure had often been considered and was now being taken forward again by Banque impériale ottomane. Otherwise, the Turkish tariff regime could be heightened to give the Ottoman Empire greater customs revenues with which to validate the railway's bonds. However, this would require the major European powers' support, particularly Britain's, although an approach to the British government could be paralleled by one to the City, whose previous participation in the Anatolian railway had been rendered stillborn by the 1890 Baring crisis. Von Gwinner had already established another basis for involving London in financing the railway's construction through an agreement made in June 1901 with J.S. Morgan & Company and Cassel.[21] Also, at some early point, the Berlin banker met Lansdowne, an encounter arranged by Lord Mount Stephens,[22] when von Gwinner and the Foreign Secretary went as far as considering the Baghdad line's extension through Persia to India.[23]

---

18   These also included a very sizeable Balkan network – the Orientbahn. See Gall, 'The Deutsche Bank from its founding to the Great War', pp. 62–4.

19   Seidenzahl, *100 Jahre Deutsche Bank*, pp. 141f.

20   R. E. Cameron, *France and the economic development of Europe 1800–1914* (Princeton NJ, 1961), p. 325, fn 131.

21   Burk, *Morgan Grenfell*, p. 124.

22   Mount Stephens was also a member of Barings's inner circle of personal clients. See Ziegler, *The sixth great power*, p. 294.

23   TBA; Bagdad Railway, I, 200248, von Gwinner to Mount Stephens, 26 May 1903, f. 309.

Lansdowne and Revelstoke: Foreign Policy and Financial Power

As Deutsche Bank developed the Bagdadbahn proposal, Lansdowne became convinced from winter 1901–2 that Britain could no longer maintain a non-committal political stance. Rather, a solution was required that accepted British dominance in the Gulf whilst not heightening Anglo-Russian tensions, especially in northern Persia.[24] The first explicit official British diplomatic involvement occurred during March 1902, when the project was briefly mentioned, although in strident terms, during conversations with the French and German ambassadors. These interchanges arose from Lansdowne having been contacted by Lord Hillingdon of Glyn, Mills and Banque impériale ottomane's London Committee to relay the likelihood of the City only having a small financial participation in the affair. In turn, his approaches to the Foreign Office stemmed from von Gwinner's feelers to the Paris capital market, following the grant of a more precise railway concession. Indeed, contacts for financing the Bagdadbahn's construction were then being made in all directions within the City and Whitehall, and Barings received a number 'from several quarters', including Lansdowne during either April or May 1902.

The Foreign Secretary sought to discover whether Barings would participate as the London lead house, thereby diminishing the role of Deutsche Bank's current City partners – J.S. Morgan & Company. With Lansdowne having taken the initiative, Revelstoke took the opportunity to discover, in turn, if the government would go as far as providing an official guarantee for the railway company's bonds. He was seeking a comparable stance from the Cabinet to that taken for preventing Cunard's absorption into J. Pierpoint Morgan's International Maritime Marine. Since the Foreign Secretary was not prepared to take this extraordinary step,[25] their conversations dwelt instead on sounding out Rothschilds over their handling of the matter in London. Rothschilds had issued Ottoman conversion loans in 1891 and 1894 and had advised Lansdowne, following the example of the Suez Canal, that the British government should take some shares in the Baghdad railway when its 1899 outline concession was granted. However, in spring 1902 this leading City merchant bank proved unresponsive to Revelstoke's approaches made on the Foreign Secretary's behalf.[26]

The need for the British government to establish a firm policy for the Bagdadbahn re-emerged in autumn 1902 when Deutsche Bank reached agreement with French financial groups. They were to participate on a 40/30 per cent basis, leaving Cassel and Dawkins, following their June 1901 agreement, to supply only 20 per cent of the funds required. With Lansdowne's backing, the City bankers successfully protested at London's minor role to gain parity – each major European power contributing

---

24 Francis, 'British withdrawal', p. 169.
25 During the nineteenth century the British government seldom gave its guarantee to a foreign issue being made on the London market. The two major exceptions were the Greek 'liberation' loan of 1824 and the 1855 Anglo-French Turkish loan that arose from the Crimean War.
26 TBA; HC 15.1.6, Revelstoke to Cromer, 4 March 1903; and Ferguson, *The world's banker*, pp. 917–18.

30 per cent of the railway's funding.[27] Von Gwinner's 1902 arrangements with French financial interests were further modified on 18 February 1903 while the Berlin banker also negotiated what proved to be the conclusive concession for the railway's construction, signed on 5 March 1903. Through the media of four per cent bonds it provided the project with state subventions of £8 500 per kilometre per annum for building and equipping the line and £180 per kilometre per annum for its operation.[28]

At first glance, von Gwinner's Parisian financial compact of February 1903 would have satisfied Lansdowne's subsequent prime objective of securing the line's internationalization. It envisaged that three-quarters of the Baghdad railway's equity would be contributed in equal parts from British, French and German sources. However, while internationalization was reflected in the proposed composition of the company's board through each major European power having eight representative directors, Deutsche Bank retained primacy in its management. Besides eight German directors, there would be one Austrian and one Swiss, both nominated by the Berlin bank, together with three from the Anatolian Railway Company, which was to take up ten per cent of the Baghdad railway's shares. The Anatolian was controlled by Deutsche Bank, and, in garnering finance for the Baghdad line's construction, its intentions regarding its future firm hand over the new line were clear.

This was understood within the Foreign Office, with Lansdowne consequently seeking from March 1903 'an international line' from the Sea of Marmara to the Gulf. Accordingly, British diplomacy required, first, the Anatolian railway's internationalization through its control passing to the Baghdad Company. Second, the envisaged national balance on the Baghdad's board was to be unaffected by any subsequent share transfers. Within the second strand of objectives, there was even later official opposition to von Gwinner's suggestion of Hillingdon as a British director. It arose because of his position within Banque impériale ottomane's tri-national management structure and so, consequently and particularly, his close relationships with the 'French group' likely to be involved.[29]

French diplomacy shared the goal of internationalizing the Bagdadbahn and therefore was 'ready and extremely anxious for English co-operation'. Like Lansdowne, its aim was equality in the line's control. Consequently, French diplomats questioned the nature of the 'so-called Swiss participation', maintaining that 'it is evidently necessary... to have a perfectly clear understanding as to the real distribution of each [share] portion'. This Gallic stance also included objections to the proposed ten per cent Turkish interest – that is, that of the Anatolian Company.[30]

However, the revised financial agreement with Banque impériale ottomane made by von Gwinner had no firm basis, with Cassel and Dawkins therefore regarding the 'business to be still very indefinite'. This was partly due to it having been signed with

27   Francis, 'British withdrawal', p. 170.
28   See A. von Gwinner, 'The Baghdad railway and the question of British co-operation', *Nineteenth Century* (June 1909), pp. 1088–94; cuttings copy in TBA; 203146.
29   TBA; 200248, draft minutes of meeting (Revelstoke, Cassel, Dawkins), 12 March 1903, f. 74; and Dawkins to von Gwinner, draft, 12 March 1903, f. 76.
30   TBA; 200248, H. Babington Smith, Constantinople, to Dawkins, 9 March 1903, f. 78; and Babington Smith, Constantinople, to Dawkins, 13 March 1903, f. 81.

reservations. The French parties required their own government's approval, which they thought would be dependent upon Russian participation.[31]

The Background to the Anglo-German Financial Alliance

The Foreign Secretary's aim of internationalizing the Bagdadbahn had been put to Cassel and Dawkins during early February 1903, and, at the close of the month, Lansdowne asked Revelstoke to approve a draft letter that the Foreign Office proposed sending to his City house, Barings. It concerned their potential participation in financing the Baghdad railway, and Revelstoke considered its contents to be 'excellent in every way'.[32] This piece of correspondence arose from Cassel and Dawkins having sought a fresh reaction from the government to their possible cooperation with Deutsche Bank. With the imminent inception of the Bagdadbahn project, Lansdowne remained unwilling for Dawkins's Anglo-American house of J.S. Morgan & Company to be the lead City institution in financing the Baghdad railway.[33]

Deutsche Bank was not only seeking funds in London during early spring 1903 but also concessions from, and the diplomatic support of, the British government to assist the Baghdad railway's viability. The latter constituted the crux of the matter for von Gwinner, who later maintained that he had never counted on the involvement of British funds. If they had been forthcoming, it would simply have meant that the line could have been completed more rapidly.[34] The diplomatic assurances had three dimensions: a mail subsidy on the Anglo-Indian post to be conveyed by the railway; the establishment of a Kuwait terminus with specific facilities for clearing Turkish customs; and the facilitation of a new Turkish tariff regime. These were required for various reasons. The subsidy would assist the railway's revenues and be justified by the line's shorter route to India than that by sea. Actually, von Gwinner wanted to go further in this regard by also obtaining official encouragement for Anglo-Indian passenger traffic on the railway. Second, a Turkish customs house at Kuwait would facilitate administrative arrangements for freight consigned to the Orient and so further aid the railway's projected revenues. Lastly, the trade duties' revision arose from the objective of augmenting Turkey's customs revenues, so enabling them to provide the required security for the state guarantees contained in the Baghdad railway's March 1903 concession.[35]

Personally, Lansdowne was favourable. His only reservations were that mails on the railway should enjoy 'substantial advantages' compared with their carriage by sea, whereas the Berlin banker needed to comprehend that the British government could not influence the particular route to, and from, the East chosen by private

31   TBA; 200248, Lansdowne to Revelstoke, 25 February 1903, ff. 7, 8.; and Dawkins to Revelstoke, 27 February 1903, f. 13.
32   Francis, 'British withdrawal', pp. 170–1; and TBA; 200248, Lansdowne to Revelstoke, 21 February 1903, f. 1; and Revelstoke to Lansdowne, 23 February 1903, f. 2.
33   TBA; HC 15.1.6, Revelstoke to Cromer, 4 March 1903.
34   TBA; 200248, von Gwinner to Mount Stephens, 26 May 1903, f. 309.
35   TBA; 200248, von Gwinner to Revelstoke, 18 March 1903.

passengers.[36] Furthermore, the Foreign Secretary had no objection to a portion of a 'reasonable increase' in the Turkish customs tariff being amongst the Baghdad railway's financial guarantees. However, the London bankers appreciated that the British government could not make a more precise commitment, such as the 75 per cent rise that von Gwinner later sought.[37] Although Lansdowne was responsive, he was equally determined to ensure that the Baghdad railway was truly an international project. No section of its route nor that of its progenitor – the Anatolian – through Asia Minor was to be solely controlled by one great power, so removing an element of likely tension from national rivalries.

During their February 1903 discussions with the Foreign Office, the railway's putative London-based financiers also indicated that they required wider support within the City. Accordingly, Cassel and Dawkins were ready to accept any official suggestions as to where this might be obtained. It was this aspect that enabled Lansdowne to contact Revelstoke again to seek once more to displace J.S. Morgan & Company's primacy in the scheme. Within 24 hours, he responded that Barings 'place[d] ourselves at the disposal of His Majesty's Government... [being] prepared to take charge of the British participation'.[38] The merchant banker's rapid acceptance went beyond fulfilling the state's wishes and meeting a friend's request. He calculated that, since there was an 'excellent market for Turkish securities on the Continent', there would be 'no difficulty in disposing' of Baghdad railway bonds. It would be 'good' and 'profitable' business which would further his house's revival from the trauma of its 1890 crisis by doing 'us good to have the leadership in what is practically an international business'.[39] Furthermore, financing the Baghdad line's construction could lead to more as there might be 'a great deal of money to be made' from the country through which it was to run.[40] Revelstoke very soon persuaded himself that there was 'no reason why the almost illimitable productive capacity... [of the Tigris valley] in the past should not be renewed' by modern irrigation.[41] The likelihood of wider gains went along with not disrupting the City's webs of personal connections since Revelstoke was on good terms with J.S. Morgan & Company, having during 1900–1 even thought of merging his merchant bank with theirs.[42]

---

36   TBA; 200248, Lansdowne to 'my dear John', 23 March, 1903, ff. 137, 139.
37   TBA; 200248, minutes of a meeting in Paris (Cassel, Dawkins, von Gwinner, Revelstoke), 24 March 1903, f. 161.
38   TBA; 200248, Foreign Office to Barings, 24 February 1903, f. 3; and Revelstoke to Sir Thomas Sanderson, Foreign Office, 25 February 1903, f. 5.
39   TBA; HC 15.1.6, Revelstoke to Cromer, 13 March 1903.
40   TBA; HC 15.1.6, Revelstoke to Cromer, 4 March 1903.
41   TBA; HC 15.1.6, Revelstoke to Cromer, 9 April 1903. See also Cromer to Lansdowne, 12 March 1903; and Cromer to Revelstoke ('John'), 14 March 1903.
42   Cassis, *City bankers*, p. 41. A different bond was Revelstoke's relationship with Ettie Grenfell, for which see ibid, p. 257 fn 52; and N. Moseley, *Julian Grenfell. His life and the times of his death 1888–1915* (New York, 1976), pp. 41–53.

Progress of the Anglo-German Compact

As the major resources for the Baghdad line's construction were to come from bond issues, with a flotation of Ff 54 million (£2.15 million) planned for spring 1903, the officially constituted London banking group quickly got down to business. They held a meeting on 12 March. Although Revelstoke was now the official London representative, he acknowledged Cassel's and Dawkins's continuing respective one-third interests in the English part of the affair. Their conclusion of the need for a wider basis within the City was reconfirmed. Revelstoke stressed that the highly speculative project required the adhesion of other banks and bankers within the 'Square Mile'. However, the discussion concentrated primarily upon gaining British policy objectives rather than the railway's financing so that the prospectus before them only received cursory examination. Cassel had seen Lansdowne again, when it was established that the critical issues were the international control of the Baghdad railway being unaffected by any share transfers following its establishment, coupled with its board governing the Anatolian railway's policy. Acting as secretary to the City banker's 12 March meeting, Dawkins pledged their involvement to von Gwinner if the Baghdad company's statutes embraced its board's permanent internationalization.[43]

That London meeting was followed within a fortnight by talks in Paris with von Gwinner. The German banker had apparently conceded internationalization in prior correspondence, beginning with the Anatolian. Nevertheless, he implied that its securement was dependent upon the British government giving the Baghdad railway the diplomatic support – what were to become known as 'the assurances' – that he sought. It seems that von Gwinner reckoned on these being forthcoming since he now required the names of British directors for the Baghdad railway, together with that of a contractor to be involved in its construction.[44] Despite early concessions from Deutsche Bank over control, later discussions were to have an entirely opposite basis. Whereas the British bankers considered that the Bagdadbahn's internationalization had been gained, they remained concerned that the Anatolian company's parallel transformation still had to be achieved.

In Paris, the London bankers were somewhat affronted on discovering that von Gwinner was conducting his various negotiations entirely independently so that their French counterparts were ignorant of his talks with them. They immediately pointed out that representatives of all putative groups in the affair 'should have complete knowledge of respective positions held'. Von Gwinner demurred, maintaining that he was anxious to obtain concessions from the French – control of the Cassaba railway – in return for surrendering a participation in the Anatolian's management. They had no objection to this objective, indeed backed him, but were very uncomfortable over the tactics being employed for its securement.

---

43 TBA; 200248, Baghdad Railway, draft minutes of meeting (Revelstoke, Cassel, Dawkins), 12 March 1903, f. 74; 201338, Minutes, 12 Mar. 1903; and 200248, Dawkins to von Gwinner, draft, 12 March 1903, f. 76.
44 TBA; 200248, von Gwinner to Revelstoke, 18 March 1903, f. 82.

To fulfil Lansdowne's aims, the London bankers stressed that the Baghdad railway's directors should be permanently drawn from each major national group involved. Cassel, with Revelstoke's firm backing, stood out for the three major European powers each having equity shares of 26 per cent so that none in collaboration with a minor financial bloc could oppose the other two important groups' wishes. Von Gwinner made no substantive response, only returning to the question after a lengthy presentation concerning the constitutions of both the railway-operating company and the construction company. Only then did he agree with the concept of permanent national directors. However, this was solely for the pan-European syndicate he was attempting to constitute for Baghdad shares and which, in his mind, would be dissolved one year after the entire line – from Konya to the Gulf – had been laid.

Rather than hammering out the issue of international control in which the London bankers had the greatest interest, von Gwinner occupied much of their time with detail. He gave his view of the financial and managerial aspects of the Baghdad railway; elaborated issuing mechanisms for the bonds to finance building its first 200 km section; and reviewed the construction company's nature and character. All was a repetition of the approach that his predecessor, Georg von Siemens, had successfully applied for the Anatolian concession but without a Swiss-based financing bank.

Von Gwinner had German directors for the Baghdad Company, together with the Anatolian Railway Company's nominations, and proposed three English directors of the eight required. He then turned to construction finance, pointing out that the bonds, if taken by a banking syndicate at 82·5, would, after preliminary expenses, provide Ff 38 million, with the institutions and houses involved making a gross turn of 10.3 per cent on their flotation. All this was followed by a detailed plan for the construction company. It, he estimated, would make a Ff 8 million gross profit, enabling the eventual refunding of its equity plus a return of five per cent. Yet again, the Berlin banker was ready with particulars of the proposed Zürich-based company's board. It was to comprise 12 members, with Britain, France, Germany, and Switzerland having equal representation. Middle management would be constituted by Holtzmann, the Anatolian's German contractor, along with his French and Austrian counterparts – Vitali and Biedermann – who von Gwinner considered almost partners. They were to be supported by Dr Frey from Switzerland, an Austrian railway director, and a still-to-be-selected English railway contractor. On all this, Cassel, Dawkins and Revelstoke could express no immediate opinion and made that clearly apparent.[45]

Within a week, Lansdowne received a verbal report from Revelstoke. He reassured the Foreign Secretary that he and his colleagues had made it evident that their government would be unable to commit itself to a precise share, such as 75 per cent, of the proposed increase in Turkish customs revenues acting as security for Baghdad railway bonds. On the other hand, they had accepted the company's internationalization in terms of each major European power involved delegating eight directors, with the board filled out by three Swiss and three from the Anatolian

---

45   TBA; 200248, minutes of a meeting in Paris (Cassel, Dawkins, von Gwinner, Revelstoke), 24 March 1903, f. 161. See also 201338, memorandum of a meeting, Paris, 24 March 1903.

Company. The meeting had produced 'an understanding' with their German and French counterparts.[46]

Of equal concern to at least the senior Baring partner was von Gwinner's business approach that secured profits from merely building the line. Revelstoke's adverse reaction was subsequently confirmed by his colleague, Gaspard Farrer, who insisted that 'we should fix our aim on the ultimate profit of the railway as a property, a going concern, and make an early stand against greed of immediate gain in construction at the property's expense'. If the basis currently put forward was accepted, then it provided 'little... for financing future and inevitable requirements'. Nonetheless, Farrer was reassuring since 'Happily your friend Ernest Cassel is built on the biggest platform and I believe will not only assist but insist on all this. Gwinner will need pressing but will do the right thing if brought to face the question squarely.'[47] Even before this concern was met, Revelstoke had begun to seek further City partners for issuing Baghdad bonds, including the Rothschilds.

Reasons for British Withdrawal

Following the 24 March meeting in Paris, there was a break of nearly a fortnight in communications between von Gwinner and his putative London partners, which proved to be important. The German banker only wrote to Revelstoke again on 3 April – in terms of the Anatolian's internationalization providing that his British associates gained their government's support for assisting both the finance of the Baghdad railway's construction and the line's subsequent revenues.[48] It was an about-turn over the issue of internationalization compared with his letters that had preceded the Paris talks. A further communication from Berlin dealt with detail while assuming that at least outline agreement had been reached. von Gwinner was convinced the affair was going forward, shown by his indicating that the banking-syndicate agreements would shortly be despatched to London. A syndicate committee meeting in Paris was arranged for the afternoon of 21 April, following an assembly of the Baghdad's European directors earlier that day.[49]

Copies of von Gwinner's letters of 3 April were soon given to Lansdowne by Revelstoke, and both he and the Foreign Secretary were concerned about the precise wording for the undertaking for the Anatolian's internationalization.[50] However, these serious qualms were being overtaken by a press campaign, led by *The National Review* and *The Spectator*, against British capital's involvement in the Baghdad railway. Subsequently, von Gwinner attributed its instigation to the activities of Russian press agents in England, particularly Wesselitzki of Associated Press and *Novoje Vremja*, and Poklewsky at the Russian Embassy but with links to *The National*

---

46 TBA; 200248, Discussion between Revelstoke and Lansdowne, 30 March 1903 am, f. 186; and HC 15.1.6, Revelstoke to Cromer, 9 April 1903.
47 TBA; 200248, Gaspard Farrer to John [Revelstoke], 9 April 1903, f. 218.
48 TBA; 200248, von Gwinner to Revelstoke, 3 April 1903, f. 178.
49 TBA; 200248, von Gwinner to Revelstoke, 3 April 1903, f. 175.
50 TBA; 200248, Revelstoke to Lansdowne, 6 April 1903, f. 180; and Draft for von Gwinner, f. 182.

Review. Their disrupting activities, in his view, were joined by those of Sir R. Blenner-Hasset of *The Spectator.*[51] Dawkins had a somewhat different opinion over the exact source of this destabilization, attributing it to activities of the Russian Embassy in Paris.[52] Nonetheless, they both perceived its aim to arise from the long-standing Russian policy of weakening the Ottoman Empire.

The likelihood of Anglo-German financial cooperation had gained some public knowledge through a lecture on the Baghdad railway, given in late March to the Berlin Geographical Society by General von der Goltz, previously a chief instructor to Turkish forces. He made reference to negotiations over a Kuwait terminus for the railway. London papers took up the story on 28 March but, according to von Gwinner, their reports were based upon the misrepresentations of Reuters' Berlin correspondent. In its brief initial coverage on 1 April, *The Times* simply pointed out that Kuwait was the principal source for Arab horses supplied to the British government, while the Sheikh of Kuwait had yet to sign the Treaty with Britain so that whether Kuwait was either 'British' or Turkish territory remained an open question.[53] Subsequently, *The Times* and *The Morning Post* joined the rapidly developing press antagonism to British finance supporting the Baghdad railway, as did *The Times of India.*[54]

The growing adverse press commentary caused Lansdowne serious concern, to the extent that he had discussions with Balfour, especially as it had given rise to a motion in the House, to be moved on 8 April. It would also appear that Joseph Chamberlain, the protagonist for greater Imperial economic unity, opposed British involvement in the Bagdadbahn and threatened to split the Cabinet.[55] Lansdowne informed Revelstoke that all this meant that no immediate reply could be given to von Gwinner over 'the assurances' and, furthermore, it would be 'impossible to proceed' should the attitude in the House on 8 April prove hostile.[56] Equally, Revelstoke had 'no wish to be gibbeted as promoting a Turkish Railroad for mere motives of cash gain'.[57]

The extent of the British bankers' own continuing commitment to the Baghdad railway is shown by their discussions with Sir John Aird on 7 April. Aird was willing in principle to be nominated as the English contractor but, understandably, would make no definite decision over his firm's involvement until he had taken his own soundings over the project. Revelstoke and Cassel also used this occasion to review the British government's now uncertain attitude in order to reply to von Gwinner's letters of 3 April. With the developing adverse British political situation, they decided to put the matter on hold, using the excuse of it being impossible to make further

---

51   TBA; 200248, von Gwinner to Mount Stephens, 25 May 1903, f. 309.
52   See Dawkins to von Gwinner, 23 April 1903, published by von Gwinner in 'Baghdad Railway' but to be read with the commentary of E.C. Grenfell to Revelstoke, 8 June 1909, in TBA; 203146, f. 12.
53   *The Times*, 1 April 1903, cuttings copy in TBA; 200248, f 174[a]. See also Francis, 'British withdrawal', pp. 171–2.
54   See TBA; 200248, von Gwinner to Baring Bros., 6 April 1903, f. 195.
55   TBA; HC 15.1.6, Revelstoke to Cromer, 9 April 1903; and Francis, 'British withdrawal', pp. 172–3.
56   TBA; 200248, memorandum of a conversation, Lansdowne House, 7 April 1903, f. 184.
57   TBA; HC 15.1.6, Revelstoke to Cromer, 9 April 1903.

progress over the two or three weeks of the imminent Easter holidays and its parliamentary recess.[58] The resultant 'holding' telegram was subsequently amplified, with blame for the situation being also put on their 'not having heard from you [von Gwinner] until now', while the Easter holidays were employed again to explain their own inabilities to attend the Paris meetings of 21 April.[59]

The bankers' reluctance to commit themselves further came as a shock to von Gwinner, who thought that 'having agreed to all your suggestions... nothing further would be required'. He also explained in this anguished telegram the situation's urgency, while he was not to be put off. He heralded the despatch of the draft syndicate agreements that had been foreshadowed in one of his 3 April letters.[60] Their practical use in London was soon put in doubt through Rothschilds declining to become involved since cousin Alphonse in Paris had informed his British relations that he did not 'take a very favourable view of the operation'.[61] This was sufficient for the leading merchant bank at New Court to take no part in the affair.

Within a week von Gwinner was also aware of the adverse British press reaction and asked his City associates on 6 April to arrange for it to be known that the 'Baghdad Railway was not hostile to British interests'. The Berlin banker excluded *The National Review* from this requested briefing since he regarded it as 'futile even to attempt reconciliation where opinion is influenced by other motives but common impartiality and common sense'.[62]

Von Gwinner's plea for press rebuttals was overtaken by the Commons debate of 8 April. Its mover was Gibson Bowles, the Member for Kings Lynn, who mounted his attack upon British involvement in the Baghdad railway utilizing a report from Vice-Consul Waugh. It had been written on 9 March and received at the Foreign Office a week later and so predated the opening of close discussions between von Gwinner and Cassel, Dawkins and Revelstoke. From it, Gibson Bowles maintained that the Baghdad railway was solely a German undertaking that would destroy two British-owned railways – the Bosphorous and the Smyrna. Furthermore, the line would place an adverse burden on British trade, based upon Waugh's assertion regarding the commercial behaviour of the German-owned Haida Pascha Port Company at the Anatolian's Bosphorous terminal. British seaborne exports would be subjected to higher charges compared with German goods conveyed over land – by rail. Somewhat later, von Gwinner dismissed this, pointing out that the only current German-Turkish overland trade comprised small parcels for which there were standard tariffs set by the international postal convention.[63] Apart from injuring British business interests, Gibson Bowles also considered that the Baghdad railway would negatively affect discussions between the British and Russian governments that promised an amicable understanding over the two countries' respective interests in Persia. This widening of the issue to even broader aspects of international relations

58   TBA; 200248, memorandum of a meeting (Revelstoke and Cassel), 7 April 1903, f. 190a (further copy at TBA; 201338).
59   TBA; 200248, letter to von Gwinner, 7 April 1903, f. 192.
60   TBA; 200248, telegram, von Gwinner to Revelstoke, 7 April 1903, f. 193.
61   TBA; 200248, Rothschild to Revelstoke, 7 April 1903, f. 194.
62   TBA; 200248, von Gwinner to Baring Bros., 6 April 1903, f. 195.
63   TBA; 200248, von Gwinner to Mount Stephens, 26 May 1903, f. 309.

led others, such as Lord Fitzmaurice, to pose the nub of the Baghdad-railway issue, namely the company's future political control. Yet they simultaneously caused this to get lost in further concerns over an Albanian rising and its consequent threat to stability in Macedonia and Serbia.[64]

Cassel subsequently explained to French associates that the hostility of British public opinion towards the Baghdad railway was a repetition of attitudes that had arisen over the Venezuela incident. In turn, they had their roots in the German press's stance towards England during the Boer War. He appreciated its force, stressing to his French colleague that 'no doubt public opinion has to be reckoned with more [in England] than anywhere else'.[65] Revelstoke separately remarked that 'it is interesting to note the virulence of the anti-German mania which is still raging in England'.[66]

As the press campaign further developed, it was also alleged that the Baghdad railway's financial guarantees, arising from the proposed augmented Turkish Customs, would reduce the security of existing Turkish bondholders. This, as with many of the charges levied against the projected Anglo-French-German financial scheme, was at best a misunderstanding since Turkish trade tariff receipts had never constituted the basis for external loans raised by the Ottoman authorities.[67] Furthermore, it was an aspect that became enmeshed in the discussion of proposals for consolidating Turkish debt, which, as von Gwinner later stressed, were not being tabled by rapacious German financiers but by Banque impériale ottomane. Moreover, the same outcome of the Turkish authorities employing 75 per cent of increased customs revenues as the security for further borrowing would arise from the debt-consolidation scheme being separately developed by Sir Henry Babington Smith, the British member of the Ottoman Debt Administration.[68]

As the parliamentary debate of 8 April took place, members of Barings reconsidered the question of the Baghdad railway's control. This arose from the recognition that the government's diplomatic support would only be forthcoming if the constitutions of both the Anatolian Company and the Baghdad Company gave rise to their internationalization. The conclusion drawn was that this would be gained in practice if three conditions were fulfilled: namely, that each of the major financial

---

64   *The Times*, 9 April 1903, 'cuttings' copy in TBA; 200248, f. 210.
65   TBA; 200248, Cassel to Noel Bardac, 21 April 1903, f. 258. Anglo-German economic rivalry had long roots, see W.E. Minchinton, 'E.E. Williams: "Made in Germany" and after', *Vierteljahrschrift für Sozial- und Wirtschaftgeschichte*, 62 (1975); P. Kennedy, *The rise of Anglo-German antagonisms 1860–1914* (London, 1980); H. Kieswetter, 'Competing for wealth and power: the growing rivalry between industrial Britain and industrial Germany, *Journal of European Economic History*, 20 (1991) and P.L. Cottrell, '"Lo, ... Nineveh and Tyre": British trading anxieties and official reactions, 1870–1929', in D.H. Aldcroft and A. Slaven (eds), *Enterprise and management: essays in honour of P.L. Payne* (Aldershot, 1995). In the particular case of banking, see P. Hertner, 'German banks abroad before 1914', in G. Jones (ed.), *Banks as multinationals 1860–1960* (London, 1990); and G.F. Young, 'British overseas banking in Latin America and the encroachment of German competition, 1887–1914', *Albion*, 23 (1991).
66   TBA; HC 15.1.6, Revelstoke to Cromer, 9 April 1903.
67   TBA; 200248, von Gwinner to Revelstoke, 23 April 1903, f. 272.
68   TBA; 200248, von Gwinner to Mount Stephens, 25 May 1903, f. 309.

groups involved – British, French and German – voted 'en bloc'; that their respective entitlements persisted for the concession's duration; and that their voting rights were independent of shareholdings' actual 'host residences'.[69] As the government became increasingly reluctant in the face of growing public hostility to give its diplomatic support to the project, there were further forays on paper within Barings to resolve the political-control question. One saw a solution in maintaining the railway company's current constitution but with the construction undertaken through three distinct national components – British, French and German. Each would have an outlet to the sea, with the British section comprising the route from Baghdad to the Gulf – 'the smallest share' but which would 'secure... where her interests predominate.'[70] This approach was later put forward by Sir Edward Fitzgerald Law. Nonetheless, von Gwinner maintained that it was untenable since Sultan Abdul Hamud had ensured through Article 29 of the 1903 Baghdad Railway Convention that no traffic was to run on the railway's section south of Baghdad until the line from Konya had been completed. Furthermore, the railway company was responsible for paying interest on its bonds until the line was generating revenues.[71]

Bankers' attempts to resolve the affair's critical issue went along with their own rising doubts over whether it had a future for British finance. Revelstoke indicated to Lansdowne the day after the 8 April parliamentary debate that it would 'be a great convenience to receive a definite idea of your wishes at an early date'. This implicit question was coupled with a succinct précis of the London bankers' position. They were 'anxious to meet the wishes of His Majesty's Government' but were not 'inclined to move further in the matter without their cordial approval'.[72]

The Retreat of British Capital

The first and only public test of whether there would an Anglo-German financial alliance for the Baghdad railway arose on 10 April 1903, when a meeting in Constantinople was legally to constitute the railway company. Its calling put the only British-nominated company director locally resident – Babington Smith – in a quandary. He asked London for directions while indicating that his French counterparts proposed to attend unless otherwise instructed.[73] Ultimately, Babington Smith absented himself, a considered reaction since he had realized that it was essential for three British directors to be present for them collectively to employ the full quota of their proxy votes. He had offered London the option of his non-attendance, with which the City bankers had concurred. As a result of this and the French directors also failing to attend, the Baghdad railway's constitution was postponed to 13 April, but when only its German directors were present. This caused divisions between the putative German and French financial groups for the line's construction. For his part,

---

69   TBA; 200248, Control of Baghdad Railway, undated but bound file position indicates 9 April 1903, f. 214. See also Babington Smith to Dawkins, 7 April 1903, f. 236.
70   TBA; 200248, memorandum, 17 April 1903, f. 247.
71   Von Gwinner, 'The Baghdad Railway'.
72   TBA; 200248, Revelstoke to Lansdowne, 9 April 1903, f. 208.
73   TBA; 200248, Dawkins to Revelstoke, 11 April 1903, f. 232.

Babington Smith sent a message to this second meeting that indicated his willingness in principle to be a director but with his active participation in the company's affairs delayed until he could be joined on its board by his fellow British members.[74]

Von Gwinner either refused to accept what this presaged for British financial involvement or failed to understand its underlying meaning. He asked Revelstoke by telegram whether the London merchant banker could come to Berlin on 25 April, when he would arrange meetings of both the railway's directors and managers of the banking syndicate. Revelstoke replied immediately, by telegram, to give the clearest possible message: 'We have already pressed as much as we can the urgency of obtaining definite opinion. Until receipt of this our hands are tied.'[75] In other words, the matter of British involvement was with the British government.

Despite Revelstoke indicating that London was now only marking time, von Gwinner attempted to push the City bankers into at least taking a stance with the British press. He pointed that, with English correspondents besieging Deutsche Bank's Berlin offices, the London bankers should issue a statement to their local newspapers.[76] All he received in reply was agreement to the 'wise' attitude that he had taken *vis-à-vis* the British press.[77] Frustrated by a *Times* leader on 15 April which, like Gibson Bowles a week earlier in the House, was based upon Vice-Consul Waugh's dated report, Gwinner maintained desperately to Revelstoke that 'the longer the clearing up of public opinion is delayed, the more difficult it will become to correct the misleading of public opinion.'[78]

Five days after *The Times* leader, von Gwinner's putative London banking partners drew up a statement explaining their views in the light of public criticism of the Baghdad railway affair.[79] It was for Lansdowne, as was a draft of a communication from Lansdowne to them. These marked the effective end of Anglo-German financial cooperation for the railway, just as Lansdowne seeking on 21 February Revelstoke's approval of a draft letter from the Foreign Office to Barings had been its beginning. Although not rejecting British involvement outright, Lansdowne's draft put forward that the affair was still too premature for the government to give the desired diplomatic assurances. These might be forthcoming once the Anatolian Railway Company had been internationalized in a comparable manner to the Baghdad and when the French and German governments had also provided similar diplomatic assurances.[80]

Two days later, the bankers discussed a draft of an answer to a parliamentary question regarding the constitution of their group.[81] They also considered the Foreign

---

74   TBA; 200248, von Gwinner to Revelstoke, 14 April 1903, f. 234; and Babington Smith to Dawkins, 13 April 1903, f. 244.
75   TBA; 200248, Revelstoke to von Gwinner, 14 April 1903, f. 235.
76   TBA; 200248, von Gwinner to Revelstoke, 15 April 1903, f. 242.
77   TBA; 200248, Revelstoke to von Gwinner, 15 April 1903, f. 242.
78   TBA; 200248, von Gwinner to Revelstoke, 16 April 1903, f. 245.
79   TBA; 200248, meeting (Cassel, Dawkins and Revelstoke), 20 April 1903, f. 296; see also 201338, minutes of a meeting, 20 April 1903.
80   TBA; 200248, draft for Lansdowne, 20 April 1903, f. 255.
81   TBA; 200248, meeting (Cassel, Dawkins and Revelstoke), 22 April 1903, f. 300. See also 201338, minutes of a meeting, 22 April 1903.

Secretary's reaction to their 20 April discussion. This led to the drawing up of a telegram for Lansdowne to send to von Gwinner that stressed the affair's immaturity preventing the British government's diplomatic support for the railway. After an interview with the Foreign Secretary that immediately followed and an ensuing further discussion amongst the City bankers, Revelstoke wrote twice to von Gwinner. The letters indicated the end of the City's cooperation in the finance of the Baghdad railway while pointing to media hostility as a major cause. The senior Baring partner maintained that 'any effort on our part to modify the virulence of the press comments would have been worse than useless'.[82]

Nonetheless, the London bankers subsequently agreed to meet their Berlin counterpart's request to provide *The Times* with details of the Baghdad Railway Company. The 'Thunderer' was continuing its campaign against British involvement for which it was now employing a pamphlet by 'Cheradam', who von Gwinner identified as someone pursuing personal claims about whom he had already been warned by English directors of the Mersina railway.[83] The German banker's exasperation became even greater when he found in late April even *The Economist* also lining up in the opposition camp, as was *The Westminster Gazette*.[84] His only allies had proven to be the British Chamber of Commerce at Constantinople and a leading English mercantile firm on the Gulf,[85] whose countervailing opinions had counted for little in the jingoism of spring 1903. Von Gwinner's defeat in the attempt to obtain the British government's diplomatic support for the Baghdad railway was sealed by further debates in the Commons, and in the Lords on 6 May, when his friend Mount Stephens participated.

The fate of officially-supported British financial involvement in the Bagdadbahn had been decided at a cabinet meeting on 24 April. Revelstoke saw Lansdowne the day before to indicate the bankers' willingness that the matter should be forgotten. He pointed out that, with the 'violent press', the 'issue of [the railway's] bonds [was now] little attractive from a City point of view' and, consequently, it was 'unwise to move without the government's cordial approval'. Revelstoke also maintained that Lansdowne should 'not press if the Cabinet was not anxious to proceed'. Being aware of Chamberlain's opposition, he reinforced this in terms of Lansdowne not allowing 'any feelings of loyalty... for us to lead him to have a tussle with other members of the Cabinet'. There was acknowledgement within the government of the difficult situation in which Cassel, Dawkins, and Revelstoke had been placed, with Balfour being 'anxious to do what he can to please us'. The Prime Minister was prepared to make reference to the Baghdad railway in a speech being drafted for 1 May but this did not chime with Revelstoke's taciturn demeanour.[86]

---

82 TBA; 200248, Revelstoke to von Gwinner, 22 April 1903, f. 265; and Revelstoke to von Gwinner, 22 April 1903, f. 266.
83 TBA; 200248, von Gwinner to Revelstoke, 21 April 1903, f. 260.
84 TBA; HC 15.1.6, Revelstoke to Cromer, 1 May 1903.
85 TBA; 200248, von Gwinner to Mount Stephens, 26 May 1903.
86 TBA; HC 15.1.6, Revelstoke to Cromer, 1 May 1903.

Consequences and Implications

Frustrated in London, von Gwinner turned to other options for the Bagdadbahn's financing. This meant cooperation solely with French institutions for which there was a basis in an 1899 agreement with Banque impériale ottomane, albeit that this understanding initially had no definite form.[87] Von Gwinner revived it to achieve the long-considered consolidation of Ottoman debt for which Banque impériale ottomane had developed fresh proposals over winter 1902–3. The required modification to the 1881 Muharram decree was finally made between mid-September 1903 and early 1904, releasing Ottoman state guarantees that could be utilized as security backing finance for the Bagdadbahn's construction.

A Deutsche Bank-headed consortium undertook the first issue of the Bagdadbahn bond series during February 1904. Although involving Banque impériale ottomane's indirect participation, it was a relatively small affair. Thereafter, despite increasing international instability, which had a major focus in the Balkans, and the 'Young Turk' revolt, Deutsche Bank directors decided in May 1909 to implement an 11-month-old agreement for taking the line to El Helif, Mesopotamia. Karl Helfferich, who had joined Deutsche Bank's managing board in 1908, played a number of parts in this, the German press pointing to his past contacts, first with the Foreign Ministry's Colonial Department and, then, the Anatolian's management.[88] Despite Helfferich's fervent nationalism, nearly half of the subscriptions for the second issue of Baghdad bonds, totalling Ff 108 million, came from abroad during an operation that ran from February 1910 to the end of the year. About a third was raised in France through the active support of French banks despite the lack of a French prospectus.

Anglo-German concord for the line's Gulf section was finally reached on 15 June 1914. Although concerning conditions for the railway's completion, its more important clauses addressed the extraction and marketing of Mesopotamian oil.[89] This pointed to the future – the new technology of the twentieth century, giving even greater importance to the Suez Canal – and did not look back to 1903. By summer 1914, much of the Ottoman railway system planned by von Pressel during the early 1870s under Sultan Abdul Hamid II had been constructed. However, the Baghdad line, its largest component, was not finally completed until 1940 – under very different circumstances, economic, financial, and political.

The retreat of the British government from the Baghdad project in April 1903 did cause the London bankers some considerable anguish, although Cassel was ultimately more upset by a *Times* leader that suggested that they had withheld important points of the matter from ministers. Being as Revelstoke put it '*plus*

---

87   J. Thobie, *Les Intérêts économiques, financiers et politiques français dans la Partie asiatique de L'Empire ottoman de 1895 à 1914*, two volumes (Lille, 1973), pp. 618f; and Seidenzahl, *100 Jahre Deutsche Bank*, pp. 148f.

88   For one appreciation of Helfferich, see J.G. Williamson, *Karl Helfferich 1872–1924* (Princeton NJ, 1971).

89   This section has been largely drawn from Barth, 'Financial history', the footnotes of which provide a clear guide to the historiography. See also Barth, *Die deutsche Hochfinanz* and, for Deutsche Bank, both Gall, 'The Deutsche Bank from its founding'; and Seidenzahl, *100 Jahre Deutsche Bank*.

Royaliste que le Roi', Cassel had wanted to send a letter of justification to the newspapers and the senior Baring partner had great difficulty in restraining him. Revelstoke continued to take the view that von Gwinner had found increasingly frustrating during the affair, namely the 'less said the better', so that it was 'especially unwise to make any comment to the public press'.[90] Ultimately, only von Gwinner of the four put his position on the public record through an article in *Nineteenth Century* in June 1909. This caused a minor frisson amongst the London bankers as he reprinted within his piece a letter from Dawkins, written when the affair was terminated. E.C. Grenfell of Morgan Grenfell & Company checked it against the copy within his merchant bank's files of his recently deceased partner's correspondence and remarked to Revelstoke that the Berlin banker had undertaken some judicious editing. Of more concern within the City was that Dawkins had clearly indicated that his letter contained 'confidential and exclusive information', a permanent stricture that von Gwinner had transgressed.[91]

The distaste that the Baghdad railway affair of spring 1903 caused within the City on a number of counts did not result in its direct participants withdrawing from Ottoman finances. Morgan Grenfell & Company, in conjunction with Parr's Bank, floated a portion of the 1909 Turkish loan for which the lead house was Banque impériale ottomane. Although the English tranche was issued at a lower price than that of the bonds marketed in Paris, it met with a poor response amongst British investors.[92] A year earlier, Cassel and Revelstoke had begun to develop plans to establish the National Bank of Turkey, an institution that would assist the British government in 'arranging for English capital to help in the present difficulties' within the Ottoman Empire following the 'Young Turk' insurrection.[93] Unlike in spring 1903, the new bank had the full and sustained, if discreet, support of the British government. Sir Edward Grey sought through it to establish an Anglo-French alliance, following the Entente of 1904, which would maintain the stability of Turkish finances and confront growing German influence within the Balkans and Asia Minor. Nonetheless, it failed in 1913, having over the previous four years neither come to constitute a conduit for the flow of British funds to Turkey nor to be an effective vehicle for the expression of British foreign policy.[94]

The nature of the Bagdadbahn affair of 1902–3 has some congruency with the concept of 'Gentlemanly Capitalism', through which Cain and Hopkins have sought to put forward explanations of British imperialism from a 'a close study of economic structure and change in Britain' that goes beyond the impact of the Industrial

---

90   TBA; HC 15.1.6, Revelstoke to Cromer, 1 May 1903.
91   TBA; 201338, E. C. Grenfell to Revelstoke, 8 June 1909.
92   TBA; Turkish Affairs, II, 200196, memorandum on recent Ottoman loan negotiations, Cassel and Babington Smith, 4 October 1910.
93   TBA; Turkish Affairs, I, 200195, Revelstoke to Hottinguer, Paris, 2 November 1908, f. 4; Revelstoke to Hottinguer (Paris), 26 January 1909 f. 14; and memorandum from Cassel, 20 July 1909, no folio but following f. 11.
94   See Mclean ' "Informal empire" ', pp. 293–7; Thane, 'Cassel', pp. 93–4; and M. Kent, 'Agent of empire? The National Bank of Turkey and British foreign policy', *Historical Journal*, 18 (1975).

Revolution.[95] This has led them to emphasize the role of finance from the mid-nineteenth century, which transmuted the previous Gentlemanly Capitalism (based primarily upon the landed interest) to one which had its nexus in the service-sector economy of the south-eastern region. They perceive its links with state power running after 1850 through the Treasury and the Bank of England.

Certainly, parts of the City were allied with organs of the state to defend British interests in the realization of the Baghdad railway. Furthermore, the affair involved actors who were at some 'distance from the everyday and demeaning world of work'; indeed, some were very wealthy members of fashionable 'High Society'. Moreover, Dawkins had been a civil servant specializing in finance until 1900, although in the Baghdad affair he was 'junior' to Cassel and Revelstoke and other business in Spain occupied much of his time following the Paris meeting of 24 March. Nonetheless, while Revelstoke was a director of the Bank of England for 30 years from 1898, this played no part in determining his role. Instead, it would appear that more important was his personal relationship with Lansdowne together with his house's regained leading role within the City. Conversely, it was not the Treasury that initiated action, although Cassel and Revelstoke like Dawkins were well known to its senior officials, but rather the Foreign Office in the further development of a policy of some seven years standing. Above all, it was not members of the City that attempted to employ political pressures for furthering their own affairs. Instead, it was the state that turned to private financial power to protect British overseas political interests. In the cases of both the Bagdadbahn in 1902–3 and, subsequently, the National Bank of Turkey, Cassel and Revelstoke responded compliantly to the Foreign Secretary's requests.

In attempting to employ the City in the defence of British interests against American and German encroachment, Lansdowne was defeated in spring 1903 not by the popular press but the newspapers, reviews, and quarterlies more likely to be read by the investing middle classes, the new life blood of Gentlemanly Capitalism during the second half of the nineteenth century. Lastly, British withdrawal from the Baghdad railway project in 1903 was not an isolated defeat (albeit one at home) in the state's attempt from the mid-1890s to employ finance in the defence of strategic political interests in specific areas abroad. It was one of a sequence of defeats for a policy that only achieved some limited gains through Imperial Bank of Persia. In this perspective, the relationship that ran from the Foreign Office to the City was unproductive, contrasting poorly with the more effective mobilization of finance by France and Germany to aid the achievement of foreign-policy goals during the period of 'Finance Capitalism' and 'New Imperialism'.

---

95 P.J. Cain and A.G. Hopkins, 'Gentlemanly capitalism and British expansion overseas, I. The old colonial system 1688–1850', *Economic History Review*, 2nd. series, 39 (1986); and idem, 'II: new imperialism, 1850–1945', *Economic History Review*, 2nd series, 50 (1987). For conception, and subsequent further development, see idem, 'The political economy of British expansion overseas, 1750–1914', *Economic History Review*, 2nd series, 33 (1980); and idem, *British imperialism: innovation and expansion, 1688–1914* (London, 1993).

SECTION II
SPIRIT, MIND AND EYE

Chapter 7

The 'Broad Gauge' and the 'Narrow Gauge': Railways and Religion in Victorian England

R.C. Richardson

'Another page of Godless legislation, another national sin invokes the displeasure of the Almighty.'
> Rev. Francis Close on the introduction of Sunday trains to Cheltenham in 1846.
> J. Goding, *Norman's History of Cheltenham* (1863 ed.), pp. 570–1, 600, quoted in J. Simmons, *The Victorian Railway* (London, 1991), p. 283.

'You know Sir, us chaps are just like them Israelites as you read of in the Bible; we goes from place to place, we pitches our tents here and there, and then goes on just like they did...'
> Quoted in D.W. Barrett, *Life and Work among the Navvies* (London, 1880), p. 140.

'While the railways have done so much to advance the material interests of men they have also afforded precious opportunities for advancing their spiritual and eternal interests.'
> *The Railway Signal* (July 1883), p. 242.

The biography of a Victorian bishop might seem an unlikely platform from which to launch a study of some of the interconnections between railways and religion in the nineteenth century. In fact it performs this function very well. *The Lancashire Life of Bishop Fraser* by the Rev. J.W. Diggle, published in London in 1889, documents the achievements of the second churchman to hold the diocese of Manchester. Gladstone's offer of the bishopric in 1870 to James Fraser (1818–85) had made clear the challenge of the appointment: 'Manchester is the centre of the modern life of the country. I cannot exaggerate the importance of the see, or the weight and force of the demands it will place on the energies of a Bishop, and on his spirit of self-sacrifice.'[1]

Gladstone's confidence in Fraser as the man for the job was not misplaced. He quickly established himself as an energetic and caring '*Citizen Bishop*',

> the prince and leader in every movement of civic improvement, civic elevation, civic righteousness.... Mayors and corporations were not less interesting to him than Archdeacons and Rural Deans. He cared for well ventilated rooms and good drainage and pure water, as

---

1   Diggle, op. cit., p. 41.

well as for church-building and lay readers and mission women. He threw the spirit of religion into every manner of good and useful work.... The newspapers soon found him out and insisted upon reporting him. The mass of newspaper cuttings which record his utterances is simply enormous – probably all the English Bench taken together did not fill the same amount of newspaper space from 1870 to 1885 as he filled singly and alone.[2]

Fraser's diocese, founded as recently as 1847, was geographically compact but densely urban and industrial. Besides the cottonopolis of Manchester itself the diocese contained at least 12 other towns with populations of more than 50 000.[3] Fraser was everywhere and into everything, conducting visitations, and confirmation and ordination ceremonies; the bishop's biographer calculated that he confirmed no fewer than 29 000 candidates in 160 separate gatherings. He preached in three-quarters of the 420 churches in his see, consecrated 26 new ones, founded a diocesan board of education, and formed a third archdeaconry. He gave hundreds of addresses on civil and religious issues to crowds of hearers. 'A Bishop of Manchester cannot, if he would [Fraser declared], as long as he has health and strength, lead the life of a recluse'.[4]

Railways transformed the functioning of dioceses in the nineteenth century, as modern historians like Owen Chadwick have shown. Fraser's biographer was quick to underline that the key to the bishop's success in Manchester was that 'systems of railways cover the diocese with a network of steel. Each parish is easily accessible to the Bishop, and the Diocesan Registry easily accessible to every churchman'.[5] Bishop Fraser's hyperactivity constantly confronted him with the juxtaposition of immense wealth and abject poverty in industrial Lancashire and with the problems inherent in the 'relations of the classes to the masses', exposed most starkly in the bitter cotton strike of 1878. 'He felt he was Bishop not merely of the worshipping few but of the non-worshipping many', to the extent that he became known as 'the working man's Bishop'.[6] Fraser threw himself wholeheartedly into the work of the mission movement of the day, addressing large gatherings of cotton workers, colliers, nightsoil workers, foundry men, postal workers, and railwaymen in their places of work. 'He would sometimes address as many as two thousand employés during their "dinner hour."'[7] Very frequently he spoke to, and corresponded with, railway workers, offering – if his biographer's examples are a reliable guide – down-to-earth spiritual instruction and moralizing which 'christianized the laws and lessons of political economy' and emphasized the benefits of thrift and sobriety. He was greatly touched when the men at the Gorton railway works in Manchester (where he had

---

2   Ibid., pp. 22–3.
3   Ibid., p. 65.
4   Ibid., p. 69.
5   O. Chadwick, *The Victorian Church I: 1829–56* (London, 1966, 2nd edition, 1971), pp. 514–15. Samuel Wilberforce is another example of a Victorian bishop constantly using railways in the conduct of his work. See R.K. Pugh (ed.), *The Letter Books of Samuel Wilberforce, 1843–68*, Oxfordshire Record Society, XLVII for 1969 (1970), passim; Diggle, op. cit., p. 68.
6   Ibid., pp. 337–8, 193.
7   Ibid., p. 330.

frequently preached) clubbed together to present him with a handsome Bible, service book and gold pencil case in 1882. 'I would a hundred times rather preach to a congregation such as I see before me [he told them] than to the most fashionable congregation of the most fashionable church.'[8]

Unusually for a Victorian bishop, he was tolerant of the Sunday opening of art galleries, museums, libraries, and workmen's clubs:

> People are not made saints, necessarily, by listening to sermons or singing songs about heaven. I would rather see men trying to do their duty and please God in that station of life, and in those relations of life, in which God's Providence has seen fit to place them.[9]

Better a Sunday afternoon visit to an exhibition, or a railway excursion, than 'drunkenness, profaneness, and grossness of conduct and language' that might well characterize working-class alternatives. Fraser held aloof from the Sabbatarian controversy about railways which aroused the feelings of many of his contemporaries (especially north of the Scottish border) and led to the emergence of the 'church interval' and 'the Railway Sunday'. ('I am told that the Scots are even more religious than the English', wrote the Frenchman, Hippolyte Taine, in the early 1860s. 'Compared with Edinburgh a Sunday in London is positively agreeable.')[10] Even the Tay Bridge disaster of 28 December 1879 – viewed by Sabbatarian extremists of the day as a stern divine judgement on a Sunday train – provoked no such moralizing from Bishop Fraser.

> The terrible railway accident at the Tay Bridge [he wrote] has been hanging like a great weight about me all day. It threatens to be the most appalling railway accident on record, and I don't expect to hear tomorrow morning that a single life has been saved.[11]

Bishop Fraser took his work with the Manchester Mission very seriously and he became a close friend of Jeremiah Chadwick, its chief evangelist. To an audience of railwaymen in 1877 Fraser proclaimed that missionary work sought only to win converts to Christianity, not to any particular religious denomination.

> There is work enough for all religious men. There is work enough and to spare for them all in Manchester, without endeavouring to draw anyone away from one religious denomination to another, except in so far as the spirit of God moves them.[12]

8   Ibid., p. 374.

9   Ibid., p. 247.

10   See J. Wigley, *The Rise and Fall of the Victorian Sunday* (Manchester, 1980); D. Brooke, 'The opposition to Sunday rail services in N.E. England, 1834–1914', *Journal of Transport History*, V (1963–4); M. Robbins, *The Railway Age* (London, 1962), pp. 48–9; J. Simmons, *The Victorian Railway* (London, 1991), pp. 282–9; J. Simmons and G. Biddle (eds), *The Oxford Companion to British Railway History* (Oxford, 1997), p. 486. The Anti-Sunday Travel Union claimed 22 500 members in 1898. E. Hyams (ed.), *Taine's Notes on England* (London, 1957), p. 283.

11   Simmons, *Victorian Railway*, p. 217; Diggle, op. cit., p. 48.

12   Diggle, op. cit., p. 76.

He was right. Railway workers were targeted and the railway itself was utilized for Christ's cause by men of many religious persuasions. Unavoidably, in an age ubiquitously dominated by the railway, Victorian religion and the new regime of transport were inextricably connected in a huge variety of ways. Even the names of leading railway stations followed ecclesiastical tradition – St Enoch's in Glasgow, St David's in Exeter, and – that great cathedral of railway architecture, beloved by Jack Simmons – St Pancras in London.

Ecclesiastical patronage exercised by railway companies was not insignificant. In the new railway towns they built churches as part of the urban fabric. Wolverton, on the London & Birmingham Railway near Stony Stratford in Buckinghamshire was the first. The small town was planted by the railway company to serve its workshops, and its directors provided at their sole charge in 1842 a church, St George's, built in the Norman style, costing £5 000, and a school.[13] The same railway company built a new church, vicarage, and school at nearby Stantonbury in 1859–60 at a cost of £6 000, two-thirds of the expenditure being borne by the company's shareholders.[14] Earlier, in 1845, the Great Western Railway Company spared no expense in erecting St Mark's Church in New Swindon at a cost of £8 000, and for the first five years also paid the £150 annual stipend to the incumbent.[15] Crewe, 'an entirely new place due to the formation of the railways', and equipped with 'all the appliances of a busy, well-contrived, improving town' acquired two railway churches early in its history. The first, Christ Church, was consecrated in 1845 – the same year as St Mark's at New Swindon – and its relatively modest £2 300 building charges were spread between the shareholders and directors. (The sabbatarians among them contributed their share of the profits derived from Sunday trains.) John Cunningham, the railway architect, was enlisted to design the church, and the resultant edifice was equipped with goods-shed style roof-vents and exterior noticeboards in company colours. A second church was added in 1869 – St Paul's, Hightown – and to this Francis Webb, head of the Crewe railway works, nominated his brother as vicar and bullied increasing numbers of railwaymen and their families into forming its congregation. Three more company Anglican churches had been founded in Crewe by 1900 in an attempt to combat the stubborn and rapid rise of Nonconformity in the railway town.[16] The railway church in Doncaster, St James's, was built in 1858 at a cost of £5 000, 'with a row of alternately round and octagonal pillars down the middle, and a spire 120 feet high'.[17] Other railway companies approached church provision from different directions. In the early days of the London & Greenwich Railway the company rented pews at St

13 J.M. Wilson, *The Imperial Gazetteer of England and Wales* (Edinburgh, 1870), VI, p. 1107.
14 Ibid., VI, p. 855.
15 *Victoria County History of Wiltshire*, IX (London, 1970), pp. 148–9.
16 Wilson, op. cit., II, p. 512; W.H. Chaloner, *The Social and Economic Development of Crewe, 1780–1923* (Manchester, 1950), p. 62; P.W. Kingsford, *Victorian Railwaymen* (London, 1970), p. 75; F. McKenna, *The Railway Workers, 1840–1970* (London, 1980), p. 48; Diane Drummond, *Crewe, Railway Town, Company and People, 1840–1914* (Aldershot, 1995), pp. 133–41.
17 Wilson, op. cit., II, p. 581.

James's, Bermondsey, for the use of its employees.[18] Other railway companies, while making no special provision of their own, pointedly exhorted their staff to attend divine worship on Sundays, and linked this to promotion prospects.[19]

As well as railway companies, individual railway leaders of one kind or another from time to time figure in Victorian religious history. David Waddington, for example, vice-chairman of the Eastern Counties Railway and vicar's warden at St Andrew's, Enfield, in 1853 asserted the power of railway capitalism by forcefully supporting the incumbent in ridding the church of its old, family-owned, box pews and in imposing new, orderly seating arrangements.[20] Later in the century, it was said of Councillor Henry Willmer, stationmaster, JP, and churchwarden in the railway town of Eastleigh, Hampshire, from 1868 to 1902, that 'Nothing in the Eastleigh of his day was done without his cognizance and approval'.[21]

More notable by far, however, was the railway contractor Sir Samuel Morton Peto (1809–89). Peto prospered in a series of family ventures, first in his uncle's building firm and then in two successive partnerships, first with his cousin Thomas Grissell, and then with his brother-in-law Edward Ladd Betts. Extensive railway building not only in Britain – the development of Lowestoft was but one of his many projects – but in Algeria, Argentina, Australia, Canada, Denmark, France, Norway, and Russia, assured his fortune. For more than 20 years he was an MP and he was knighted for his (railway) services to the Crimean War effort. A devout Baptist, he built chapels in Bloomsbury and Regent's Park. The twin-spired Bloomsbury Chapel, opened in 1848, made Baptist history on account of its architectural style and prominent metropolitan location – earlier meeting-houses had been tucked away in back alleys and upper rooms – and inaugurated a new era in Baptist witness in central London. Positioned as it was between the fashionable Bloomsbury squares to the north and the slums of St Giles to the south, it endeavoured to meet the spiritual needs of all classes. From his parliamentary constituency of Norwich, Peto invited the Rev. William Brock to be the first pastor of the imposing new church. His work was complemented by that of George M'Cree who led a domestic mission in the Seven Dials area.[22] Peto also came to the rescue of the heavily indebted Baptist chapel in Banbury.[23] Proving that he was no narrow denominationalist, however, Peto rebuilt the parish church in

18 McKenna, op. cit., p. 46.
19 For examples see P.W. Kingsford, 'Labour relations on the railways, 1835–75', *Journal of Transport History*, I (1953–4), p. 73.
20 G. Dalling, *Enfield's Railway King. David Waddington and the Great Pew Controversy*, Edmonton Hundred Historical Society, 38 (Enfield, 1978), passim.
21 A. Drewitt, *Eastleigh's Yesterdays* (Eastleigh, 1935), p. 97.
22 *DNB*; *Bloomsbury Central Baptist Church. A Brief History* (London, n.d.), pp. 1–2. I am indebted to Professor P.S. Bagwell for this reference.
23 B. Trinder, *Victorian Banbury* (Chichester, 1982), p. 111. Banbury provides another example of a railway philanthropist in the person of William Mewburn, stockbroker and chairman of the South Eastern Railway and Star Life Insurance Companies. As steward of the Wesleyan Methodist circuit in Banbury from 1866 he made huge contributions to the costs of both the Marlborough Road and Grimsbury chapels. Chapels of other religious denominations, however, benefited from his generosity. Thanks largely to him, no Nonconformist chapel in Banbury in the 1870s was in debt (Trinder, op. cit., p. 117).

Somerleyton, Suffolk, where he resided in style in a mock Jacobean hall surrounded by a newly built model village.[24]

In 1850, with 14 000 men on his payroll, Peto was one of the largest employers of labour in the country, and renowned as one of the most caring. To Edwin Chadwick, Peto represented the untypically moral face of railway capitalism, and Peto's ventures were noticeably free from the unfeeling exploitation which increasingly attracted notice in the mid-Victorian period.[25] Peto not only provided housing (where necessary) but practical Christanity in the form of schools and schoolbooks, chaplains and scripture readers, preferably working-class men themselves to whom the navvies could relate more easily. (He employed ten of them on the Ely–Peterborough line.) Weekly, not monthly, wages were given. Tommy-shops – a notorious abuse practised by most other railway contractors – were banned. The sale of beer was prohibited on his workings. Compensation for injury and death (small, but at least something) was provided.[26]

> Give [the railway worker] legitimate occupation [said Peto in the House of Commons in 1847], and remuneration for his services, show him you appreciate those services, and you may be sure you put an end to all agitation. He will be your faithful servant.[27]

The Dean of Ely confirmed the fact. The conduct of Peto's labour force during two years in the Fens, he declared, had been 'an example to the district'.[28] William Wordsworth, no railway lover, was struck by the respectful demeanour of other navvies in June 1845 taking their 'noontide rest' amidst the ruins of Furness Abbey next to the line they were constructing.

> All seem to feel the spirit of the place
> And by the general reverence God is praised.[29]

Rather later, in *Self Help* (1859), Samuel Smiles retailed the story of two English navvies – *en route* for foreign employment – being moved to follow a lonely, unknown funeral cortège to a burial in a Paris cemetery.[30]

---

24   N. Pevsner, *The Buildings of England. Suffolk* (2nd edition, Harmondsworth, 1974), pp. 421–3. Somerleyton, of course, also had its own railway station, complete with Peto's crest above the entrance porch (G. Biddle, *Victorian Stations* (Newton Abbot, 1973), p. 206). Sadly, Peto was bankrupted by the Overend & Gurney crash of 1866, sold his estate, resigned from Parliament, and lived out the rest of his life in provincial obscurity.

25   R.A. Lewis, 'Edwin Chadwick and the Railway Labourers', *Economic History Review*, 2nd series, III (1950), pp. 107–16. T.A. Walker and, later, Messrs Lucas & Aird, were other Christian-spirited railway contractors.

26   R.J. Joby, *The Railway Builders. Lives and Works of the Victorian Railway Contractors* (Newton Abbot, 1983), pp. 58–62; T. Coleman, *The Railway Navvies* (Harmondsworth, 1968), pp. 29, 58, 74, 172, 174; D. Brooke, *The Railway Navvy* (Newton Abbot, 1983), pp. 151, 153, 154.

27   Quoted in Coleman, op. cit., p. 70.

28   Quoted in Brooke, op. cit., p. 158.

29   J. Morley (ed.), *Complete Poetical Works of William Wordsworth* (London, 1909), p. 787.

30   S. Smiles, *Self Help* (London, 1859, rp. 1905), pp. 472–3.

Such instances notwithstanding, railway construction workers in the nineteenth century – whose numbers at times exceeded the combined strength of the armed forces – were not renowned for piety and churchgoing.[31] Devout contractors like Morton Peto tried to win them to the gospel. Railway companies sometimes donated funds for the same purpose.[32] The Church Pastoral Aid Society, founded in 1836, and individual local initiatives also made well-intentioned but often inept and ineffectual contributions.

> It is no use a preacher going occasionally on a fine Sunday [said the no-nonsense Mrs Garnett]. Navvies have learnt, practically, to disbelieve in Christians. They judge men by their fruits. If we are to do them good we must deny ourselves. Establish regular services, day schools, night schools, reading rooms, house-to-house visitation etc... The work, to do anything, must be constant.[33]

Missionary efforts of this intensive kind attracted more publicity. Thomas Fayers's *Labour among the Navvies* (London, 1862) was based on his personal experience of missionary work on two sections of the railway line under construction in Westmorland. Another such account was the anonymous *Death or Life, or the Story of my Experience on 'the Line'* (London, 1864) which documented the trials, tribulations, and rejoicings of a female evangelist who devoted herself to tract distribution and school teaching. She told of one of her converts being taunted 'There's a navvy! A navvy going to church!', adding that his spiritual progress was 'in some respects the greatest encouragement I ever had'.[34] D.W. Barrett had been curate in charge of the Bishop of Peterborough's mission to railway navvies in 1876–8. *Life and Work among the Navvies* (London, 1880) was his report on the growth of mission chapels and schools, savings banks, and libraries, and on the full and active encouragement given to all these endeavours by the railway contractors, Messrs Lucas & Aird. Barrett's book, in fact, was dedicated to them and offered the writer's 'grateful acknowledgement of your keen desire and generous efforts to ameliorate the social condition, and still better, to further the spiritual interests of the men in your employ'. Not only the Bishop of Peterborough but also the Bishop of Lincoln came to preach in the mission chapels. And what more appropriate text could the inspired second prelate have chosen than Isaiah, XL, 3?

> Prepare ye the way of the Lord!
> Make straight in the desert
> A highway for our God.[35]

---

31   Their reputation for disorder and violence, however, has been exaggerated. See Brooke, *The Railway Navvy*, and D. Brooke, 'The "lawless" navvy: a study of the crime associated with railway building', *Journal of Transport History*, 3rd series, X (1989), pp. 145–65.

32   For examples from the years 1845–9 see P.W. Kingsford, 'Labour relations on the railways, 1835–75', *Journal of Transport History*, I (1953–4), p. 72.

33   Brooke, *The Railway Navvy*, p. 132; Chadwick, *The Victorian Church*, I, pp. 449–50; E. Garnett, *Our Navvies, A Dozen Years Ago and Today* (London, 1885), p. 20. On Mrs Garnett see p. 108 below.

34   *Death or Life*, pp. 136–37.

35   Barrett, op. cit., p. 113.

As ever in missionary activity of this kind, much effort was devoted to the distribution of religious tracts and other improving literature. Barrett strongly recommended titles by Samuel Smiles – *Self Help*, *Thrift*, and the *Life of George Stephenson* – as books calculated to stir the latent capacities of navvy readers. Mr Morley, navvy missionary on the Hull & Barnsley line, gave out 16 000 books and tracts in 1882 alone! *Death or Life* (1864) had earlier listed railway tracts then in circulation among navvies; *Light for the Line*, *The Bar of Iron*, *Sunset on the Line*, *The Railroadman's Last Year*, *A Dying Navvy* were some of them.[36]

By the time Barrett's book had been published in 1880, local missions to railway workers, such as that initiated in the diocese of Peterborough, were being supplemented by the activities of the Christian Excavators' Union and the Navvy Mission Society, founded in 1875 and 1877 respectively. Associated with the redoubtable campaigner Mrs Elizabeth Garnett (1839–1921), clergyman's daughter and clergyman's widow, the second society had a long and vigorous life and published a regular *Quarterly Letter to Navvies* which combined hot gospelling with practical advice and danger warnings, hints on first aid and savings banks, and up-to-date news of job vacancies, marriages, and deaths.[37]

The Navvy Mission Society was quintessentially a temperance movement. 'Our settlements are soaked in drink', thundered Mrs Garnett.[38] D.W. Barrett took the same line. 312 000 gallons of beer sold in a year on the Peterborough line was his estimate. 'The quantity consumed was something fearful.'[39] Mrs Garnett promoted tea as a wholesome alternative to the demon drink and hymn singing – especially hymns with uplifting choruses – as a counter to alehouse rowdyism.

> Yes, I am an English navvy, but oh, not an English sot,
> I have run my pick through alcohol in bottle, glass, or pot.
> And with the spade of abstinence and all the power I can
> I am spreading out a better road for every working man.[40]

Who could resist the call?

The majority of navvies, of course, probably did. But the temperance and mission movements – usually inseparably bound up together – targeted company railway servants and not just the armies of men involved in the construction of the lines. In so many ways, in fact, the Victorian temperance campaign and the railway went hand in hand, as Brian Harrison has convincingly shown. As rail took over from long-distance coach travel old-style inns declined. Some railway companies refused to sell intoxicating liquor at their stations and were keen to recruit total abstainers to their staff. In the railway towns the controlling companies exercised strict control over the numbers of public houses. Temperance reformers invested in railways. It was no accident that Preston, Lancashire – a railway junction – became a major hub of teetotalism as well. Thomas Cook (1808–92) and his son, John Mason Cook

---

36   Garnett, op. cit., p. 83; *Death or Life*, pp. 187–8, 253, 254.
37   For Mrs Garnett see Coleman, *The Railway Navvies*, pp. 176–86.
38   Garnett, op. cit., p. 141.
39   Barrett, op. cit., p. 17.
40   Garnett, op. cit., pp. 110–11.

(1834–99), architects in so many ways of the growth of the nineteenth-century tourist trade, were both ardent temperance reformers. (Cook's first organized railway excursion in 1841 from Leicester to Loughborough, as is well known, was to a temperance convention.[41]) 'The railway probably did more for temperance in the nineteenth century', says Harrison, 'than either the Temperance Movement or the Vice Society.'[42]

Always stronger among railway company servants than among navvies, whose shorter contracts and greater mobility made them intrinsically less susceptible to discipline of all kinds, the Temperance Movement established key bastions in the great railway works at Crewe and Swindon. The Railway Mission in London was a major centre for meetings and publications. Its principal mouthpiece was *The Railway Signal, or Lights along the Line*, a monthly journal launched in 1882 'advocating Christian life and Christian work on the railways of Great Britain'. It carried regular news items and reports, testimonies, and tracts – such as one on the awesome contrast between the destinations reached at the end of the broad gauge railway which accommodated sin and the narrow gauge to heaven which eschewed it. There were also many publishers' advertisements, and announcements concerning mission services for railwaymen.

Issue number 6 (January, 1883), for example, carried advance notice of prayer meetings and a short address every Tuesday and Friday 'at the breakfast hour' in Bristol in the Midland locomotive shed. Details were also given of regular Bible classes on Sunday afternoons in the second-class ladies' waiting-room at Portsmouth Town station, and a Sunday gospel service at 11 a.m. in the luxury of the first-class waiting-room at Lowestoft GER station. The same issue carried the itinerary for the Sankey and Moody crusade from 30 December 1882 to 14 April 1883. Publisher's advertisements for Sankey and Moody hymnbooks, tracts, and books appeared regularly in *The Railway Signal*. (Morgan & Scott were offering the 104th thousand printing of Moody's addresses in June 1883.) The following month's issue paraded a prominent advertisement for a special 15-day series of services at the Railway Mission Hall and Kensal Hall led by William Groove (Happy Bill), William R. Lane, and George Clarke, secretary of the Railway Mission.

The issue for February 1885 carried a report on a daily Bible class for working men on the 5.57 a.m. train from Enfield to London, and exhorted others to follow the example and become spiritual commuters. Interspersed in this and most other issues were poems and hymns, many of them by the prolific Samuel Peach, a kind of aspiring William McGonagall. One example of his *oeuvre* will suffice to capture its prevailing spirit:

> Lord of Glory still watch o'er us
> Thou our pilot be
> On the line of life before us
> Guide this company.

---

41   *DNB*.
42   B. Harrison, *Drink and the Victorians. The Temperance Question in England 1815–1872* (London, 1971), pp. 334, 335, 117, 336.

Chorus
In thy service faithful keep us
Guard us night and day.
Fit us for the final signal
Ready, right away!

...
Keep us watching and re-firing
With full pressure on
For promotion still aspiring
Till the prize is won.

Chorus
In thy service faithful keep us, etc.[43]

Evangelists were quick to exploit the potential of the railway as a religious metaphor, a new way of depicting pilgrims' progress. An inscription on a tombstone in the south porch of Ely Cathedral offered a shortened version – six verses out of eighteen – of a railway poem circulating in the early 1840s:

The Spiritual Railway

The Line to heaven by Christ was made
With heavenly truth the Rails are laid.
From Earth to Heaven the Line extends
To Life Eternal where it ends.

Repentance is the Station then
Where passengers are taken in.
No fee for them is there to pay
For Jesus is himself the way.

God's word is the first Engineer.
It points the way to Heaven so dear,
Through tunnels dark and dreary here
It does the way to Glory steer.

God's Love the Fire, his Truth the Steam
Which drives the Engine and the Train.
All you who would to Glory ride
Must come to Christ, in him abide.

In First and Second, and Third Class,
Repentance, Faith and Holiness
You must the way to Glory gain
Or you with Christ will not remain.

---

43   *Railway Signal* (February 1885), p. 89.

Come then poor Sinners, now's the time
At any station on the Line
If you'll repent and turn from Sin
The Train will stop and take you in.[44]

Examples abound of railway literature of this type, both prose and poetry. A broadside ballad, *The Railroad to Heaven*, appeared in 1845. In the same year, but in a quite different vein, a forceful pamphlet entitled *The Railroad Considered in a Moral and Religious Point of View*, inveighed against the new mode of communication as 'the masterpiece of Satan', and identified railways with Sabbath desecration, violence, fraud, and injustice. J. Wright's *Christianity and Commerce* (London, 1851) and *The Latter Days: Railways, Steam and Emigration* (Dublin, 1854) were more positive. In 1872, the SPCK brought out *The Railway Ticket: a tract on infant baptism*. Rather later in the century, in 1883 the Rev. F.J. Bird published a forthright tract, uncompromisingly entitled *SHUNTED. A Sermon to Railwaymen*. First preached at the Baptist Chapel, Walsworth Road, Hitchin, and with all profits generously donated to the Railway Orphan Fund, the sermon squeezed every last drop out of its text (Galatians, V, 7: 'You did run well; who did hinder you?') and its chosen keyword. The different causes, courses, and consequences of shunting were vigorously rehearsed, as was the spiritual resonance of the railway wagon labels 'Not to go' and 'Home with all speed'. The preacher was clearly panting with ecstasy as he reached his bright finale.

> Then, with slackened speed, it may be, and wasted steam, but at nonetheless sure pace, may you finally draw into the platform of the Great City – and stop, your journey safely accomplished, its dangers passed, at home with God, safe in Heaven at last![45]

The issue of *The Railway Signal* for September of the same year (1883) included the no-less-edifying poem, *Life: a Railway* (anon).

> Birthdays, like stations, fill the space between
> And darksome vales and spots of cheerful green
> Alternate rise, while mortals hurrying on
> Behold life's many stations past and gone;
> The warning sounds, the end approaches fast
> And then the terminus is reached at last.

Some of this literature was graphic as well as figurative and translated railway maps into spiritual messages. *The New Railway Guide* (London, 1848) is one such publication, with its fold-out map depicting two clusters of railway routes, separated by Neutrality-shire and leading either to perdition and eternal death or to Glory and eternal life. The first group of lines (intemperance was one of them) passed through Unbelief-shire and Evilhabit-shire while the other groups of lines went by way of

---

44   Martha Vicinus is clearly wrong to claim the spiritual railway metaphor as an import to Britain which came with American evangelists (*The Industrial Muse* (London, 1974), pp. 38–9). The British Library has a copy of the poem, described as a song, dated c.1840.

45   Bird, op. cit., p. 16.

Self-renunciation-shire and Sanctification-shire, and crossed the sea at Victory Point. The map and accompanying text briefly noted some unfrequented or unfinished lines. 'It is said that there is one line called Unitarian line which reaches as far as Enquiry-shire but no further.'[46] As for the Popery line, much of it was said to be no more than a branch line, 'very zig-zag and tedious'. Amazingly, however,

> many nervous persons have given the preference to this line, on account of its antiquity and the dictatorial manner of its Director (for there is only one) who makes so many boasts of his infallibility that they believe him at last... It is hoped, however [said the frank and impartial author], that the line will be closed 'ere long since it is evidently an imposition on the public, and then there will be a better prospect of all the lines that lead to Glory being amalgamated and becoming one grand line with an altogether new name.[47]

Another example, from the Railway Tract Depot and advertised in *The Railway Signal* in January 1895, was a map only of *The Up and Down Lines*. It was described as 'size 24 × 20, beautifully coloured. Also mounted on rollers' – clearly envisaged as something that should adorn every railway worker's home. It was, said the accompanying text, 'a most interesting and instructive picture showing the Down Grade of Sin from Eden and the Upward Course of the Believer from the City of Destruction to Glory'. The down line made its descent from paradise, through wilderness and captivity, by-passing Calvary, and with alluring stopping points offering sport, gambling, dancing, and the demon drink. The final stop – Pit station – was shown as being open day and night and selling only one-way tickets. Not far beyond it, all trains, bearing their unfortunate human freight, crashed down eternity rapids. The Celestial Route station, by contrast, had placards offering refreshments, life insurance, grace, pardon, and peace. The main line went over the bridge of mercy, through Grace Station, and on through Calvary Junction to Glory.

Missions to railway navvies and to railway company servants, of course, represented only some of the connections between railways and Victorian evangelism. W.H. Smith (1825–91), nicknamed 'the Northwestern Missionary' and 'Old Morality', cleansed the railway bookstalls of unwholesome reading matter and acted as treasurer of the SPCK.[48] 'At the [London] railway stations', observed the stunned Frenchman Hippolyte Taine in 1862, 'there are large chained Bibles for travellers to read while waiting'.[49] The railway network facilitated the preaching itineraries of such great preachers as Charles Haddon Spurgeon (1834–92), the last major Calvinist leader, and brought great crowds into special services held at mission halls and cathedrals throughout the country.[50] Near Aberdeen in the early 1860s Taine encountered railway excursionists heading for a mass gathering which 20 000 were expected to attend.[51]

---

46 *New Railway Guide*, p. 21.
47 Ibid., pp. 12, 13, 14.
48 *DNB*.
49 *Taine's Notes on England*, p. 13.
50 On Spurgeon see Chadwick, *The Victorian Church*, I, pp. 417–21 and E.W. Bacon, *Spurgeon, Heir of the Puritans* (London, 1967), passim.
51 *Taine's Notes on England*, pp. 281–2.

Much more mundanely and routinely railways came to underpin both itinerancy and the circuit system, the twin pillars on which Wesleyan Methodism rested. In the 1870s when a great revival quickened the pulse of religious life in England, it was the railways which enabled the American evangelists, Ira D. Sankey and D.L. Moody, to stomp the country. (They returned in the early 1880s with a gruelling itinerary which took in Dublin, Birmingham, Belfast, Leicester, Nottingham, Manchester, Leeds, Liverpool, and London within the space of a few weeks.) It was railways also that, after 1875, brought throngs of people to the annual Keswick Convention.[52] It was a conjunction of tub-thumping preaching, brass bands, and the railway that propelled the Salvation Army into its early successes. A railway navvy and John Lawley, 'the saved railway guard', were two of William Booth's earliest helpers. (Lawley's arresting technique was 'to tear his song book to shreds (in the manner of the devil) and dive from the platform making swimming motions to demonstrate the sea of God's love and pardon'.) The age of the 'Hallelujah railway ticket' (a propaganda device of the Salvationist, William Corbridge), issued free to all seekers, had dawned. The format was exactly the same as a conventional railway ticket, but this one was issued by the Salvation Railway and guaranteed to convey its holder from Sin-land 'where they are in danger of perishing eternally' to Glory-land 'where they shall praise the Lamb for ever'.[53]

As a kind of epitaph to this study we may note that railway provision did not fail to take account of the dead as well as the living, as the Necropolis Railway amply demonstrates. The London Necropolis & National Mausoleum Society, to give it its full title, was incorporated in 1852 as a contribution to a solution to London's vast interment problem. With the cholera epidemic of 1848–9 a recent and ghastly memory and with metropolitan burials even in normal years running at about 55 000, this scheme for conducting multiple funerals at the same time at an out-of-London site seemed eminently appealing. From a dignified-looking private railway station at Waterloo, designed by William Tite and equipped with coffin lifts and mortuary space beneath the railway arches, the line extended to a 2 000-acre cemetery, cleared from the Surrey heathland at Brookwood, near Woking, and reached by way of Necropolis Junction. Bishop Sumner of Winchester officiated at the consecration ceremony on 7 November 1854. A North and South station, to which different portions of the daily funeral train were directed, gave access to Nonconformists (plus Roman Catholics and Jews) and Anglicans respectively.[54]

> This is the most peaceful railway station in the kingdom [boasted the *Railway Magazine* in 1904], this station of the dead. Even the quiet, subdued puffing of the engine seems almost sympathetic with the sorrow of its living freight. But this is a sad station... For every time it is used means an occasion of grief and pain to those who tread its platforms... One is forcibly reminded of the last great station on the railroad of life; of the final platform; of the completion of this world's journey.

52   L.E. Elliott-Binns, *Religion in the Victorian Era* (London, 1936), pp. 216–21, 223, 334, 373.

53   K.S. Inglis, *Churches and the Working Classes in Victorian England* (London, 1963), pp. 179, 187, 186.

54   J.M. Clarke, *The Brookwood Necropolis Railway* (London, 1983), pp. 8, 10, 13, 19, 33.

The Necropolis Company employed its own chaplains. Entirely in line with Victorian society and its railways, class distinctions were preserved in its day-to-day operations – both for the dead as well as for the living – in the London station and on board the trains. First-class funerals enjoyed luxury all the way; third-class funerals had to make do with communal facilities – and, indeed, with a communal interment service at Brookwood. (Even after the simplification of the passenger fare structure in 1918, three classes of coffin tickets – single fares only! – continued to be issued.)[55] Extra railway carriages, or even trains, could be provided for grand occasions or emergencies.[56]

With cruel irony, the Brookwood funeral which seems to have attracted the largest attendance was that of the Victorian freethinker, Charles Bradlaugh. A crowd of 5 000 or so gathered for the interment of this renowned opponent of Christianity and conventional respectability.[57] The Necropolis Railway was pressured to dispense with its usual elaborate funerary rituals for the occasion, which took on a different tone altogether and had some of the gaiety of a seaside railway excursion; it seemed more like Brighton than Brookwood cemetery. Only Mrs Annie Besant – Bradlaugh's co-worker – wore black.

Secularism, however, formed part of the connections between religion and railways in the Victorian age and added to the many cross-currents and contrasts. Seaside resorts, the direct product of railway expansion, dispensed both secular diversions (music halls, bandstands, piers, bathing, donkey rides, and all the rest) and spiritual refreshment from their abundant supply of churches. Perhaps the supreme paradox was that the railway, which in so many ways placed itself at the service of Victorian religion, could easily develop a quasi-religious aura of its own. How else should we read Charles Mackay's poem of 1846?

> Lay down your rails, ye nations near and far
> Yoke your full trains to steam's triumphal car;
> Link town to town; unite in iron bands
> The long-estranged and oft-embattled lands.
> Peace, mild-eyed seraph – knowledge, light divine,
> Shall send their messengers by every line.
> Men join'd in amity shall wonder long
> That hate had power to lead their fathers wrong;
> Or that false Glory lured their hearts astray
> And made it virtuous and sublime to slay...
> Blessings on science and her handmaid Steam!
> They make Utopia only half a dream.[58]

---

55   Ibid., pp. 25, 27, 28, 30, 42–4.
56   Seven thousand corpses from the burial ground of St Mary Lambeth, which was cleared to make way for the railway extension to the new Charing Cross terminus in the 1860s, were re-interred at Brookwood (A.A. Jackson, *London's Termini* (Newton Abbot, 1969), p. 244). The brutal displacement of the old St Pancras burial ground aroused a great public outcry in the 1860s (J. Simmons, *St Pancras Station* (London, 1968), pp. 38–41).
57   Clarke, op. cit., p. 33. On Bradlaugh see E. Royle, *Radicals, Secularists and Republicans. Popular Freethought in Britain, 1866–1915* (Manchester, 1980), pp. 88–92.
58   Charles Mackay, 'Railways 1846' in *Poetical Works of Charles Mackay* (1876) quoted in S. Legg (ed.), *The Railway Book. An Anthology* (London, 1952), p. 125.

An article in *Building News* in 1875 urged – with the neo-Gothic of St Pancras chiefly in mind – that the great railway termini and their hotels should be viewed as the nineteenth-century equivalent of medieval cathedrals and monasteries. 'They are truly the only representative buildings we possess.'[59] More than a century later and shunning reverence and nostalgia, Jack Simmons has wisely taught us to recognize the many layers of meaning that such statements contain.

---

59 *Building News*, 29 (1875), p. 133, quoted in J. Richards and J. Mackenzie, *The Railway Station. A Social History* (Oxford, 1986), p. 20.

Chapter 8

Railways, their Builders, and the Environment

Gordon Biddle

Stepping into the twenty-first century e-world has interesting parallels with the beginning of the steam railway 200 years ago. As a means of transport the railway was already over a century old, but primitive and slow. Steam transformed it. Like the internet, the transformation came suddenly. An American wrote, 'it burst rather than stole upon the world. Its advent was in the highest degree dramatic. It was even more so than the discovery of America.'[1] Steam power was the world's first truly great technological breakthrough, leading by various routes to today's technology. Writing about it, Asa Briggs said, 'It was steam locomotion which was, and still is, usually taken to be the chief of the *triumphs of steam*.'[2] He was referring, of course, to the railway.

Like the internet, railways gained adherents who saw in them a means of combating mankind's social ills.

> We see, in this magnificent invention, the well-spring of intellectual, moral, and political benefits, beyond all measurement and price – the source of a better physical distribution of our population – a check to the alarming growth of cities, especially of manufacturing towns, and of this Babylon in which we write – and the source, above all, of... a diffusion of intelligence over the whole country.[3]

The writer of those words could have been living today. In fact, the opposite happened, and it is the purpose here to examine the environmental effects of railways that were so optimistically forecast. But to do so one must first briefly look at the aspirations and motives of some representative figures who played dominant, sometimes conflicting, roles which brought those effects about.

1   C.F. Adams, jr, *Railroads: their Origin and Problems* (1886 edition).
2   Asa Briggs, *The Power of Steam* (1982).
3   *Quarterly Review*, 42 (1830).

Railway Visionaries

The system developed piecemeal, purely on a commercial basis, with no overall plan of any kind, although two early contemporaries stand out for their vision of a national system designed to serve the rapidly changing needs of an emerging industrial economy. Thomas Gray (1787–1848) passionately advocated a network built and owned by the government. He was regarded as a crank, but he lived long enough to see the framework of a system, built by private interests with no thought of planning.[4] Gray lacked the skills and resources to back his arguments, unlike William James (1771–1837) who had both. He was a wealthy engineer and surveyor turned coal owner and canal proprietor who in 1800–20 unsuccessfully tried to promote many railways at his own expense but who overreached himself and became bankrupt. With greater patience he could have been more convincing and would have profited from his enterprise. Like Gray, he died a poor man.[5]

The successful visionaries were realists, who knew that turning ideas into reality needed ingenuity, adaptability, hard work, and, above all, money. There were practical men around who had the needed technical skills, or who were able to learn them, but in Britain the private money market was the only source of capital to finance their work. The financiers and entrepreneurs also had vision, of a different kind, seeing that money spent on building railways could transform the economy and thereby make more money.

The earliest promoters, however, had little thought of the wider benefits of railways. They were motivated by more immediate needs. These were men like the Duke of Portland, whose Kilmarnock & Troon Railway of 1807, the first in Scotland to use locomotives, was built to take coal from his pits to the nearest harbour for shipment, or the Leicestershire coal owners who promoted the Leicester & Swannington Railway of 1832 to connect their collieries to the nearest market. For a wider business vision one has to look to north-eastern England, and the Quaker father and son, Edward and Joseph Pease, who led the projection of the Stockton & Darlington Railway (1825). They marshalled investment from the Quaker network and built up a regional industrial empire with the North Eastern Railway as its hub; a railway that up to 1860 was the most profitable in the country.[6]

The success of the Stockton & Darlington acted as a catalyst. Five years later the Liverpool & Manchester Railway was opened, the first 'main line' in the modern sense. Its promoters were led by Joseph Sandars of Liverpool, another Quaker, whose connections, like the Peases', created a Liverpool group of investors who built up a capital base that had more influence on the country's railways than any other, including London. Their vision was shared by others, like the Bristol merchants who saw that a railway to London could arrest their city's declining fortunes in maritime activities. There were also those who stepped beyond acceptable business ethics, sometimes beyond the law, notably George Hudson, the 'Railway King', although

---

4   W.T. Jackman, *The Development of Transportation in Modern England* (1916).

5   E.M.S. [Paine], *The Two James and the Two Stephensons* (1961 edition).

6   M.W. Kirby, *The Origins of Railway Enterprise: A History of the Stockton & Darlington Railway, 1821–1863* (1993).

even he had a vision of a national railway system, albeit one under his own control centred on his power-base at York. Manipulators like Hudson were observed by lesser men, fraudsters ready to make a fast buck by deluding naive investors into losing their savings in the most impracticable speculative schemes; investors who were mindless or ignorant of the South Sea Bubble a century before. Those dishonest dealings were the negative side of railway promotion. Yet despite the fervour for railways there was no grand plan. The longest of the early trunk routes, the Great Western from London to Bristol, was only 116 miles, and the 1 000-odd miles of trunk line opened by 1845 were operated by 17 companies.[7] It is significant that many European countries adopted a completely different way of developing a railway system.

The early engineers could see the advantages of a nationally-planned network, but they could only work within the prevailing order. George Stephenson had the sagacity to combine opportunism with his much sought-after talent. For instance, he did not let his friendship with William James prevent him from taking over James's surveys of the Liverpool & Manchester, and eventually its construction, followed by the Canterbury & Whitstable. James bitterly averred that it was done with Stephenson's connivance. Neither did Stephenson hesitate to claim credit for the work of his assistants, nor to pass on blame for mistakes. He probably made more money from railway share-dealing and the coal and mineral interests he acquired during railway work than from building the lines themselves, and unlike his younger colleagues he retired to enjoy the proceeds, living until he was 67.

His son Robert was more innovative and built up the highest reputation as the foremost and most prolific railway engineer of his day. But he worked himself to death, dying at the age of 51. His contemporary, I.K. Brunel, was an enigma. Possessed of an inventive vision that extended beyond railways to steamships and much else, his single-mindedness, some say stubbornness, led him up blind alleys like the abortive atmospheric propulsion system and, with far-reaching consequences for the Great Western Railway, the broad gauge. He refused to admit defeat even in the face of 1 901 route miles of 4 ft 8½ in. gauge – the 'standard gauge' – against only 274 miles of his 7 ft gauge. As his biographer commented, 'It may have been commercial folly, but then like oil and water, ideals and commerce do not mix.'[8] Brunel, too, died young of overwork, aged only 47.

The last of the three great contemporaries, Joseph Locke, may be said to have had a consistent mission. Of all the many lines he built, most of those that came to form the West Coast Main Line were made as the result of his constant advocacy of a continuous route from London to Scotland. When it finally reached Aberdeen in 1850 it had been built by no less than 18 different companies, the greater part of it under Locke's supervision. Although rivals, these three engineers were good friends, and Locke outlived the others by only a year, dying aged 55.

Engineers who followed the pioneers needed more condensed vision in building on their predecessors' work; men like Joseph Mitchell who pushed railways into the Highlands of Scotland, and J.S. Crossley, a company man who built the Settle &

---

7   J. Simmons & G. Biddle, *The Oxford Companion to British Railway History* (1997), article 'system, development of the'.

8   L.T.C. Rolt, *Isambard Kingdom Brunel* (1957), p. 159.

Carlisle line for the Midland Railway. Although their tasks were no less demanding, their sense of purpose was of necessity different, tuned to the highly competitive nature of railway expansion, akin to the great company chairmen and managers who turned amalgamations of often antagonistic companies into large territorial corporations with jealously guarded spheres of operation. Intent on increasing what today is called market share, they were concerned to weld diverse attitudes and loyalties into a single corporate ethos, which they did with a determination that was at times ruthless. The careers of the redoubtable Mark Huish of the London & North Western and Richard Hodgson of the North British were among those that ended in disgrace. Others gained honours. James Allport, who guided the provincial Midland Railway through its great period of expansion to London and to Scotland, received a well-deserved knighthood; the less-scrupulous Edward Watkin's was less so. Chairman of three companies and the first, unsuccessful, Channel Tunnel company, he was a man before his time. His dream was to use his railways as the English part of a continuous route from Manchester to Paris.

All these men had tremendous corporate vision, and if there is emphasis on civil engineers and managers at the seeming neglect of often equally far-sighted designers of locomotives and carriages, or the men responsible for signalling and the running of trains, it is because it was they who, primarily, created the environmental impact of railways with which we are concerned; an impact still prominent, but now largely taken for granted. In the early days of railways, just as today, there were many in the old world outside the new technology who were apprehensive of the changes that were happening with such speed, unwelcome changes in society that aroused social consciences, inbuilt prejudice, self-interest, or simply fear of new processes that were little understood. Their influence was significant. The visionaries, engineers, and power-brokers did not by any means have it all their own way.

Railways in the Landscape

Railways affected the environment in three main ways: predominantly on the landscape; then by atmospheric pollution; and thirdly on the ear. They were criticized for the way in which they were thought to despoil town and country, far more than the canals had done, and they were also dirty and noisy. Early opposition was greatest from landowners, mainly driven by self-interest, whose families had spent large sums in landscaping parks and improving agriculture. Some opposed railways through an altruism derived from the romantic movement and the concept of the picturesque in natural scenery. Consequently, many demanded high prices for their land, in cash and in kind in the shape of deviations of route, unnecessary tunnels to hide the railway from view, and ornamental bridges. The Earl of Lichfield, for instance, secured all three at Shugborough in Staffordshire, together with decorated portals to the tunnel and a station away from sight and sound but near enough for his convenience. Yet who today can deny that Shugborough, owned by the National Trust, is not the better for them?

It took all the persuasion of Joseph Mitchell, the celebrated Scottish engineer, to gain the Duke of Atholl's reluctant assent to a line through 24 miles of Highland scenery that included Birnam and Killiecrankie passes, and Blair Castle, at the price of carefully-executed engineering works that soon rendered the railway inconspicuous – unlike the widened A9 trunk road today – or, where intrusion was unavoidable, such as river bridges, by bold designs that provoked admiration. Government agencies were equally defensive of Crown property, keeping two railways out of Windsor Home Park, while the Southampton & Dorchester Railway was forced into a roundabout route through the New Forest. The Admiralty and the War Office would only sanction railways on their own terms, the former often to secure improved sea defences at the railways' expense, the latter to prevent imagined impairment of military defences.[9]

Writers and poets were vocal critics. John Clare was one of the earliest, complaining in 1825 about railway surveyors invading a favourite wood.[10] John Stuart Mill, in 1836, cited proposed routes for the London & Brighton Railway running close to Box Hill when he declared 'in this country the sense of beauty, as a national characteristic, scarcely exists.'[11] William Wordsworth and John Ruskin were particularly vociferous. In 1833 Wordsworth thought that the benefits of railways outweighed the marring of 'the loveliness of Nature',[12] but quickly changed his attitude when the Lake District was threatened by the Kendal & Windermere Railway in 1844, composing his well-known sonnet that began 'Is then no nook of English ground secure / From rash assault?'[13] He failed to stop it, but with support from Ruskin and others prevented an extension to Keswick, although successive schemes had to be fought until 1870. A few years later proposed lines through the Newlands Valley to Buttermere, and into Ennerdale, were defeated. Ruskin was particularly vitriolic in his opposition to railways in the Derbyshire Peak District, producing his famous blast against the line through Monsal Dale whereby 'every fool in Buxton can be in Bakewell in half-an-hour, and every fool in Bakewell in Buxton.'[14] Henry James, too, in 1879 bitterly decried the despoliation of the Isle of Wight by the 'hideous embankments and tunnels' of 'the detestable little railway.'[15]

Others saw social benefits outweighing environmental damage. In 1855 Harriet Martineau considered that the biggest change ever to have affected the Lake District, the railway, was also the best, by introducing moral, intellectual, and economic improvements.[16] T.H. Huxley, in 1886, suggested that railways made 'the beautiful scenery accessible to all the world'. To him, the single line along Bassenthwaite Lake did nothing to interfere with his 'keen enjoyment of the lake any more than the tarmacadamised road did'.[17] He might not have said the same about the broad tarmac highway that has taken the railway's place.

---

9   See G. Biddle, *The Railway Surveyors* (1950), pp. 84–8, for landowners' attitudes.
10   J.W. & A. Tibble, *The Prose of John Clare* (1951).
11   *Collected Works* (1982), vi, pp. 327–8.
12   *Steamboats, Viaducts and Railways*, in *Shorter Poems* (Everyman edition), p. 573.
13   *On the Projected Kendal & Windermere Railway*, in K. Hopkins, *The Poetry of Railways* (1966), p. 72.
14   E.T. Cook & A. Wedderburn, *Works of John Ruskin*, xvii (1907), p. 86.
15   *English Hours* (1905), p. 232.
16   *Guide to the English Lakes* (1855), pp. 141–4.
17   L. Huxley, *Life and Letters of Thomas Henry Huxley* (1913 edition), ii, pp. 454–5.

What was the reality during the 30 years that followed the opening of the Liverpool & Manchester Railway, in 1830? In the countryside, raw earthworks were quickly covered with vegetation, and, contrary to the predictions of the railways' opponents, provided new natural habitats. W. Warde Fowler, in 1886, noted how around Kingham in the Cotswolds, where four lines met, 'this large mileage of railway within a small radius acts beneficially on our bird life'.[18] Again, Beatrix Potter observed that game grazed unconcernedly alongside the line the Duke of Atholl had opposed between Perth and Inverness.[19]

Arched bridges sat comfortably along the permanent way, while viaducts introduced elements of dramatic grandeur, even elegance, into the natural scene. Tunnel portals were usually hidden deep in cuttings, but those that were prominent, and some that were less so, received architectural treatment that lessened their impact and, by their solidity, reassured the timid traveller. As Michael Robbins has pointed out, country people were well-used to upheaval: first the enclosures, then the building of turnpike roads, followed by canals which, despite earthworks and structures on an unprecedented scale, were quickly assimilated into the landscape. On a smaller scale, they showed how the rural scene could absorb the railway.[20]

This absorption process contained two dominant ingredients: good design and the use of local building materials. Red brick in the Midlands soon mellowed. Local stone, like golden limestone in the Cotswolds, pale grey in Derbyshire and Westmorland, gritstone in the Pennines, or dark red sandstone in Dumfriesshire, to name a few, quickly weathered. Country stations, too, received care in designs that put them at home in their surroundings, looking like cottages, gate lodges, farmhouses, and small manor houses, in accordance with their importance. In them, also, local vernacular styles predominated, whether in brick, stone, flint, or half-timbering, according to district, and in the south-east, traditional clapboarding.[21]

Here again there were critics. Ruskin, shifting his ground, asserted that stations were over-elaborate: 'Railroad architecture has, or would have, a dignity of its own if it were only left to its work'.[22] The Gothic purist A.W.N. Pugin, who disliked railways, criticized stations like Euston that were built to look like something else.[23] Francis Whishaw, whose penetrating description of railways in 1842 is a classic, deplored 'the growing evil of expending large sums of money on railway appendages... instead of cottage buildings which... would have been amply sufficient.'[24]

It must also be said that in some stations there was not the consistent quality required of civil engineering works, where the stability of the line was paramount, and which could absorb most of a company's capital. Between the Tees and the Tyne, for instance, and in industrial Lancashire, parts of south-east England, south Wales, and Cornwall, there were some dreadfully cheap and nasty stations. Eventually most of them were rebuilt, but some lasted for a long time. Scotland's record was much better.

---

18   *A Year with the Birds* (1886), pp. 144–51.
19   L. Linder (ed.), *The Journal of Beatrix Potter* (1966), p. 260.
20   *The Railway Age* (1962), pp. 65–6.
21   For fuller treatment see G. Biddle, *Victorian Stations* (1973).
22   Cook & Wedderburn, viii, pp. 159–60.
23   *An Apology for the Revival of Christian Architecture in England* (1843), p. 11.
24   *Railways in Great Britain and Ireland* (1842), 1969 reprint, pp. 367–8.

The Urban Environment

Civic attitudes to railways in towns varied, although there appears to have been little active discouragement. Large towns, especially those with corporations, were usually powerful enough to admit railways on their own terms. Liverpool contributed to a suitable facade for the first Lime Street station, designed by its own surveyor;[25] Southampton sold land on a site of its own choosing at a knockdown price;[26] Carlisle also, through a specially set up railway committee; and Derby got its way over a single station for three separate railways, for which it sold land and provided improved road access.[27] Several towns, like Brighton, Chester, and Norwich, built new streets to the station of their own volition, for the benefits the railway would bring. Others made the railway do it. At Perth, the citizens successfully resisted a government-supported plan for a station on the South Inch, a large riverside public space. Parliament upheld them.[28] But at many towns the railway at first chose to stay outside, where land was cheaper. Only later did lines penetrate further in, at considerable cost.[29] London was a special case. A government commission of 1846 advised against railways entering the central area, which is why the capital still has 13 termini.[30]

In those early days, care was often taken over a railway's approach lines. The descent of the London & Birmingham Railway from Camden to Euston was through a newly developing area of villas. It was put into a cutting of some elegance, of which a fragment remains. Where the Great Western cut through Sydney Gardens in Bath, a handsome stone-walled cutting and graceful bridges were built, while the viaduct skirting the city was given Tudor detailing. The London Road viaduct which dominates the north side of Brighton is a noble structure with classical elements, while the great Wicker Arch spanning a main road into Sheffield is evidence of an early railway's good manners in an industrial city.

By 1863 such courtesies were coming to be disregarded. That year, in Leeds, the North Eastern Railway planned to connect its separated lines by a viaduct across the city, where new streets had been laid out. Widespread indignation forced the railway to adopt a revised route devised by the city's surveyor. Even then it separated the parish church from the town, and required two public buildings to be demolished.[31] In Edinburgh, fighting between a railway and the forces of local opinion went on intermittently for more than 50 years over the line through Princes Street Gardens. Each time the railway won, costing the city tunnels under the National Gallery on The Mound, loss of the fifteenth-century Trinity Chapel, and a pall of smoke until the coming of diesel trains in the 1960s. Neither of the railways that served Edinburgh built stations with faces worthy of Scotland's capital. After two reconstructions the

---

25   G. Biddle, *Great Railway Stations of Britain* (1986), p. 46.
26   R.A. Williams, *The London & South Western Railway*, i (1968), p. 26.
27   J. Simmons, *The Railway in Town and Country, 1830–1914* (1986), p. 188.
28   Biddle, *Great Railway Stations*, pp. 41–2.
29   For detailed examination of the whole subject see J.R. Kellett, *The Impact of Railways on Victorian Cities* (1969).
30   J. Simmons, *The Railway in England and Wales, 1830–1914* (1978), pp. 116–18.
31   Biddle, *Great Railway Stations*, p. 43.

North British Railway's Waverley station still had no sort of frontage. Neither had the Caledonian's Princes Street station, which originally was wooden. Instead, each became dominated by a railway hotel.[32]

The quality of provision in cities, therefore, was mixed. Although in the first years nothing quite compared with the Doric portico at Euston and its Ionic counterpart in Birmingham, other railways emulated them by building a variety of impressive station facades: Brighton, Bristol, Chester, Glasgow Bridge Street, Huddersfield, Hull, Newcastle upon Tyne, and Stoke-on-Trent were amongst the best. But rarely did they form the focus of a central square, as happened on the continent and in North America, because there was little large-scale town planning in Britain. Euston Square, Stoke, Hull, Blackburn, and Leeds were exceptions, the first four quite small. At Leeds the wooden station was hidden behind a railway hotel, which in the circumstances was just as well.[33] A sizeable minority of important city stations were downright wretched, often regarded as temporary to reduce the first cost, but frequently becoming permanent, doing no credit to the town. Glasgow's other early stations, Coventry, the entrance buildings at Manchester Central, and Brunel's Great Western wooden barns such as Cardiff, Swansea and Plymouth, were examples. All three Leeds stations had grave shortcomings, while Oxford's wooden station was a disgrace for over a century.

Railways were also guilty of damaging or destroying historic monuments, rarely accompanied by protest. York, for instance, raised no objection to arches being made in the city wall to enable a station to be built inside, although there the politics were strongly influential. Chester allowed a corner of the city wall to be pierced – admittedly a slight intrusion. But when a national monument was threatened at Conway the Commissioners of Woods & Forests stepped in. Telford's road bridge had been required to harmonize with the castle, and the railway was required to pay respect to both. So Robert Stephenson's tubular bridge was given crenellated portals, and alongside the castle a Gothic arch was cut through the town wall, with a Tudor-style station immediately beyond it. The damage was minimal, unlike at Shrewsbury which suffered grievous harm. Three railways devised an unusually collaborative scheme for a station alongside the castle in the neck of the River Severn's horseshoe loop around the town. The townspeople rejected it, but were unable to agree on two alternatives, so the railways won by the expedient of divide and rule. They did try to make amends by erecting a handsome Tudor station, but also built a viaduct which completely obstructed the fine view up the river. Their successors compounded the injury by extending the station on to the viaduct, which they widened with a hideous iron structure, and by building a loud red-brick retaining wall under the castle rampart.

Other historic sites were allowed to suffer – Lewes lost most of the medieval St Pancras Priory in return for a modestly classical station. Berwick-on-Tweed castle lost its keep, as did Northampton. Lesser damage was done to castles at Huntingdon and Penrith. The proposed railway through Furness Abbey, on its approach to Barrow in 1844, was deflected to one side, but only Wordsworth condemned the desecration

---

32   Biddle, *Great Railway Stations*, p. 42.
33   G. Biddle, *The Railway Surveyors* (1990), pp. 155–73.

of the secluded valley. Later on Nathaniel Hawthorne concurred, and later still Ruskin refused a Royal Institute of British Architects gold medal as a protest against four public environmental outrages: three in Italy and one in Britain – Furness Abbey. Of course, they were too late in shutting the stable door. Perhaps the worst example is at Newcastle, where the lines sliced right through the castle bailey, separating the keep, accompanied by widespread destruction of other historic property and part of a public open space. It was part of a deal for a much-needed new road as part of a railway bridge across the Tyne. There were other, aesthetic, compensations. Newcastle gained what in many ways is the finest station in Britain, and the bridge is an undoubted masterpiece.

Rural antiquities were also threatened. Maumbury Rings, near Dorchester, were saved; Spettisbury Rings in the same county, and a Roman camp at Normanton, Yorkshire, were not, among many more.[34]

Urban viaducts brought urban blight, especially in South London, in Glasgow, and in Manchester, where the centre is almost ringed by viaducts. Some were built in slum areas where property was cheap, while in others it became devalued, descending into slums. Paradoxically, railways were applauded for demolishing slums to make way for new stations – in London, Birmingham, and Glasgow, for instance – while simultaneously they were creating new ones. No thought was given to where the inhabitants would go. They simply created more overcrowding elsewhere.

From the 1850s railway construction coarsened. Local materials were often no longer used; the railway itself made it cheaper to import from other places. Ugly plate-girder bridges cost less than brick or stone arches, and where brick was used it was often the blue engineering kind, chosen for its durability, even for patching up red brick or masonry. For example, new viaducts replacing timber ones at Blakedown and near Kidderminster in Worcestershire – a red-brick and sandstone county – were severe in blue brick. Station designs to a common pattern, often using standard components, enabled a railway to be identified by the style of its stations, regardless of locality.[35]

Where the heart of a town was penetrated, fine street vistas could be ruined. The main street of Leamington Spa, lined with Regency buildings, was terminated by a pair of railway bridges. The slightly better of the two was removed in the 1960s, revealing the stark bow-string girders of its companion. Glasgow's Tolbooth Steeple of 1626 is squarely blocked by a lattice-girder bridge, and handsome streets in Worcester and Derby are severed by bridges only partly ameliorated by decorative treatment which today has merited listing.[36] In 1863 the engineer Sir John Hawkshaw explained his dilemma when faced with a statutory requirement to build bridges at a uniform height over streets between London Bridge and Charing Cross with a single wall-to-wall span and no intermediate supports, which made him abandon any attempt at architectural effect. 'A girder so restricted cannot be made architecturally beautiful'. He went on to denounce the railway practice of using attractively-designed

---

34 For full account see J. Simmons, *The Victorian Railway* (1991), pp. 155–73.

35 Biddle, *Victorian Stations*, pp. 144–62.

36 For more extensive treatment, see G. Biddle, 'Railways in Towns', *Journal of the Railway & Canal Historical Society*, 31, p. 156 (November 1993).

bridges for advertising.[37] The new Hungerford Bridge over the Thames at Charing Cross was damned by an observer as 'a horizontal line of huge gratings', while at London Bridge station 'No words are strong enough to condemn the scandalous and irretrievable ugliness [of the bridge] which has spoilt the old station and the entrance to the Borough.'[38] The greatest insult was the bridge that completely spoiled the view of St Paul's up Ludgate Hill, until it was demolished in 1991.

Smoke

Almost from the beginning of steam railways, smoke was a nuisance. Railway Acts required engines to be constructed so as to consume their own smoke, codified in the Railways' Clauses Consolidation Act, 1845.[39] In order to comply, coke was used until the invention of the brick-arched firebox enabled coal to be burned cleanly from 1860. Even so, atmospheric pollution was severe around stations, yards and, above all, engine sheds. Standing locomotives, engines being lit up, being badly fired, or using poor coal, inevitably emitted smoke, creating a permanent sooty pall. Camden in London was a case in point, with a three-fold infliction: from an important engine shed; a large goods depot; and heavy trains climbing the bank from Euston where the engines had started 'cold'. Regular complaints were made for years, right until the end of steam.

Railway smoke also affected country locations, such as a junction like Woodford Halse in rural Northamptonshire which had an important locomotive establishment and a large marshalling yard, or Rowsley in the Peak District, where there was also a shed and a marshalling yard. A letter to *The Times* in 1864 complained bitterly that 'between Slough and Wycombe the country is poisoned and the passengers asphyxiated by the foulest, and the blackest, and the most sulphurous pest of smoke', calling on the Board of Trade to see that the railways obeyed the law.[40] Unfortunately, the law was specific only to the construction of locomotives. It said nothing about the actual emission of smoke caused by human agency.

Birmingham's Street Commissioners gained provisions in the 1846 Act[41] for the building of New Street station which attempted to conceal it from sight and sound. The station, which was below ground level, had to be entirely roofed over (which caused problems in design); construction of a new approach street was specified in detail; bridges at each end were required to have parapets 15 feet high (a section still remains); goods trains were not permitted (there were avoiding lines anyway); and engines were prohibited from standing in the station any longer than necessary. This last provision was doubtless an attempt to reduce the smoke nuisance, although it was impossible to enforce.

---

37 *Parliamentary Papers* (1863), viii, p. 149.
38 D. Hudson, *Munby: Man of Two Worlds* (1972), p. 175.
39 *The Railways' Clauses Consolidation Act, 1845*, 8 Vict., cap. xx, s. 114.
40 J. Simmons, *Railways: an Anthology* (1991), p. 67.
41 *The Birmingham, Wolverhampton and Stour Valley Railway Act, 1846*, 9 & 10 Vict., cap. cccxxviii.

Noise

Occasionally efforts were made at abating noise. Influential landowners succeeded in having the line into Victoria station in London enclosed by a glass roof and brick walls, with the track cushioned with rubber.[42] The 1865 Act[43] for the new line across Leeds contained a similar requirement for cushioning, and for semaphore signalling to reduce the noise of engines whistling. This could indeed be a torment: 'the enormous evil... of peace disturbed day and night by the shrieks of railway whistles.'[44] An association of residents in Primrose Hill, London, was formed in 1860 to try to have it reduced, and in 1871 the whistling of trains on the North London Railway on Sundays was said to have stopped church services. One resident in 1903 threatened 'to shoot either the driver or the engine.'[45] The north tunnel at Harecastle in Staffordshire was built solely to prevent trains disturbing worshippers in Kidsgrove church on Sundays, but was opened out as a cutting in 1966 when the line was electrified.

An Assessment

What conclusions may be drawn from this brief survey? Until the 1850s railways took a fairly responsible attitude to the natural and built environment; sometimes it was praiseworthy. But receding memories of the age of romanticism were accompanied by rapid industrialization that produced progressive degeneration of whole areas, to which the concomitant expansion of railways greatly contributed, creating entire 'railway districts' in some places, hand-in-hand with a steady decline in taste as the century progressed. There were exceptions, notably in some large public, corporate, or private buildings, including on the railways. Some work was exemplary, like the widening of the Wharncliffe and Maidenhead viaducts, the enlargement of the front of Shrewsbury station, the building of St Pancras in 1868–73, and the West Highland line in 1894. But good practice was not consistent, as Shrewsbury showed. There was no worse an example than the progressively deplorable treatment of the Euston portico, culminating in its demolition in 1961. Cost was always the primary factor, influenced to a varying degree by external forces. The physical evidence is still there, most of it still in use.

There is less readily-available evidence on smoke and noise. When locomotives started burning coal instead of coke, factory chimneys had already turned industrial areas black, helped by domestic smoke which also afflicted residential areas and smaller towns. Smoke generally, to which the railways contributed only a share, was unquestioned as a feature of life, and regarded as unavoidable.

Noise was probably considered to be the more objectionable. Outside the large areas of heavy industry such as engineering, steel-making, and ship-building,

---

42   A.A. Jackson, *London's Termini* (1985 edition), p. 272.
43   *The North Eastern Railway (Leeds Extension) Act, 1865*, 28 & 29 Vict., cap. ccli.
44   D. Duncan, *Life and Letters of Herbert Spencer* (1908), p. 314.
45   Simmons, *The Victorian Railway*, p. 370.

railways were the greatest single generator of noise. The sounds of locomotives shunting and the clanging of buffers from only a medium-sized goods yard was continuous and widespread, more than from, say, a textile mill, a brick works, or a brewery. Yet in country districts some people liked to hear the sound of a train. Somehow it was reassuring.

To summarize, therefore, the effect of the railway on the environment was mixed, and considerably dependent on the locality. Detrimental effects excited comment mainly from social reformers and, if they were exceptionally severe, perhaps from local bodies and inhabitants. Most of the time the railways ignored them, as well they might, because to the Victorians, railways and industry were the mark of progress; their by-products and what they left behind them were unquestioningly accepted as part of everyday existence.

At the beginning of the twenty-first century it is now the road and aviation which pollute, while the modern railway is clean and a friend of the environment. Moreover, after more than 200 years of continuous development, a device that pre-dates steam power now embraces twenty-first century technology. The metal wheel on the metal rail is still turning, on an infrastructure which those early visionaries so confidently built to last.

Chapter 9

Ruskin and the Railway

J. Mordaunt Crook

Like the proverbial parson pontificating on sin, Ruskin had only one attitude to the railway: he was against it. The steam engine itself, however, filled him with amazement. In a passage worthy of Thomas Carlyle, he marvelled at its sophisticated mechanism, its imagery of limitless power:

> I cannot express the amazed awe, the crushed humility, with which I sometimes watch a locomotive take its breath at a railway station, and think what work there is in its bars and wheels, and what manner of men they must be who dig brown iron-stone out of the ground, and forge it into THAT! What assemblage of accurate and mighty faculties in them; more than fleshly power over melting crag and coiling fire, fettered and finessed at last into the precision of watchmaking; Titanic hammer-strokes beating, out of lava, those glittering cylinders and timely-respondent valves, and fine-ribbed rods, which touch each other as a serpent writhes, in noiseless gliding, and omnipotence of grasp; infinitely complex anatomy of active steel, compared with which the skeleton of a living creature would seem, to a careless observer, clumsy and vile – a mere morbid secretion and phosphatous prop of flesh![1]

So it was not the steam engine, but its misuse, which aroused Ruskin's anger. In effect, he came to see the railway as an instrument of the devil; an agent of modernism, disruptive of the peace, the beauty, the civility, and the natural harmony of the world. Railroads, he informed *The Times* in 1887, 'are to me the loathsomest form of devilry now extant, animated and deliberate earthquakes, destructive of all wise social habit or possible natural beauty, carriages of damned souls on the ridges of their own graves'.[2] Throughout England, Ruskin saw the railroad's tentacles strangling town and countryside; eroding the contours of familiar landscapes, and sundering a multitude of social relationships built up seamlessly over centuries. And it was a European phenomenon, not just an English one. From Conway Castle to the Castle of Chillon; from Furness Abbey to the Falls of Schaffhousen, the progress of the iron monster seemed irresistible. What alarmed Ruskin particularly was the cultural chasm which appeared to be opening up between past and present. With Venice,

NOTE: References throughout are to *The Works of John Ruskin*, ed. E.T. Cook and A. Wedderburn (Library Edition, 39 volumes, 1903–12).
1   *Works* XIX, 60–61: *The Cestus of Aglaia* (1865–6).
2   *Works* XXXIV, 604: *The Times*, 3 March 1887.

Florence, Paris, and Birmingham all reduced to the equality of mere platforms on a gigantic network of railway lines, Ruskin sensed the impending disintegration of all traditional cultural forms. 'The life of the Middle Ages', he concluded, was surely 'dying'; and with it 'the warm mingling of past and present' on which human civility was based.[3]

Ultimately, the development of the railway system was for Ruskin a question of morality. The railroad was the embodiment of the industrial process, a potent metaphor of the capitalist principle. Its construction was financed by speculation; its accomplishment carried through by exploitation. Its very existence was the product of misapplied science and multiplied usury. 'Railways', he writes in 1874, 'should no more pay dividends than carriage roads or field paths'. For 'a railroad dividend is a tax on its servants – ultimately, a tax on the traveller'; its finances are based on the payment of interest, and 'interest [itself] is... a tax by the idle on the busy and by the rogue on the honest man'. 'Interest [indeed] is always either Usury on Loan, or Tax on Industry'. And 'there is no reason why a railroad should [ever] pay a dividend [any] more than the pavement of Fleet Street'.[4]

> Neither the roads nor the railways of any nation [he concludes] should belong to any private person. All means of public transit should be provided at public expense... and the public should be its own 'shareholder'... All dividends are simply a tax on the traveller and the goods, levied by the person to whom the road or canal [or railway] belongs, for the right of passing over his property... [Therefore] this right should at once be purchased by the nation.[5]

Such demands for state action – what the twentieth century learned to call nationalization – have often led scholars to hail Ruskin as a pioneer Socialist. Nothing could be further from the truth. 'I am', he assured the *Daily Telegraph* in 1871, 'an old and thoroughbred Tory'.[6] By that he meant a radical paternalist, a 'violent Tory of the old school (Walter Scott's... and Homer's).'[7] If Socialism is about equality, then Ruskin was a thoroughgoing anti-Socialist. 'So far from wishing to give votes to women', he explained in 1870, 'I would fain take them away from most men'.[8] Ruskin was neither a Socialist nor a Liberal. 'I hate all Liberalism', he told the students of Glasgow in 1880; 'I hate all Liberalism as I do Beelzebub'.[9] Spurning the delusions of both liberty and equality, he placed his faith in the older values of community and trust. Revisions of the franchise, he believed, were worthless unless accompanied by a fundamental revision of morality based on interpersonal ethics and ecological conscience. In short, his thinking was closer to the communitarian

---

3   *Works* VII, 423: *Modern Painters* V (1860); *Works* XII, 314–15: 'Samuel Prout' (1849); *Works* XXXIII, 404–5: *The Art of England* (1883).

4   *Works* XXVIII, 201: *Fors Clavigera*, November 1874; *Works* XXIX, 570–73: 'Interest and Railways'.

5   *Works* XVII, 528–35: letter to the *Daily Telegraph*, 6 August 1868, reprinted in *Arrows of the Chace* ii (1880).

6   *Works* XXXIV, 506–7: letter to the *Daily Telegraph*, 20 December 1871, reprinted in *Arrows of the Chace* i (1880).

7   *Works* XXVII, 167: *Fors Clavigera*, October 1871.

8   *Works* XXXIV, 499: 29 May 1870.

9   *Works* XXXIV, 549: September 1880.

philosophy of medieval Christianity than to the atomistic assumptions of modern industrial democracy. 'You will see', he told Coventry Patmore in 1876, 'in [the] next [number of] *Fors* [*Clavigera*], something of a Catholic Faith wider than yours'.[10] What he meant by that, however, was Catholic social theory shorn of its doctrinal metaphysics. 'It is *only* their doctrine of the Mass', he told Kathleen Oglander in 1888, 'that separates me from them... [otherwise] I should be a Roman Catholic at once'.[11]

At the root of Ruskin's rejection of liberal economics – and hence his detestation of the railway system – lay a fundamental contempt for the moral vacuity of *laissez faire*.

> The Science of Political Economy [he wrote in 1862] *is* a Lie, wholly and to the very root... To this 'science' and to this alone (the professed and organised pursuit of Money) is owing *All* the evil of modern days. I say All. The Monastic Theory is at an end. It is now the Money theory which corrupts the Church, corrupts the household life, destroys honour, beauty and life... [Economists] don't know what they are talking about. They don't even know what Money is, but tacitly assume that money is desirable as a sign of wealth, without defining Wealth itself.[12]

The pursuit of money through the mechanism of the market was but a higher form of 'cannibalism'; 'the first principle of economy is employment'; and ultimately, 'there is no wealth but life'.[13]

Ruskin looked at the Victorian railway and saw there not an index of multiplied prosperity, but the cloven hoof of *laissez faire* economics. 'Your railroad mounds,' he wrote in 1866, 'vaster than the walls of Babylon; your railroad stations, vaster than the temple of Ephesus, and innumerable; your chimneys, how much more mighty and costly than cathedral spires! your harbour piers; your warehouses; your exchanges! – all these are built to your great goddess [of] "Getting on"'[14] Ornament such buildings as we may, their debasement of soul will always filter through. 'Millions upon millions', he pointed out in 1867, 'have... been spent within the last twenty years, on ornamental arrangements of zigzag bricks, black and blue tiles, cast iron foliage and the like... and (trust me for this!) *all that architecture is bad*.'[15] Attempts to disguise the ugliness of railway structures with a mere mask of decoration will always be futile. A truly Gothic architect might have sublimated industrial design by employing the imagery of strength. Modern architects were doomed merely to embody in their work the symbolism of profit. Medieval masons 'would have put some life into those iron tenons [at Blackfriars Bridge]... Whereas, now, the entire invention of the

---

10   *Works* XXXVII, 191: February 1876.
11   *The Gulf Years: Letters of John Ruskin to Kathleen Oglander*, ed. R. Unwin (1953), 15–16.
12   *Works* XVII, lxxxii and XXXVI, 417–8: letter to Dr John Brown, August 1862; *Fraser's Magazine* (1863), reprinted in *Munera Pulveris* (1872).
13   *Works* XXVIII, 669–74: *Fors Clavigera*, August 1876; *Works* XVII, 528–35: *Arrows of the Chace* (1880); *Works* XVII, 105: *Unto This Last* (1862).
14   *Works* XVIII, 448: *The Crown of Wild Olive* (1866).
15   *Works* XVII, 389–90: *Time and Tide* (1867).

designer seems to have exhausted itself in exaggerating to an enormous size a weak form of iron nut, and in conveying the information upon it, in large letters, that it belongs to the London, Chatham and Dover Railway Company.'[16]

Such images haunted Ruskin's travels, particularly in places of solitude and wilderness. 'It always puts me in a passion to think of Derbyshire', he noted in 1873, 'the whole county is spoilt with "works" and railroads; there are no more trout in the Dove... and you have got embankments and tunnels where there were rocks and caves'.[17] Two years before, that had been the theme of one of his more ferocious denunciations of Progress:

> There was a rocky valley between Buxton and Bakewell, once upon a time, divine as the Vale of Tempe; you might have seen the Gods there morning and evening – Apollo and all the sweet Muses of the light – walking in fair procession on the lawns of it, to and fro among the pinnacles of its crags. You cared neither for Gods nor grass, but for cash... You Enterprised a Railroad through the valley – you blasted its rocks away, heaped thousands of tons of shale upon its lovely stream. The valley is gone, and the Gods with it; and now every fool in Buxton can be in Bakewell in half an hour, and every fool in Bakewell at Buxton; which you think a lucrative process of exchange – you Fools Everywhere.'[18]

By 1884, with the planning of the Dore & Chinley Railway – opening up the North Peak district around Castleton – Ruskin's worst fears seemed to be coming true:

> Derbyshire is a lovely child's alphabet; an alluring first lesson in all that's admirable... the grace of it all!... And half a day's work of half a dozen navvies, and a snuff-box full of dynamite, may blow it all to Erebus and diabolic Night, for ever and ever... In Derbyshire the whole gift of the country is in its glens... The rocks are not big enough to be tunnelled, they are simply blasted away; the brook is not wide enough to be bridged, it is covered in, and is thenceforward a drain; and the only scenery left for you in the once delicious valley is alternation of embankments of slag with pools of slime... Commerce has filched the Earth, and Science shut the Sky.[19]

And where the Peak District had gone the Lake District threatened to follow. In 1875 a railway extension was planned from Windermere to Ambleside, and thence by Rydal and Grasmere, over Dunmail Raise to Keswick. One resident, Robert Somervell, drew up a protest, and Ruskin penned a memorable preface. Ruskin knew all about the impact of railways in the Lakes. The old journey from Coniston to Ulverston – an idyllic 12-mile walk – had been turned into a 24-mile rail journey, costing two shillings a time: 'the results, absolute loss and demoralisation to the poor... and iniquitous gains to the rich'.[20] Even so, this preface was not easy to compose. 'It will

---

16   *Works* XIX, 25–6: *The Study of Architecture in Schools* (1865).
17   *Works* XXXIV, 511–12: letter to Derby School of Art, December 1873, reprinted in *Arrows of the Chace* (1880).
18   *Works* XXVII, 86: *Fors Clavigera*, May 1871.
19   *Works* XXXIV, 568–72: letters to *Manchester City News*, 5, 12, 19 April 1884. The Dore & Chinley line was completed in 1894. For a parody of Ruskin's letter, see 'On All Fours Clavigera', *Punch*, 23 August 1884, p. 85.
20   *Works* XXVIII, 129–30: *Fors Clavigera*, August 1874.

not come right', Ruskin complained in June 1876; but eventually it was done. 'I've done the Preface at last', he told Somervell, 'and I think it stunning. It came to me all of a heap as I was shaving.'[21]

He began by taking up Wordsworth's environmental protest against the 'false, utilitarian lure' of accessibility and profit. There were insufficient minerals in the hills to justify mining, and casual travellers would merely destroy what they came to see. 'The stupid herds of modern tourists,' he protested, would simply 'be emptied, like coals from a sack, at Windermere and Keswick'; the new railway would merely 'shovel those who have come to Keswick to Windermere, and... those who have come to Windermere to Keswick... I [certainly] don't want them on Helvellyn while they are drunk.' As for Grasmere, it would soon be 'nothing but a pool of drainage, with a beach of broken gingerbeer bottles', defiled by 'taverns and skittle grounds', all too reminiscent of Blackpool. The only profits would be those accruing to those 'tribes of louts and scoundrels', the developers; profits 'stolen between the Jerusalem and Jericho of Keswick and Ambleside, out of the poor, drunken traveller's pocket.'[22]

On this occasion, Ruskin had his way. The scheme was postponed. But when it was revived in 1884–7, he returned to the fray with even greater vehemence. Opening up the northern lakes, he explained, would merely pollute the waters of Rydal and Grasmere with sewage; the lakeland solitude of Helvellyn and Ullswater would be scattered with suburban gaming houses, 'decorated in the ultimate exquisiteness of Parisian taste'.[23] Fortunately for Ruskin's sanity, the scheme was rejected at Westminster and eventually abandoned.

Of course Ruskin, like his chief disciple William Morris, was himself a capitalist and a railway user. His family business produced profits on the import of wine; his investments in Bank of England stock brought in returns of ten to fifteen per cent. But at least – so he assures us in 1874 – he was 'a well-meaning usurer';[24] and – so he tells us in 1880 – he remained a reluctant rail-traveller all his life: 'I have never held a rag of railroad scrip..., nor ever willingly travelled behind an engine where a horse could pull me'.[25] When he did use the railway, he was always 'looking... hopefully forward to the day when [its] embankments will be ploughed down again, like the camps of Rome, into our English fields.'[26] Occasionally he did enjoy the views opened up by railway travel, particularly in France and Italy.[27] But he continued to abuse the system as fundamentally antisocial, a prime contributor to 'the indolence, ill-health, discomfort, thoughtlessness, selfishness, sin, and misery of this life.'[28] In essence, it involved the misuse of science in pursuit of that vainest of all ends: sensation. 'The [very] clouds', he wrote in 1856, have been 'packed into iron cylinders' in a foolish

21   *Works* XXXIV, xxxi: 22 June 1876.
22   *Works* XXXIV, xxx–xxxi, 135–43: 'Railways in the Lake District' (1876).
23   *Works* XXXIV, 568–72: letters to *Manchester City News*, 5, 12, 19 April 1884.
24   *Works* XXIX, 570–73: September 1874.
25   *Works* XVII, 528–35: letter to *Daily Telegraph*, 6 August 1868, reprinted in *Arrows of the Chace* ii (1880).
26   *Works* XXVII, xxviii and XXVIII, 247: *Fors Clavigera*, January 1875.
27   *Works* VII, xlvi and XXXIII, xliii.
28   *Works* XXXVI, 62: letter to Dr John Brown, 27 June 1846.

attempt 'to go as fast as the clouds...'[29] If only 'we English, who are so fond of travelling in the body, would also travel a little in the soul!'[30]

In other words, the railways, and all they stood for, were for Ruskin mere machinery. Unless radically redirected, they would surely destroy our ecological inheritance and multiply the misery of mankind. Far from embodying the 'natural law' of economics, the railroad supplied overwhelming evidence of a need for social control. Something of Ruskin's programme came eventually to fruition in the shape of Gladstone's legislation on railway safety.[31] But on the main issue – the railway as a symbol of market economics – Ruskin railed in vain. 'There is no such [thing as a] natural law [of economics]', he wrote in 1868; 'prices can be, and ought to be, regulated by laws of expediency and justice.'[32] 'Political economy is neither an art nor a science; but a system of conduct and legislature founded on the sciences, including the arts, and impossible except under certain conditions of moral culture.'[33] It was the corruption of that 'moral culture' which Ruskin set out to heal. The medicine he prescribed proved too strong for most Victorian stomachs; even so, its prescription still makes salutary reading.

---

29   *Works* V, 383: *Modern Painters* iii (1856).
30   *Works* XVI, 74: *A Joy For Ever* (1857).
31   For Ruskin on railway accidents, the result of 'miserable greed... gross negligence and... criminal recklessness', see *Works* XXVII, 664–6.
32   *Works* XVII, 528–35: letter to *Daily Telegraph*, 10 August 1868, reprinted in *Arrows of the Chace* ii (1880).
33   *Works* XVII, 532: *Fraser's Magazine*, June 1862, 784, reprinted in *Munera Pulveris* (1872).

Chapter 10

Philip Larkin's Railways

Roger Craik

Had he lived, Philip Larkin would have been invited to contribute to this volume. To conjecture about the essay he would have written in affection for Jack Simmons (with whom he worked at Leicester University, then a University College, from 1946 to 1950) is a wistful pleasure, for so many aspects of Larkin's personality would surely have vivified it: the wide range of reading, the enjoyment of the Englishness of railway travel, the love of the everyday, the sheep and fields and hedgerows and Odeons and level crossings and cricket matches, the railway stations with their crackling announcements in received pronunciation, and possibly also a nostalgia for the times when lunch and dinner were announced by a steward passing along the train. There might also be, however, that Larkinian waspishness towards train travel in the 1980s and after, with soccer-shirted children and adults, mobile phones, and Sony Walkmans, and dry, air-conditioned air and windows that do not open, all of which Larkin would have abominated and which his death in 1985 spared him. And, beyond all these, informing a voice which is conservative in tone yet universal in its appeal, would be Larkin's own lifetime of rail travel.

Born in 1922, well before the heyday of the motor car and the long-distance bus, Philip Larkin did all his travelling by train until 1964 (when he passed his driving test and bought his first car), and even then he relied on the train for his longer journeys. These had started with 'all those annual hols' as he puts it with self-mocking schoolboy pomposity in 'I Remember, I Remember', with his parents to Bigbury-on-Sea, Folkestone, and Caton Bay in Yorkshire, and which continued from his native Coventry to Oxford from 1940 to 1943 when he was an undergraduate. From 1943 to 1946 Larkin would regularly take the local train for the 12-mile trip from Wellington, where he held his first job as a librarian, to see his college friend, Bruce Montgomery, in Shropshire. Larkin appears to have travelled less in his third post, when he worked as a librarian at Queen's University, Belfast, (1950–55) than in his second, at Leicester University (1946–50), but his return to England in 1955 to work at Hull University Library saw him using the train more than ever, not just to visit his companion Monica Jones in Leicester and his mother in Loughborough, but also to attend library conferences and to collect literary awards which he had won for his first major collection of poems, *The Less Deceived*.

It is true that in his later years in particular, Larkin was given to producing a grand display of irritation towards any journey for the disruption it would cause to his routine and his writing: 'I am not a natural traveller: a place has to be pretty intolerable before it enters my head that somewhere else might be nicer' (quoted in Chambers, p. 48); 'I wouldn't mind seeing China if I could come back the same day... Generally speaking, the further one gets from home, the greater the misery' (*RW*, p. 55). Such remarks, though, are in keeping with his persona of an intransigent homebound recluse which he enjoyed presenting to his friends, and are in any case directed against being shifted against his will. He never once complained about any rail journey he took, and indeed there is much to suggest that he enjoyed his travelling by train.

But if to muse about the essay that Larkin would have written for this Festschrift is to miss him afresh, it is also to be grateful for what he has given us, for Philip Larkin stands unrivalled as the English railway poet of the twentieth century. Outdistancing John Betjeman who loved the architecture of railway stations, or Auden who observed the *Night Mail* crossing the Border, Larkin in some of his very best poems wrote of how it feels to travel through the English railway landscape in the decades of the mid-century. Over 24 years, starting in 1940, Larkin's train journeys are variously long, brief, taken singly or in company; and the speaker may gaze at the passing scenery or at his fellow travellers if there are any, or may even fall asleep in the middle of a poem. This is a special kind of poetry because by its very nature rail travel induces a particular kind of thought: the musing and observation of a mind at ease allow the poet to think his way towards conclusions while the landscape passes undemandingly by. My thesis is that in Larkin's very best poems, of which none, significantly, was written after *The Whitsun Weddings* (1964), a railway journey is more than either a setting or a background but is an experience which determines the changes of mood and the poetic thought-process itself. This is truly a poetry of mind and eye, and eye and mind.

Although Larkin's best-known railway poems were those he wrote in the 1950s and 1960s, he started writing about trains in his schooldays and continued through his undergraduate years. Virtually unknown until the publication of *Collected Poems* in 1988, these early pieces collectively announce the themes on which Larkin was to dwell in his prime. More particularly, they show him attracted enough by trains, the railway landscape, and the interior of railway carriages to make them settings for two poems which foreshadow his later famous train journeys. One poem, 'So you have been, despite parental ban', written in 1940, when Larkin was 18, opens with the speaker envying a 17-year-old contemporary who has broken free of parental strictures and stultifying provincialism and who 'through rain to empty station ran / And bought a ticket for the early train' but whose return years later fulfils neither others' projected hopes nor his own:

Today your journey home is nearly done:
That bag above your head, the one you had
When seventeen, when you were a son,
Is labelled now with names we do not know;
The gloved hands hang beneath the static knees
And show no glee at closing evening's glow –
Are you possessor of the sought-for ease? (*CP*, p. 236)

This disillusioning scene in the railway compartment finds counterparts in many of Larkin's later poems; the unnoticed exotically-labelled bag reappears in 'The local snivels through the fields', while the 'closing evening's glow' anticipates many other evenings, generally associated with the approach of death. More importantly, the poem shows Larkin suggesting that the stride towards independence, enviable though it may seem, may not only fail to establish identity ('That name for which you fought – does it quite fit?' he asks pointedly) but also may bring diminution; the wanderer returns crestfallen and his father has died during his absence. The train which bore him away flamboyant brings him back years later, inert and broken down by circumstances. For his part, the speaker remains where he is, both attracted by the thought of adventure and conservatively opposed to it.

Four years later, in 'One man walking a deserted platform', Larkin's traveller does not return:

> One man walking along a deserted platform,
> Dawn coming, and rain
> Driving across a darkening autumn;
> One man restlessly waiting a train
> While round the streets the wind runs wild,
> Beating each shuttered house, that seems
> Folded full of the dark silk of dreams,
> A shell of sleep cradling a wife or child.
>
> Who can this ambition trace,
> To be each dawn perpetually journeying?
> To trick this hour when lovers re-embrace
> With the unguessed-at heart riding
> The winds as gulls do? What lips said
> Starset and cockcrow call the dispossessed
> On to the next desert, lest
> Love sink a grave round the still-sleeping head? (*CP*, p. 293)

Larkin's original title was 'Getaway'. The thought of escape is no less attractive because the speaker knows he cannot manage it whereas his 'one man' can. 'One man walking' appears to ally itself with 'So you have been, despite parental ban' through the exhilaration of breaking free, but in fact the speaker sees only the man and the platform, never the train's arrival or departure, bearing the passenger away. The *idea* of leaving rather than leaving itself is the clue to the real preoccupation: fear of entrapment. In his wondering envy of the man who can escape the enclosed domesticity of 'Each shuttered house, that seems / Folded full of the dark silk of dreams, / A shell of sleep cradling a wife or child' and who can sidestep the 'hour when lovers re-embrace' (that is, re-establish their commitment to each other) and who can depart before love has the chance to 'sink a grave round the still-sleeping head', the speaker by implication roots himself in the very circumstances that the wanderer shuns, leaving himself to reconcile the conflicting claims of independence, solitude and love. This dilemma is central to Larkin's work as a whole, and 'One man walking' tentatively approaches it.

It is worth remembering that during the 1940s, when he was writing the poems which were to make their way into his first collection, *The North Ship* (1956), Larkin saw himself as a novelist rather than a poet. In the first chapter of his strongly autobiographical first novel, *Jill*, written from 1943 to 1944, Larkin placed his 18-year-old central character, John Kemp, on a train travelling the last stretch of line from Coventry to Oxford:

Clear of the gasometers, the wagons and blackened bridges of Banbury, he looked out over the fields, noticing the clumps of trees that sped by, whose dying leaves each had an individual colour... The windows of the carriages were bluish with the swirls of the cleaner's leather still showing on the glass, and he confined his eyes to the compartment. It was a third-class carriage, and the crimson seats smelt of dust and engines and tobacco, but the air was warm...

As if the train knew his destination was near it seemed to quicken speed, plunging on with a regular pattern of beats. He looked from the window and saw a man with a gun entering a field, two horses by a gate, and presently the railway was joined by a canal, and rows of homes appeared. He got to his feet and stared at the approaching city across allotments, back gardens and piles of coal covered with fallen leaves...

The train clattered by iron bridges, cabbages, and a factory painted with huge white letters he did not bother to read; smoke dirtied the sky; the train swung violently over set after set of points. Their speed seemed to increase as they swept round a long curve of line through much rolling-stock... (pp. 21–4)

Here the landscape and interior detail give a strong sense of 'The Whitsun Weddings', and with this scene from *Jill* Larkin's writing about trains takes as it were a curious loop line, as if Larkin felt that he could write of trains in detail in prose but not in verse.

Two years later Larkin made a radical change in approach and style. It was Thomas Hardy, he writes, who allowed him to

simply relapse back into one's own life and write from it. Hardy taught me to feel rather than to write – of course one has to use one's own language and one's own jargon and one's own situations and he taught one as well to have confidence in what one felt. (RW, pp. 175–6)

His poetry after 1946 catches up with his manner in *Jill*. By the early 1950s, rather than dealing with abstractions (as he had done previously under Yeats's influence), Larkin was concentrating his energies on everyday situations, events, and objects, and trains take their place alongside churches, tombs, racehorses, apple cores, vases, ambulances, and photograph albums. His poetry becomes less forced, more relaxed, and altogether more accomplished, and nowhere is this difference more marked than in his poems involving trains. Here Larkin ever more deftly explores the possibilities of the railway setting as a vehicle, as it were, for the major themes of his poetry: these themes are especially effective when they stem from the interplay of scenery and thought which rail journeys uniquely encourage.

However, the most obvious of the differences between the trains of Larkin's tentative early poems and those of the 'mature' poems of *The Less Deceived* (published in 1955) is that they are no longer awaited and boarded by anonymous figures. Rather, the poet himself, writing in the first person, is no longer the onlooker but the traveller:

> The local snivels through the fields:
> I sit between felt-hatted mums
> Whose weekly day-excursion yields
> Baby-sized parcels, bags of plums,
> And bones of gossip good to clack
> Past all the seven stations back. (*CP*, p. 59)

It is not necessary to know (although some might suspect it) that this was intended as self-parody (Larkin, *Letters*, p. 266) to sense Larkin's confidence in using 'one's own language and one's own jargon and one's own situations.' His self-portrait as a slightly absurd, silent bachelor is one which with variations is replicated in poem after poem of *The Less Deceived*, *The Whitsun Weddings* and *High Windows*: in particular this poem anticipates 'The Whitsun Weddings' and, to an extent, 'Here', in its affectionate mockery (just this side of condescension) of people whom he sees from his train seat, along with an understated recognition of the differences between him and them. But if the first verse reads as self-parody throughout, the second is ruefully pensive:

> Strange that my own elaborate spree
> Should after fourteen days run out
> In torpid rural company
> Ignoring what my labels shout.
> Death will be such another thing,
> All we have done not mattering. (*CP*, p. 59)

Rail travel by its very nature often induces our minds to wander and to settle, finally, on our own most pressing concerns; this is what is happening here. During the artful pause between verses, the speaker, now that he has exhausted the interest of what is in sight and with no book to divert him (unlike the traveller of 'The Whitsun Weddings'), finds his thoughts turning inward. It is interesting to compare this poem with 'Dockery and Son' in this respect. There the speaker is actually musing, whereas in 'The local' the speaker *has* mused in the break between verses to yield up the lugubrious comparison between his own 'elaborate spree' of 'fourteen days', which presumably make up his annual holiday, with the 'weekly day-excursions' of his fellow travellers. Like 'So you have been', 'The local' concerns a traveller who finds himself shunted back, his adventure a failure, into a provincialism which takes no notice of the labels on his suitcases. This dismal realization, though, does not prepare the reader for the extremely abrupt association with death, shocking because of its abruptness, at which all of life's accomplishments will count for nothing.

Larkin did not select this poem for publication in *The Less Deceived*, possibly considering the change in registers too jarring, but more probably judging that he had better poems to hand. But over the next 13 years, during which the poems published in *The Less Deceived* and *The Whitsun Weddings* were written, he made trains the setting for several poems, some of which rank as his very best. It would be a slight overstatement to characterize these poems as 'dramatic' – after all, a traveller only sits, and at most changes trains or leans out of an open window – but they are nevertheless thought out on a stage which is largely static (the interior of a railway compartment) yet which is also moving (the landscape passing by the window).

These poems, in which the travelling is more important than the arrival, are popular because they present to us 'simple, everyday experiences' (*RW*, p. 83) so that we recognize the particular details with pleasure and empathize with the speaker. Like us, he may observe his immediate surroundings, or look out of the window, or simply allow his mind to wander. Like us, on board a train he can relish a special kind of solitude, special because it allows him to observe life in the presence of others, and at the same time to be undisturbed by them. The interplay of passing scenes and the speaker's musings guards against highly descriptive travel narratives on the one hand, and excessive introspection on the other: Larkin's train poems are never purely abstract, as they are, for example, when the speaker is alone in his room sitting under a lamp, or drinking.

Larkin's insights, though, are seldom comforting. In 'Reference Back' the speaker opines that 'though our element is time, / We are not suited to the long perspectives / Open at each instant of our lives' (*CP*, p. 106). From the early 1950s onwards, all Larkin's train poems have the speaker contemplating these 'long perspectives', whether his own or others'. Whereas 'The local snivels through the fields' ends by looking towards the speaker's death, a better-known and in every sense more ambitious poem, 'I Remember, I Remember', moves in the opposite direction:

> Coming up England by a different line
> For once, early in the cold new year,
> We stopped, and, watching men with number-plates
> Sprint down the platform to familiar gates,
> 'Why, Coventry!' I exclaimed. 'I was born here.'
>
> I leant far out, and squinnied for a sign
> That this was still the town that had been 'mine'
> So long, but found I wasn't even clear
> Which side was which. From where those cycle-crates
> Were standing, had we annually departed
>
> For all those family hols?... A whistle went:
> Things moved. I sat back, staring at my boots.
> 'Was that', my friend smiled, 'where you "have your roots"?'
> No, only where my childhood was unspent,
> I wanted to retort, just where I started:
>
> By now I've got the whole place clearly charted. (*CP*, p. 81)

The tremendous onslaught which follows on an 'unspent childhood' arises from the way that what is recognized ('familiar gates'), albeit from an unusual approach, is disconcertingly situated among the unrecognizable. During the second verse the speaker spends all of the train's stopping time (several minutes: Coventry is a major city) craning his neck out of the window and 'squinnying for a sign', but so much has changed that he is hard put to recognize anything. During the pause after the bewildered question about 'family hols' he is still searching for landmarks and casting around for associations. As the train pulls out ('A whistle went: / Things moved') the speaker sits back in his seat staring at his boots in what I take to be disillusionment, sullenness, and resentment. Out of politeness he just manages to hold back a cuttingly true rejoinder when his friend asks in pseudo-literary vein, 'Was that... where you "have your roots"?' ('Was that' rather than 'Is this' makes clear that the train has left Coventry), but after the break between the third and fourth verses, and with the train gathering speed all the time, he lets fly so vehement a tirade against D.H. Lawrence's and Dylan Thomas's presentations of childhood (his indignation rising at 'I'll show you, come to that', and continuing to rise until his friend's comment interrupts it) that one forgets that it is all thought, not uttered. 'I Remember, I Remember' is a poem of long silences punctuated by scraps of speech; the two men are silent not only as the train nears Coventry but also after it has left it, for the speaker's friend would surely make no answer to 'Nothing, like something, happens anywhere' which brings the poem to a bleak end.

An even grimmer poem, 'Dockery and Son', is for the most part set on a train from Oxford to Sheffield. According to Larkin's biographer Andrew Motion, the poem 'describes a visit Larkin had made to his old college at Oxford, St John's, on his way back from the funeral of Agnes Cuming, his predecessor as librarian at Hull' (p. 333). The funeral took place on 12 March 1962. Interestingly, timetables show that there was only one evening train directly from Oxford to Sheffield, which left in the dark at 8.27 p.m., made the complete run in darkness, and arrived at Sheffield Victoria at the ungodly hour of 12.46 a.m.[1] What is important about the poem, however, is not so much its circumstances or Larkin's changing of times, but the two 'long perspectives' and their relation to the railway setting. Before the train journey the speaker, rather than attending to the Dean of his former college at Oxford telling him that his contemporary Dockery now has a son at the same college, lets his mind run back sarcastically to undergraduate peccadilloes, ' "Our version" of "these incidents" last night.' But it is only when he is on the train, free at last from company and as 'Canal and clouds and colleges subside / Slowly from view' that his thoughts of Dockery, and what he now knows of him, begin to take shape:

> But Dockery, good Lord,
> Anyone up here today must have been born
> In '43, when I was twenty-one.
> If he was younger, did he get this son
> At nineteen, twenty? Was he that withdrawn
>
> High-collared public schoolboy, sharing rooms
> With Cartwright who was killed? (*CP*, p. 152)

Whereas the speaker in 'I Remember, I Remember' had only to be a few miles out of Coventry to have 'the whole place clearly charted', the speaker here manages only

---

1   I am indebted to John Gough for this detail and for all other details of timetables referred to in this essay.

two uncertain questions. He is tired. The day with its associations and company has wearied him, and he is glad to relax on his own in the train, as the alliterative languor of 'Canal and clouds and colleges / Subside slowly from view' makes clear. Too close to the day's events to form any conclusions about them, he falls asleep while rather half-heartedly trying to do so:

> Well, it just shows
> How much ... How little ... Yawning, I suppose
> I fell asleep, waking at the fumes
> And furnace glares of Sheffield. (*CP*, p. 152)

After a long refreshing sleep, the speaker awakens in altogether different circumstances: he is in the harsh industrial north, and it is night. He has to get up to change trains, he eats 'an awful pie', and strolls in the night air to the end of the platform where he observes the 'ranged / Joining and parting lines reflect a strong / Unhindered moon.' The 'awful pie', which as a jibe at British Railways' catering has become one of Larkin's most memorable phrases, actually fortifies the speaker and, like his sleep, the distance from Oxford and his brief walk, gathers him back to himself, undistracted, in a state he once called 'uncontradicting solitude' ('Best Society' *CP*, pp. 56–7). With the railway lines and the moon acting both as themselves and as symbols to quicken his thoughts, he now has the energy to articulate the differences between Dockery and himself:

> To have no son, no wife,
> No house or land still seemed quite natural.
> Only a numbness registered the shock
> Of finding out how much had gone of life,
> How widely from the others. Dockery, now:
> Only nineteen, he must have taken stock
> Of what he wanted, and been capable
> Of... No, that's not the difference: rather, how
>
> Convinced he was he should be added to!
> Why did he think adding meant increase?
> To me it was dilution. (*CP*, pp. 152–3)

In contrast to the earlier tired musings, this is concentrated thinking. The speaker's confidence finds voice in the ambitious syntax ('To have no son, no wife'), in his refusal to be thwarted by his mistaken assumptions as he thinks his way forwards, and in his providing answers to his own questions rather than falling back on platitudes. The determination to discover the differences between Dockery and himself and then to arrive at a philosophy of life stems from the train journey's physically restoring effects which in turn invigorate his way of thinking.

The position at the end of 'Dockery and Son' ('Life is first boredom, then fear' and 'age, and then the only end of age') is the starting point for his little-known poem of the previous year, 'A slight relax of air where cold was':

A slight relax of air where cold was
And water trickles; dark ruinous light,
Scratched like old film, above wet slates withdraws.
At garden ends, on railway banks, sad white
Shrinkage of snow shows clearer than the net
Stiffened like ectoplasm in front windows.

Shielded, what sorts of life are stirring yet:
Legs lagged like drains, slippers soft as fungus,
The gas and grate, the old cold sour grey bed.
Some ajar face, corpse-stubbled, bends round
To see the sky over the aerials–
Sky, absent paleness across which the gulls
Wing to the Corporation rubbish ground.
A slight relax of air. All is not dead. (*CP*, p. 142)

This sonnet differs from all of Larkin's other train poems in that it is entirely exterior. Perceived from a slowly moving train or one which has stopped between stations, this *tableau mourant* reminds me of how, when I used to commute to London, the train would pass very close to, and in most cases higher than, squalid houses surprisingly close to the track. Like many people, I imagine, I used to wonder what it must be like to live that way, and at the same time I felt relieved I was not doing so. Larkin's stance is more intense: again he is struck by the differences between people's lives, but here the dominant elements are fear of age, particularly in poverty, and revulsion at its appearance. The speaker in his railway carriage is close enough to the 'ajar face' to notice that it is 'corpse-stubbled', and his imagination has little difficulty in picturing the house's interior as damp, distasteful, and sexually arid. The reminder that 'All is not dead' purposely leaves the reader feeling that it would be better if it were. Death was already an established theme in Larkin, but the savaging of the most distressing aspects of age at its very closest to death and in the worst circumstances is a new note in 'The Whitsun Weddings' and it is heard most distinctly in this poem, which he wrote when he was 40.

Unlike 'A slight relax of air' with its dismal gaze, 'Here' and 'The Whitsun Weddings' enact lengthy journeys which, for the first time, include arrival. Although well apart from each other in *The Whitsun Weddings* collection, geographically the poems are counterparts, 'Here' being a railway journey northwards through Yorkshire to Hull, and 'The Whitsun Weddings' from Hull to London. The speaker travels the same stretch of track in opposite directions, noting in 'Here' 'the widening river's slow presence, / The piled gold clouds, the shining gull-marked mud', while in 'The Whitsun Weddings' he can see not just 'the river's level drifting breadth' once it has widened, but traces its appearance on the horizon to 'where sky and Lincolnshire and water meet', which is the remote area, the Holderness peninsula, contemplated at the end of 'Here'. And whereas the train of 'Here' is 'swerving east' (that is, north-east, from Doncaster), that of 'The Whitsun Weddings' keeps 'a slow and stopping curve southwards' (that is, south-west). Both poems mention Hull's fish dock.

Despite the 'common ground', 'Here' and 'The Whitsun Weddings' are very different poems. 'Here' ushers in *The Whitsun Weddings* both as a tribute to Hull as the city in which Larkin wrote the collection, and as a statement of his personality: Hull and its surroundings, which are 'clearly charted', suit his temperament. In this poem of recognition rather than surprise, only the occasional 'Harsh-named halt' (of which, incidentally, there are none between Doncaster and Hull) indicates that the swerving journey towards the city is by rail at all rather than by car, and even here the emphasis falls not on the swerving but on the speaker's relieved knowledge that it is away from 'rich industrial shadows / And traffic all night north' and towards 'solitude / Of skies and scarecrows, haystacks, hares and pheasants'. This appealing openness like the background in a Breughel painting gives way to a bustling foreground once the train arrives at its destination, Hull, where

> Residents from raw estates, brought down
> The dead straight miles by stealing flat-faced trolleys,
> Push through plate-glass swing doors to their desires–
> Cheap suits, red kitchen ware, sharp shoes, iced lollies,
> Electric mixers, toasters, washers, driers–
>
> A cut-price crowd, urban but simple. (*CP*, p. 136)

Beyond this envy-tinged distaste, tempered with reasonableness, stands the speaker's awareness of the difference between his life and others' lives; the 'cut-price crowd' are gregarious, married with children, secure in their domesticity, and unwaveringly certain of their wants. For his solace, the speaker veers away from lives with which he cannot connect, and towards the thought of the landscape beyond Hull:

> Here silence stands
> Like heat. Here leaves unnoticed thicken,
> Hidden weeds flower, neglected waters quicken,
> Luminously-peopled air ascends;
> And past the poppies bluish neutral distance
> Ends the land suddenly beyond a beach
> Of shapes and shingle. Here is unfenced existence:
> Facing the sun, untalkative, out of reach. (*CP*, pp. 136–7)

The threefold repetition of 'here', dwarfing its single appearance at Hull, identifies the word's most vehemently-felt meaning as this particular scene which so emphatically rejects the interference of others: 'unfenced', 'untalkative', and 'out of reach' pointedly deny their opposites. The furthest the speaker (and of such a personal poem one is tempted to write 'Larkin') allows the reader to travel is Hull, where the train stops for the last time, leaving the speaker to describe tenderly and longingly a land he knows well and from which the reader is excluded. The absence of a sense of travelling in the last verse moves the poem, almost without the reader's noticing it, from narrative to deeply-enjoyed self-revelation. Despite its prominent position in *The Whitsun Weddings* and the accessible appearance of rail travel, 'Here' is profoundly personal, private, and concerned with solitude.

'The Whitsun Weddings', by contrast, is openly public while being distinctively personal. Larkin now places his speaker amongst other people, interesting in their own right, yet presented in a way that is uniquely his own. Without ever giving the impression, as 'Here' and 'A slight relax' both do, of trying too hard or sounding strident, 'The Whitsun Weddings' is poised between the details of the journey and the speaker's wonder at the newly-weds boarding at each station, between indoors and outside, between landscape and people, between singleness and marriage – and all this in the course of a complete rail journey which is itself such an appealingly recognizable feature of English life. What sets it above all the other poems discussed so far is the weight it achieves while giving the impression of spontaneity. Composed over three years and passing through almost 30 pages of drafts, 'The Whitsun Weddings' negotiates Yeats's caveats in 'Adam's Curse':

> A line will take us hours maybe;
> Yet if it does not seem a moment's thought,
> Our stitching and unstitching has been nought. (Yeats, p. 78)

while satisfying Larkin's own desiderata:

> I write poems to preserve things I have seen/thought/felt... both for myself and for others, though I feel that my prime responsibility is to the experience itself... Generally my poems are related, therefore, to my personal life. (quoted in Enright, p. 79)

In the case of 'The Whitsun Weddings', the 'experience itself' took place on WhitSaturday, 28 May 1955, when he took the Pullman from Hull, which stopped at Brough, Goole, Thorne (North), Doncaster, Retford, Newark Northgate, Grantham, and Peterborough before terminating at London King's Cross some four-and-a-half hours after it had left Hull. This was, in Larkin's words,

> a very slow train that stopped at every station and I hadn't realized that, of course, this was the train that all the wedding couples would go on and go to London; it was an eye-opener to me. Every part was different but the same somehow. They all looked different but they were all doing the same things and sort of feeling the same things. I suppose the train stopped at about four, five, six stations between Hull and London and there was a sense of gathering emotional momentum. Every time you stopped fresh emotion climbed aboard. And finally between Peterborough and London when you hurtle on, you felt the whole thing was being aimed like a bullet – at the heart of things, you know. All of this fresh, open life. Incredible experience. I've never forgotten it (quoted in Hartley, p. 100).

The diction ('somehow', 'sort of') and the splintering of sentences into fragments show Larkin striving to convey precisely what was so striking about this experience, and attest to the power it still possesses: it was, in a phrase of Hardy's which Larkin appreciated, a 'moment of vision'. Interviewed on television in 1981 Larkin modestly shrugged off his own skill and sought to foist all the poem's success on to the circumstances of the particular day: 'There's hardly anything of me in it at all. It's just life as it happened' (quoted in Motion, p. 287).

'Just life as it happened' is Larkin's modesty, for having an experience and then creating a fine poem from that experience are by no means one and the same. Here, by contrast, are two verses of another writer's poem about a train journey:

I am in a long train gliding through England,
Gliding past green fields and gentle grey willows,
Past large dark elms and meadows full of buttercups,
And old farms dreaming among mossy apple trees.

Now we are in a dingy town full of small ugly houses
And tin advertisements for cocoa and Sunlight Soap,
Now we are in a dreary station built of coffee-coloured wood
Where barmaids in black stand in empty Refreshment Rooms,
And shabby old women sit on benches with suitcases...

(quoted in Everett, pp. 135–36)

Thus wrote Vivian de Sola Pinto, editor of D.H. Lawrence and former Professor of English at Nottingham University. And here are the opening verses of 'The Whitsun Weddings':

That Whitsun, I was late getting away:
Not till about
One-twenty on the sunlit Saturday
Did my three-quarters-empty train pull out,
All windows down, all cushions hot, all sense
Of being in a hurry gone. We ran
Behind the backs of houses, crossed a street
Of blinding windscreens, smelt the fish-dock; thence
The river's level drifting breadth began,
Where sky and Lincolnshire and water meet.

All afternoon, through the tall heat that slept
For miles inland,
A slow and stopping curve southwards we kept.
Wide farms went by, short-shadowed cattle, and
Canals with floatings of industrial froth;
A hothouse flashed uniquely: hedges dipped
And rose: and now and then a smell of grass
Displaced the reek of buttoned carriage-cloth
Until the next town, new and nondescript,
Approached with acres of dismantled cars. (*CP*, p. 114)

The comparisons make themselves. Nevertheless, it is a paradox of Larkin's work that its very ease and truth to experience can obscure its agility of tone and nuance and the perceptive intelligence in making poetry out of things we have all seen. Despite Larkin's disclaimer, 'The Whitsun Weddings' abounds with personality which expresses itself first of all in a matter-of-fact casualness as he tells the reader 'That Whitsun I was late getting away', and then in the languor of a man on holiday with a long undemanding journey ahead of him ('all sense / Of being in a hurry gone,') which in turn eases itself into the relaxed contemplation of 'The river's drifting level breadth began' and then rises to the dry humour, which makes a geographical point, of 'Where sky and Lincolnshire and water meet'. Of course, the first two verses ensconce the reader in the speaker's railway carriage in order to experience, as the poet does, the dipping and rising of the hedges, and 'a smell of grass' which so refreshingly 'Displaced the reek of buttoned carriage-cloth'. But at the same time the narrative framework is being meticulously built. The unobtrusive 'we', which is how any rail traveller would speak, quietly acknowledges the speaker's fellow travellers on his 'three-quarters-empty train' even though he never mentions them, but also prepares for the very different 'we', the newly-weds, with whom he does feel involved. The 'short-shadowed cattle' are not merely observed but serve to make the contrast to the 'long shadows' cast by the poplars outside London as the day comes to an end. Throughout the second verse the speaker gazes out of his carriage window on and off all afternoon, taking in the details 'Until the next town, new and nondescript, / Approached with acres of dismantled cars', whereupon he very reasonably assumes that there will be nothing further of interest to be seen, and turns to his book as the train pulls into the station.

At this most unpromising stage, with the speaker reading (as he has done at other stations too), and, ironically, missing something more interesting than anything he has yet seen, the narrative begins to gain momentum. Even so, it is briefly delayed: 'Sun destroys / The interest of what's happening in the shade' is a remark which strikes the reader as self-evident, but only once Larkin has articulated it. Yet unlike 'Dockery and Son' or 'I Remember, I Remember', which having once embarked on generalities dwell on them, 'The Whitsun Weddings' promptly returns to the local situation with the speaker's eye caught, as the train pulls out, by

> girls
> In parodies of fashion, hats and veils
> All posed irresolutely, watching us go
> As if out on the end of an event
>   Waving goodbye
> To something that survived it. (*CP*, pp. 114–15)

The delicate metonymy of the end-of-platform image first of all locates the wedding guests and contrasts with the speaker's arrival at the next station where 'Struck, I leant / More promptly out... more curiously'. Therefore he *did* lean out of his window at the previous station but only once the train had passed the wedding guests whom he saw at the end of the platform, waving goodbye. This time, because he leans out and stares while the train is stationary, he has time to 'see it all again in different terms':

> The fathers with broad belts under their suits
> And seamy foreheads; mothers loud and fat;
> An uncle shouting smut; and then the perms,
> The nylon gloves and jewellery-substitutes,
> The lemons, mauves and olive-ochres that

> Marked off the girls unreally from the rest. (*CP*, p. 115)

These 'different terms' come from the different order; as the stress on 'then' makes clear, he sees members of immediate family first and in detail rather than, as before, just the bride's girlfriends and the bridesmaids. Yet the 'different terms' also display the speaker's increasing involvement: earlier he had perceived 'curiously' (the implication of the later 'more curiously') the girls 'grinning and pomaded... / In parodies of fashion', but now leaning out in an unambiguously nosy gesture he finds his sense of the girls' tastelessness supplanted by a fascination. Unlike 'Here', which suffers from the over-obvious rigidity of the speaker's stance, known well beforehand, 'The Whitsun Weddings' succeeds because it so convincingly takes its direction from its speaker's changing apprehensions in the course of the journey itself. As couple after couple board the train and once 'The last confetti and advice were thrown' (with the adroit suggestion of each wedding party *and* the very last one) and the train moves off again, the speaker, probably hanging even further out of his window, is rapt to the point of illumination:

> each face seemed to define
> Just what it saw departing: children frowned
> At something dull; fathers had never known
>
> Success so huge and wholly farcical;
>   The women shared
> The secret like a happy funeral;
> While girls, gripping their handbags tighter, stared
> At a religious wounding. (*CP*, p. 115)

This, held back until the train moves off, is the last of the 'different terms'. Next, all the wedding parties are left behind as the 'hurtling' begins:

> Free at last,
> And loaded with the sum of all they saw,
> We hurried towards London, shuffling gouts of steam. (*CP*, p. 115)

These lines make very precise distinctions. If 'we', the subject of the sentence, still sounds slightly detached because 'loaded with the sum of all *they* saw' (my italics), it is also qualified by 'free', implying that speaker and newly-weds alike share a sense of freedom now that there are no more stops. Thereafter everyone watches the same landscape, itself precisely observed ('An Odeon went past, a cooling tower, / And someone running up to bowl': the train is going so quickly that no one sees the cricket ball delivered); and as the destination approaches, the passengers see the 'walls of blackened moss' at which the train begins to slow down.

The poem's memorable climax has to be quoted in full:

> I thought of London spread out in the sun,
> Its postal districts packed like squares of wheat:
>
> There we were aimed. As we raced across
>   Bright knots of rail
> Past standing Pullmans, walls of blackened moss
> Came close, and it was nearly done, this frail
> Travelling coincidence; and what it held
> Stood ready to be loosed with all the power
> That being changed can give. We slowed again,

And as the tightened brakes took hold, there swelled
A sense of falling, like an arrow-shower
Sent out of sight, somewhere becoming rain. (*CP*, p. 116)

Before one comes to these lines one wonders how Larkin can possibly avoid a somewhat disappointing conclusion, for the journey has to end. A lesser poet would have brought the poem to a lame halt by re-emphasizing how memorable the day was. And indeed, although the train is at full speed or thereabouts as it 'race[s] across / Bright knots of rail', it slows down when corridored by the blackened-moss walls and changes from being a 'frail / Travelling coincidence' which is 'nearly done' to one which *is* done as it enters the station and comes to a standstill with the 'tightened brakes' taking hold and jerking the passengers towards the engine. By writing 'there *swelled* / A sense of falling' (my italics) Larkin associates the idea of the train's stopping with the potential to release its contained energy on 'London spread out in the sun, / Its postal districts packed like squares of wheat', and with outward movement, thus permitting the easy connection to 'an arrow-shower/Sent out of sight' (connoting spreading, dispersal, and, ingeniously, height) resulting in increase for the honeymoon couples ('somewhere becoming rain'). Larkin also anticipates the paradoxically releasing effect of the brakes *tightening* when he stresses that the couples' energy 'stood ready to be *loosed* with all the power / That being changed can give' (my italics). Larkin's metaphors are at once true to the experience's power and to the journey's actual circumstances; the everyday is not transformed into the uplifting but partakes of it while remaining itself.

After the publication of *The Whitsun Weddings* Larkin never wrote about trains again. This has far less to do with his passing his driving test in 1964 than with the attitudes governing *High Windows*, his final work. With the voice of a man who knows what he thinks and has made his mind up, Larkin writes bitterly and testily of age, death, the irrecoverable nature of youth, and his alienation from the young. In this climate there is no place for trains because they involve transience, narrative, and, frequently, the unexpected. Fine as *High Windows* is, what is lacking is the engaging sense of the poet experiencing a situation, or a state of mind, for the first time or by accident, and then thinking his way forward.

Indeed, what is important in Larkin's best writing, which I take to be *The Less Deceived* and *The Whitsun Weddings*, is the intelligent thinking which tacks, often erratically, sometimes to the speaker's surprise, sometimes in brief pursuit of an *ignis fatuus*, often through unrewarding digression, towards conclusions. It is this natural yet illuminating way of thinking which receives particularly effective expression when Larkin is writing about trains, with the speaker observing life and yet remaining uncontactable outside its mainstream. Broadly speaking, rail travel both takes one away from one's concerns yet allows the mind, in a peculiar state of unforced receptiveness, to return unhurriedly to those very concerns. (One wonders how many poems, not necessarily having trains as their setting, were silently gestated while Larkin was travelling.) In Larkin rail travel creates a poetry in which very frequently 'his moods are our moods' (Davie, p. 64), but even when they initially are not, the recognizable thought processes, being our own, place us in the poet's frame of mind. Furthermore Larkin, in his train poems alone, brings back the narrative skill of his novels *Jill* and *A Girl in Winter*, novels which for all their craftsmanship never stir the emotions, yet narrates in a way that hardly seems narrative at all, just the speaker travelling, observing his fellow travellers when there are any, and noting, with none of John Betjeman's nostalgia, the sights and details which pass by his window.

Larkin has produced a poetry in which the interplay of company, of a mind musing on its own concerns yet open to impressions, and of the moving landscape, is so natural that it draws no attention to the means by which it articulates itself, still less to Larkin's unfaltering control over the balance of the conditions out of which the poems seem to arise. These qualities in Betjeman Larkin singled out for the highest praise, but the judgement applies equally to his own poetry: '[he] is serious because he has produced an original poetry of persons and surroundings in which neither predominates: each sustains the other, and the poetry is in the sustaining' (*RW,* p. 216).

Works Cited

Abbreviations

Two of Philip Larkin's books are abbreviated as follows:
CP   *Collected Poems*, ed. Anthony Thwaite, London, The Marvel Press and Faber, 1988.
RW   *Required Writing: Miscellaneous Pieces 1955–1982*, London, Faber, 1982.

Other works

Chambers, Harry, ed., *An Enormous Yes: in memoriam Philip Larkin (1922–1985)*, Colstock, Peterloo Poets, 1986.
Davie, Donald, *Thomas Hardy and British Poetry*, London, Routledge, 1973.
Enright, D.J. ed., *Poets of the 1950s*, Tokyo, Kenkyaska, 1956.
Everett, Barbara, 'Art and Larkin', in *Philip Larkin: The Man and His Work*, ed. Dale Salwak, London, Macmillan, 1989.
Hartley, Jean, *Philip Larkin, The Marvel Press and Me*, Manchester, Carcanet, 1993.
Larkin, Philip, *Jill*, London, Faber, 1985.
——, *Selected Letters of Philip Larkin*, ed. Anthony Thwaite, London, Faber, 1992.
Motion, Andrew, *Philip Larkin: A Writer's Life*, London, Faber, 1993.
Singer, Burns, *The Collected Poems of Burns Singer*, London, Secker & Warburg, 1970.
Yeats, W.B., *Collected Poems*, New York, Macmillan, 1956.

Chapter 11

BEWARE OF THE TRAINS:
Reflections and a Few Footnotes on the Railways of Suffolk

Norman Scarfe

There is a well-attested story of the celebrated Dr Routh, who lived from 1755 to 1854. His boyhood was passed at Beccles in Suffolk and he was President of Magdalen College at Oxford for 63 years. In about 1850, an undergraduate from Suffolk casually arrived in college a fortnight after term had begun. His tutor assumed the President would want him sent down. 'Ah, sir, the roads in Suffolk are very bad at this time of the year', was the kindly old man's reply. 'But Mr President, he didn't come by road.' 'The Suffolk roads in winter I do assure you are very bad for travelling.' 'But he didn't come by the road, sir, he came by the "rail".' 'Eh, sir? The "what" did you say? I don't know anything about "that".'

The story illustrates the nonagenarian President's steady refusal to accommodate the dangerous and intrusive forces of mechanized transport into the quiet countryside of his early recollection, or anywhere else. Had he read William and Hugh Raynbird's delightful *Report on the Agriculture of Suffolk* in 1849, I doubt if he would have been pleased to be told: 'The Suffolk drill is now the kind in the most general use throughout the kingdom, and is adapted either for drilling corn, on level land or on ridges, and on all descriptions of soil.' He might have been seriously disquieted to know that the prototype of this drill had been invented by the Revd Mr Cooke, a clergyman, of Heaton Norris in Lancashire: its general appearance, 'the seed-box, coulters, with tin funnels, the delivering the seed from cups, and the mode of action by cog-wheels', was very similar to the Suffolk machine. It was in about 1790 that Henry Baldwin, a farmer of Mendham, in the Waveney valley, only a short distance above Beccles, Bungay, and South Elmham (where Routh's father held two livings), improved upon the drill known as Cooke's Drill. It greatly improved seed-sowing as practised since Biblical times. At least it was not engine-drawn and did not involve people careering through the landscape in carriages headed by noisy, smoky steam-engines. The train service from Oxford to London was never experienced by Magdalen's Suffolk President. He went by road, and that but once after the railway came.

By the time I went to read history at Magdalen, I had come to share his disinclination to travel by rail. There had been alarming journeys to school from

Felixstowe Town station to Canterbury East, which, despite its name, stood at the true south edge of the walled town, so that one walked past the Dane John Mound and all the way up St Margaret's Street to Mercery Lane and Palace Street, where my House stood on remains of the enormous thirteenth-century hall of the archbishops. Those journeys to school involved carrying a great case bulging with clothes, books, and a plum cake, and boarding an Inner Circle train on the Underground, from Liverpool Street to Cannon Street, in time to catch a particular train to Canterbury. I still find the directions on the Inner Circle at Liverpool Street baffling: because its position in the 'circle' comes at the east end of an elipse, the invisible directors seem not to know when to stop calling westbound trains 'eastbound' and when to start admitting that they are westbound. One learnt that familiar sequence of stops: Aldgate–Mark Lane (for Fenchurch Street)–Monument. One wonders what Mark Lane has done to be erased from the only alliterative part of the sequence, and replaced now by Tower Hill? The alarm which I gradually overcame sprang from what must be fairly common fears: that one will miss a connection, get into the wrong train, or fail to get out at the right station. On car journeys, I suppose a comparable anxiety might be about losing the way, especially in a big town without an adequate map.

My lack of enthusiasm about rail travel from Suffolk was increased by experiences in the war, though a single journey in 1940 nearly changed my attitude. At Easter, we were evacuated from a relatively front-line position at Canterbury to Cornwall. In terms of the 'Baedecker' air raids, the school looked vulnerable, clustered in the Precincts all along the north side of the cathedral. (As to a German invasion, we now know that they had planned to come by the shortest crossings to Kent.) At Easter in 1940, southern Cornwall seemed reassuringly far west, but in May and June the Netherlands and France fell, and we were suddenly in the front line, facing a Brittany occupied by German forces. What seemed more real was that we arrived at St Austell from Paddington on the *Cornish Riviera Express*. For me it was both a real and a magic journey, something altogether unimaginable on the railways of Suffolk (though I can believe that Suffolk's branch lines, on, alas, none of which I travelled, had a magic of a very different kind). There was an excitement, very hard to define, as you stepped on to the *Cornish Riviera Express* in 1940: doubtless something to do with the word *Riviera*, and the attention of the staff, and the comfort of even the third-class carriages; and there was an excitement that increased as the train made its rapid and efficient way west, where the engineering of the route conveyed its own majestic design. With the devastating bombing of Plymouth, and the wartime overworking of the railways, the overcrowding, the undercleaning, and the half-lit, blacked-out carriages by night, the enjoyment of that great express gradually gave way to discomfort.

By nature Jack Simmons was a devoted West Countryman, and of course he has set into the clearest perspective the kinds of effect the *Cornish Riviera Express* had, not merely on its travellers, at least down to 1940, but, like the remarkable railway of which it was part, on the economic well-being of the West Country. His essay on 'The End of the Great Western Railway', in his book *Parish and Empire* (1952), managed with artful ease to convey, in 14 pages, how Bristol brought it into being, how physically 'Great' it was in the area it served, what Isambard Kingdom Brunel contributed and Daniel Gooch, Charles Russell, and C.A. Saunders (two of them, including Brunel, still in their twenties: choosing middle-aged men of established reputation 'was not the habit of the early Victorians'), how stolid it had nevertheless become when N.J. Burlinson became Superintendent of the Line in 1888 and transformed it. In 1904, the *Cornish Riviera Express* began to run non-stop between London and Plymouth – 'easily the longest non-stop journey in the world, made at an average speed of 55 miles per hour... In the 1930s, its express trains were better than they had ever been before, and better than they ever seemed likely to be again.' (His book of elegant essays is supplied with a businesslike Appendix, setting out the shortest times between Paddington and some principal stations between 1888 and 1951.) He added: 'If you travelled third you were well looked after. But there always remained an opulence about a Great Western first-class carriage... You travelled on the Great Western in comfort: even in the pinched and harassing 1940s it would give you the curious illusion that you were really living in 1910.'

That was just the kind of illusion I had in 1940, and, more fleetingly, in the later 1940s. But wartime confirmed my almost automatic association of railway journeys not with Edwardian comfort but with a shared, socially unifying discomfort, and with something much less bearable: the saying of melancholy farewells on platforms or at the lowered windows of carriage doors – to friends – and after the touching hopelessness of brief encounters – and to my parents, who knew we might never meet again, as I set off for the Divisional Concentration Area of tented camps surrounded by barbed wire and screened from the air by the woods in that hinterland above Portsmouth where we waited to sail, the day before D-Day, for Sword Beach. Before that, there were other dour recollections: joining, for instance, the crowded last train of the evening from Kings Cross to Darlington, and discovering in the early hours that it had broken down, so that we should be late by at least an hour for parade at our officer-cadet training course at Catterick. The penalty was a week of parading in full kit, before dawn every morning for a week, because we had not, 'as any seriously responsible potential officers would, caught the train "before" the last one scheduled to get us to parade on time'. I still, for different reasons, find myself catching the train before the one I really need. Nor do I seem able to forget journeys north to Glasgow (with a taxi-less walk between Central and Buchanan Street), and on up to Inverness, to take part in amphibious exercises from Fort George; cold though they were, those journeys were of almost Egyptian or Indian congestion, with various servicemen reduced to sharing with bulging kitbags the hard surfaces of tables between the packed seats in each compartment. Such experiences, scarcely suitable for cattle, are related in a civilized way by Evelyn Waugh at the opening of *Brideshead Revisited*.

Mention of *Brideshead* brings me back to the essay on 'The End of the Great Western Railway' because, in it, Jack Simmons opened up an idea, always present in his formidably marvellous literary memory. 'The Great Western', he claimed, 'alone among English railways has cut some figure in literature.' He thought of it as 'William Morris's railway', for it served the country of *News from Nowhere*. He remembered Beerbohm's famous opening of *Zuleika Dobson* at Oxford station, and Thomas Hardy's moving poem 'Midnight on the Great Western', and Edward Thomas's 'Adlestrop', and how the Great Western 'appears, time and again, in Richard Jefferies'. Whether, since 1952, anyone has answered his challenge seriously on behalf of other railways, I do not know.

In 1997 he edited with Gordon Biddle the magnificent *Oxford Companion to British Railway History*. In it, they and their contributors have come closer than I would have dreamt to turning me into an enthusiast for at least the history of railways. Simmons's own contributions are manifold. They range alphabetically, from 'Abandonment of railways' and 'Accidents' to the 'York, Newcastle and Berwick Railway: its central station at Newcastle remains today one of the finest in England: the bridges built for it at Newcastle and Berwick are among the outstanding achievements of British railway engineering'. The *Oxford Companion* includes his own essay on 'Railways in English literature'. It is that which has encouraged me to write these 'Festfussnoten' for his Festschrift (perhaps not least because, like many other writers, I have pained him by not liking railways, for a variety of reasons, including President Routh's; compare particularly J.M. Crook's piece in this section, on Ruskin.

Hugh Moffat, in his book on *East Anglia's First Railways*, 1987, reproduces a copy of the first edition of the one-inch OS map showing John Braithwaite's proposed great 70-foot-high viaduct crossing the Stour astride Constable's father's watermill, Flatford Mill. That proposed viaduct is dated the year after the painter's death. It was mercifully superseded by Peter Bruff's line that took more flexible account of the contours: the Flatford viaduct was succeeded by his 'substantial and scientifically constructed' timber viaducts at Cattawade (Figure 11.1). Betjeman alone seemed to relish everything about railways, as he did almost everything Victorian. Simmons notes wryly: the railways 'alone made possible the enormous increase in the dissemination of English literature during the Victorian age. Like so many of the railways' services to the community, this one excited no remark.'

When this Festschrift was taking shape, it was thought that, though I am no railway historian, I might be able to contribute something on 'railways in Suffolk'. But Jack himself had done the job already, in 1961, a compact but flawless essay on 'The Railways of Suffolk' in his book *The Railways of Britain: An Historical Introduction*. It comes in his chapter on 'Railways on the Ground', which begins with a beguiling journey from Fenchurch Street to Rochester, including the crossing from Tilbury to Gravesend by one of those ferries 'with their tall slender funnels' and with the memory, at this very point on the river, of the climax of *Great Expectations*. Following this, Suffolk's railways seem a bit tame: 'only one main line passes through Suffolk, and that not quite of the first class'. It certainly is not. Nor do Suffolk's lines get much patronage from first-class writers. I only hope there may be enough to supply a footnote or two, if not an Appendix, to Jack Simmons on 'The Railways of Suffolk'.

First, his essay is accompanied by a valuable map of them, showing you, at a glance, which bit of the network was made in which year. You see that Beccles, for instance, was not reached by rail till the year of President Routh's death, 1854, which may partly excuse his determined unawareness of their presence in Suffolk. (We are not told by his biographer, Middleton, what part of Suffolk the unpunctual undergraduate inhabited.) After noting the exceptions of the seaside resorts, the fishing and the bus/coachbuilding needs of Lowestoft (the demands of Felixstowe's great container terminal have arisen entirely since the essay was written), Jack Simmons emphasized the plain truth that, apart from the one main London–Ipswich– Norwich line, Suffolk's railways came into being to serve the farmers in their best mid-nineteenth-century years. His map shows the rather wavering course of the branch line intended to link up as many as possible of the more inaccessible parts of

Figure 11.1   Peter Bruff's timber viaduct at Cattawade on his main Colchester–Norwich line of the Eastern Union Railway, 1846 (*Illustrated London News*, 1846 / Hugh Moffat)

**Plate 11.2   Worlingworth station, on the Mid-Suffolk Light Railway, opened
           1904, closed 1952** (Jack Simmons, 1953)

arable High Suffolk; it came too late to get established in the prosperous years before
1879. It finally branched out from the main line at Haughley, wandering through
Kenton into the fields beyond Laxfield, carrying freight from 1904 and passengers
from 1908. This was *The Mid-Suffolk Light Railway*, whose story was vividly
described by N.A. Comfort (1963, with two new editions). How haphazard the project
was may be judged from the fact that in May 1902 HRH the Duke of Cambridge cut
the first sod at Westerfield, on the edge of Ipswich and many miles from Haughley,
which was the final starting point of the line as it headed for Kenton, Laxfield, and
through to Halesworth: the Duke's spadework seems to have been the only
contribution to the abandoned southern end of this branch line. F.S. Stevenson, MP
for Eye, who lived at Playford, not far from Westerfield, was by nature an antiquary,
and took more serious interest in the controversy about the Anglo-Saxon minsters at
North and South Elmham than in the company accounts: he had lost nearly £100 000
of his wife's money in this railway by the time he was bankrupted in 1906. Boosted
briefly by the Railways Act, 1921, the Mid-Suffolk Railway managed to puff on till
1952.

A survival of it that now puzzles the younger natives and newcomers alike is the
side walls of a shapely little brick bridge that never carried anything over the road
between Debenham and Aspall. Simmons's own photograph of the line shows
Worlingworth station in 1953, which tells its own story (see Plate 11.2). I was with
him on that expedition, and acquired an abandoned cast-iron plaque bearing the
considerate words, in bold sans-serif lettering:

BEWARE
OF THE
TRAINS

I used it as a door-step outside my cottage on the beach at Shingle Street, where even motor traffic peters out.

Another melancholy photograph (Plate 11.3) illustrating Simmons's Suffolk essay shows Capel station, where the Hadleigh branch line has just left its two-way junction with the main London–Ipswich line at Bentley. It is about to risk the crossing of the Norwich–Colchester Roman road between Copdock and Capel St Mary, on its own eight-mile journey to Hadleigh, where the train arrived by a gracefully curving track. The dramatic story of the opening of this line is told in Hugh Moffat's book, *East Anglia's First Railways: Peter Bruff and the Eastern Union Railway* (Lavenham, 1987). Moffat's chief passion is our local maritime history (he is my near neighbour, just off Woodbridge's Market Hill). But his book on Bruff and the Eastern Union Railway is the most serious study of the establishment of Suffolk's railways to appear since Simmons published his essay in 1961. Bruff is sometimes referred to as 'the Brunel of the Eastern Counties'. He clearly does not rank with the trio of Brunel, Joseph Locke, and Robert Stephenson, whose junior he was. His dates were 1812–1900. But he got his engineering training under Locke, and his achievement was, essentially, the Eastern Union Railway, with its main line from Colchester to Norwich. Admittedly not of the first class, it was a proud achievement of very great value to thousands of people in East Anglia.

At the ceremonial sod-cutting of the Hadleigh branch in 1846, Bruff was warmly cheered, and referred to as 'Engineer in Chief'. But he was too busy with mainline

Plate 11.3   Capel station, on the Hadleigh branch line, opened 1849, demolished 1973 (Jack Simmons, 1953)

schemes and parliamentary business, and deputed the Hadleigh work to his 'oldest assistant', Robert Richardson, 'with the assistant engineers, Charles Russell and Edward Sheppard, previously at Ipswich tunnel'. An early excursion to the Ipswich regatta in 1847 was marred by a violent gale, demolishing a wall on Hadleigh station and causing many serious injuries. However, I associate the line with more agreeable occasions. Branch lines, by definition, are relatively little-used, and correspondingly difficult to furnish in retrospect with life. I hope this may excuse my introducing Capel station's indispensable role in my own genealogy. My paternal grandfather, Samuel Scarfe, born at Capel St Mary, moved to Trimley St Martin near Felixstowe, to the Big Mill, one of East Suffolk's postmills, described by Rex Wailes as 'the finest of their type in the world'. My grandfather worked it by wind, and later by steam. When the going got tough, and Ranks started to squeeze out the windmillers, whose craft went back before Chaucer, he sold his great central oak post to the military in 1914, as his two sons went off to the war, and took to being 'corn merchant and small farmer'. In the 1880s, he had courted my grandmother, a Capel girl from his native village. Without two branch railway lines, the Hadleigh one and the one Colonel Tomline had built between Ipswich and Felixstowe, his courtship would have been almost impossible: as it was, there was a four-mile cross-country walk to Felixstowe Beach station (Trimley station came in 1891); at Ipswich a change to the London train; and, at the first stop, Bentley, another change on to the Hadleigh line, where Emma would meet him at Capel, the first stop. *Love on a Branch Line*, 1959, is the title of a comic little novel, set in Suffolk, by the late John Hadfield: nothing could have been more serious or more sensible than my grandfather's pursuit of love: no pair of branch lines ever performed a better service.

'The Hadleigh Express' occupies a chapter in one of the most engaging Suffolk books, *A Suffolk Childhood*, 1959, 1985, by Simon Dewes, the pen name of John Muriel. It describes very freshly the life of that beautiful and completely self-contained old town as he saw it in the years 1914–18, when he was a young boy, born in 1909. His father was the Hadleigh doctor, so the family knew pretty well everything that went on. In the middle of the war, a few bombs fell on the town, and the doctor packed his wife and family off to live, briefly, and very surprisingly, in Margate. This was the occasion of young John's experience of a lifetime, travelling the eight miles from Hadleigh to Bentley on the footplate with the driver and his fireman.

They had arrived in two cabs at Hadleigh station. The stationmaster escorted Mrs Muriel to the ticket-office, where the clerk already had their tickets, for 'who was there in Hadleigh who didn't know of this momentous journey?' Ever since he had become an extremely junior member of the Hadleigh Fur and Feather Society, young John Muriel had known Mr Thorpe, the guard on the Hadleigh Express, for he was a prime fancier of Belgian Hares and also kept pigeons. There he was, with almost more gold braid than the stationmaster's, and in full-length double-breasted coat, with brass buttons, and this was only a branch line! Mr Thorpe made the sudden astounding suggestion to the stationmaster that John might like to travel to Bentley junction in the cab of the engine. His mother gave reluctant consent, so he was handed up and seated on a little bench. At once the driver, guard and stationmaster were consulting great turnip watches, Mr Thorpe with a furled green flag under his armpit. He finally disposed of his watch, raised his flag and put his whistle into his mouth. At this critical moment, the booking-clerk came galloping on to the platform with news of a passenger puffing up the hill. Amiably, Mr Thorpe took his whistle from his mouth and shouted to the driver: 'You'll have to hold on a minute, booy', going to greet the late arrival. The driver noticed John's disappointment and said: 'There, tha's a bit o' luck for you: now we'll be able to show you a bit o' speed, makin' up time.'

At last we were off. The green flag was dropped. The whistle blew. An enormous belch of feathery-grey smoke was spurted out of the tall narrow chimney. The driver pulled a most inadequate piece of string so that the engine gave a deafening shriek, and we were off. Looking forward, I could see the straight single line of track that ran the eight miles to Bentley junction. Looking back I could see in the valley the tapering spire of Hadleigh church, the broad straight thread that was the High Street and the narrow winding thread that was the sleepy river Brett. The engine gave another prolonged whistle, and the driver told me he always did that when he passed the top of Cranworth Road, to let his wife know he'd got away safely. Looking out, sure enough I saw what looked like a white flag being waved in reply.

It was between Raydon Wood and Capel that we did 'the bit of speed' he had promised me, for here there were between two and three miles without any level-crossings, and here the driver announced he would 'let her goo'. He did. The little engine seemed to take the bit between her teeth. She shuddered. She hiccupped and then, by God, she bolted, and the dear fat driver lifted me from my seat, yelled in my ear 'take a hold o' that', and thrust the piece of string into my hand. I took hold of it. The engine's whistle screamed and shrieked away. 'That'll let 'em know we're comin' said the driver. 'We don't want no-one in our road the

Figure 11.4   Reg Carter's drawing in the Sorrows of Southwold postcard series No. 2 (Southwold Museum)

THE SOUTHWOLD EXPRESS - THE BRIDGE IS UNABLE TO STAND THE STRAIN OF THE ANNUAL
EXCURSION TRAIN - LUCKILY THE LIFEBOAT IS NEAR AND ALL ARE SAVED

**Figure 11.5   Reg Carter's drawing in the Sorrows of Southwold postcard
series No. 1** (Southwold Museum)

rate we're a-goin'. So I held on to the string, and the whistle continued and the little train
bucketed along like a mad thing till I was told to let go. The noise stopped, and we came to
rest, sedately, at Capel station.*

There are no adequate illustrations of the working of the Hadleigh Line, but as Jack
Simmons notices in his Suffolk Railways essay, things were different on the narrow
gauge (three-foot) Southwold Railway (1879–1929). As Simmons says: 'The line
was held in special affection by Southwold and its neighbourhood for the erratic and
leisurely conduct of its traffic. Caricatures were drawn of it and sold as postcards. It
remained staunchly independent of all amalgamations throughout its fifty years of
service; and, most surprisingly, its shares came to pay a dividend.'

Alan R. Taylor and Eric S. Tonks, in their book on *The Southwold Railway*, 1979,
write of those caricatures by Reg Carter which were sold in two series of six
postcards, and claim that the employees depicted were 'identifiable caricatures'. Jack
Simmons reproduced one of them. Here, we add two more. Figure 11.4 shows the
driver 'doing a roaring trade in hot taters owing to a delay caused by the porter over-
estimating his strength'. It illustrates that the officials at Hadleigh were by no
means alone in formality of attire. Figure 11.5 shows the 'new' swing bridge of 1907
across the Blyth river under strain during an annual excursion. It replaced the original
bow-string girder swing-bridge which had something in common with Peter Bruff's

* I am grateful to Barbara Hopkinson for permission to use this extract.

1846 bow-string truss over the navigable channel at Cattawade, just discernible towards the middle of that romantic contemporary engraving from the *Illustrated London News* we have produced on p. 155 (the thin vertical line indicating the original two-page spread).

Chapter 12

The Train in the Landscape:
Dovey Junction c. 1932[1]

Gwyn Briwnant-Jones

Few of the engineers and contractors who built railways through the predominantly rural districts of Wales were blessed with sufficient capital to realize their ambitions. The central region, for example, failed to attract the support of either of the major English companies then just across the border at Shrewsbury. Rails were eventually laid between the Marches and the coast of Cardigan Bay only through the efforts of prominent Montgomeryshire men like the engineer Benjamin Piercy, the redoubtable David Davies of Llandinam, and the rather more cavalier Savin brothers of Oswestry.

The Oswestry–Newtown section was constructed jointly by the Davies & Savin partnership; a double-track line was planned but although bridge abutments were optimistically built for a second line, a single track had to suffice. Similar constraints applied when Davies continued alone with the following section to Machynlleth. It was left to Savin (with Piercy's able support) to develop the fascinating north-south line along the Cambrian coast from Pwllheli to Aberystwyth, but, here again, limited capital influenced the final outcome.

The original scheme called for the coast line to be carried directly across two major estuaries, the Dyfi and the Mawddach. The viaduct across the latter, at Barmouth, is well known for its historic, engineering, and scenic interest, but Piercy's scheme for crossing the Dyfi would hardly have been any less spectacular; his plan for a double-tracked bridge included a roadway on a separate deck beneath the rails and featured an opening (swing) span. Unlike the Mawddach estuary, however, the Dyfi had powerful shipping interests up river. There was much opposition to the idea of a bridge and for many years lack of an adequate foundation was advanced as the principal reason for the scheme's delay and final abandonment – despite contemporary newspaper reports that 'one of the most experienced borers in England' claimed to have located 'a capital bottom'.[2]

---

1   When Jack Simmons unveiled a Railways of Wales Exhibition (at the Welsh Industrial and Maritime Museum), he urged the invited guests to view one railway painting which appealed to him particularly because it portrayed the train within the landscape. The painting featured Dovey Junction c.1950; it was the first of several inspired by this location.

2   *Oswestry Advertiser*, 15 January 1862.

Plate 12.1   Dovey Junction c.1932. Painting by Gwyn Briwnant-Jones, which appears in colour on the jacket. Reproduction by courtesy of the National Library of Wales.

Although the proposal for a bridge was first mooted in 1861, the eventual solution, a four-mile deviation-line along the north shore of the estuary to a more modest river crossing and a new junction on the Machynlleth–Aberystwyth route, was not completed until 1867.

With the benefit of hindsight, we now appreciate that Glandovey Junction – so named until 1904 – was hurriedly located, as it lies at a particularly low point on the marsh, a factor regularly demonstrated whenever the water table begins to rise. Over the decades, succeeding engineers would have been grateful had Piercy and Savin chosen a slightly more elevated site some quarter or half a mile to the east near Morben (the original transhipment point of the erstwhile Corris Tramway).

During steam days, locomotives could be relied upon to cope with flood water up to 18 inches above the top of the rail, although progress could be hindered with less water than this if the flood carried debris and driftwood to impede the signalling and point operation. In the days of the diesel multiple-units the problem was accentuated, as the DMUs were often troubled by just four inches of water above the rail.

The painting in Plate 12.1 depicts Dovey Junction during the early 1930s. The infrastructure is Cambrian, but the trains are thoroughbred Great Western Railway, hauled by examples of the 'Duke' class 4-4-0s introduced to mid-Wales after the Grouping. In days when up and down services crossed here, the normal practice was to dispatch the up train first, followed usually by the Aberystwyth service (visible in the distance). The coast train was nearly always the last to depart, leaving platform staff with little to do other than find solace in a hot mug of tea in the refreshment room.

SECTION III
THE OPENING UP OF BRITAIN

Chapter 13

The London Railway Suburb 1850–1914

Alan A. Jackson

The Railways have set us all moving far away from London – that is to say the special middle class of Londoners... people with incomes ranging from three to five hundred a year. The upper ten thousand and the abject poor still live and sleep in the metropolis. ... The middle classes betake themselves to far off spots, like Richmond, Watford, Croydon or Slough. True it is that the smaller fry content themselves with semi-detached boxes at Putney, Kilburn, New Cross or Ealing; but the wealthy take a more extended scope, and are found to go daily to and from the capital as far as even Reading or Brighton.

('An Unbeneficed District Surveyor', *Architect* x (1873), 170.)

Demand and market forces were usually the major factors in determining suburban development in the 60 years up to 1914 but good train services were always of great importance and occasionally a very powerful stimulant. For the capitalist, the best profits lay in attracting the middle classes, most of whom would walk to their trains at each end. This meant that success would depend on suitable building land being available at acceptable prices near a well-served station, preferably within a half-mile, although where stations were fed by electric trams (after 1901) this distance could be greater.[1] For the majority of potential occupiers, it was also important that fares should be moderate and the total journey to work should not take much longer than an hour.[2] Those with more comfortable incomes tended to reckon distance in terms of time rather than miles and if rail services were fast they would be very content to live well outside the city, yes even at Brighton.

Landowners' attitudes were also crucial; if they refused to break up their estates, the hopes of developers and railway promoters could be cruelly frustrated. This happened with the Hayes branch in south-east London, holding up development for many years.[3] Active property sales could raise hopes but if there was a good supply of suitable housing nearer London new stations would remain lightly-used and speculators could be left holding land on which they lost money should they start to build. This happened at some points along the 1895 line from Blackheath through Bexley Heath.[4] Also, however much their boards might wish, railway companies

---

1   F. Howkins, *An Introduction to the Development of Private Housing Estates and Town Planning*, 1926, pp. 12 and 14–15.
2   Howkins, op. cit. (1) p. 14.
3   Alan A. Jackson, *London's Local Railways*, 2nd edition, 1999, pp. 58–9.
4   E. Course, *The Bexley Heath Line*, 1980, and Jackson (3), p. 28.

were usually unable to ensure that residential settlement in a particular area was of a quality favourable to their profitable First and Second class revenue. There was one exception to this: the Metropolitan Railway, working from a base of special legislation, was able to manipulate the nature of development in its catchment areas by owning and developing land through subsidiaries.[5]

In projecting new lines, railway managers and chairmen seem sometimes to have been badly advised or perhaps did not seek professional guidance. Thus the dominance of local employment could be overlooked, the imminence of electric tramway competition not foreseen, and oversupply of new housing nearer London ignored. This happened with the Great Eastern Railway's Edmonton–Cheshunt Loop of 1891 which proved such a flop that it became one of the first railways in the metropolitan area to be closed to passenger traffic (1909).[6] The same company also came to grief north of Ilford in 1903, judging wrongly that the northward expansion of that suburb would continue at a fair pace and building at great expense a lavishly-stationed loop-line through the heavy clays of Hainault and Chigwell. Such traffic as might have been on offer at the south end was sucked away by electric trams which were in action just before the railway was ready.[7]

Railway suburbs were by no means entirely middle-class. For those immediately below in the social scale, the burden of daily travelling costs was significantly lightened by the introduction of very cheap workmen's fares from the mid 1860s, a facility which in time would create another species of suburban growth. The London County Council (LCC), formed in 1889, worked long and hard to widen the provision of workmen's trains and secure equality of treatment between the private railway companies.[8] From 1903 the LCC published a 'quarterly return' of all cheap rail and tram facilities available and by 1914 this was a quarter-inch-thick booklet listing over 460 stations offering cheap early-morning rail fares to central London.[9] Many of the more distant stations shown were however inaccessible to the average manual worker, since the return fares were 8d a day or above. It is more realistic to consider 4d as the absolute affordable daily maximum for the majority but nevertheless, given a readiness to rise early, there was a surprisingly varied group of places all round the capital between eight and eleven miles out from which journeys to work in the central area might be made for an outlay of no more than two shillings a week.[10]

---

5   Alan A. Jackson, *London's Metropolitan Railway*, 1986, pp. 140–2, 238–44, 249–50.
6   See Jackson (3), p. 366.
7   Ibid., pp. 397–8.
8   *Royal Commission on London Traffic*, 1905, vol. II, paras. 5135 et seq., 5149; vol. III pp. 24 and 125.
9   London County Council, *Quarterly Return of Workmen's Trains and other Cheap Morning Trains... and of Workmen's Trams...*, no. 43, 2 February 1914.
10   The outer limit of Cheap Trains for 4d or less return daily in 1914, clockwise from the north was: Enfield Chase (9¼ rail miles from the central area), Enfield Town (10¾), Walthamstow Wood Street (7), Leytonstone (6), Manor Park (6¼), Barking (7½), Silvertown (8), North Greenwich (4½), Plumstead (8¾), Blackheath (5), Woodside (10), Norwood Junction (8½), Selhurst (9¾), Tooting (9¼), Earlsfield (5½), East Putney (5¾), Putney (6), Acton Town (9¼), Hanwell (7¼: to Paddington only), Greenford (7¾: to Paddington only), Harrow on the Hill (9½), Hendon (9), Golders Green (6), Woodside Park (8¾). Source (9) above.

This encouraged a strong outward migration, and by 1912 something like a quarter of London's suburban rail passengers were travelling on workmen's trains; in the inner zone of six to eight miles from the centre, the proportion was as high as 40 per cent. Largely due to the efforts of the LCC, the number of workmen's trains daily was increased from 257 in 1890 to 1 806 in 1913.[11]

There were however considerable disparities between what was available in different directions around London and these had a marked effect on the nature and pace of suburban development. In 1914 a 2d workmen's ticket in south-west and north London bought eight miles return travel but in the north-east and east, 10½ and 21½ miles could be traversed for the same outlay.[12]

It was of course those railways which had no strong flow of freight or long-distance traffic that sought to build up the stable all-year-round 'residential' business. Thus whilst the London Brighton & South Coast Railway (LBSCR) offered first-class season tickets between London and Brighton for £50 a year as early as 1845 (and had reduced the cost to £33 by 1912[13]), the Great Western Railway (GWR) issued no seasons at all until 1851 and only after pressure from the LCC was it offering by 1914 a 4d workmen's return to London from Hanwell or Greenford and a 3d one from Ealing.

The companies made little if any profit from the workmen's trains. Their suburban revenue was at its highest from the top two of the three classes of carriage and it was fortunate that social pressures ensured that when the families of the first- and second-class season-ticket holders travelled, they would seek the same accommodation as paterfamilias. Victorian and Edwardian middle-class suburbanites went almost everywhere by train, making use of rail for journeys to school, to the large shopping centres and to leisure activities, notably to the London theatres. Even the most impecunious members of the middle classes, who might have tolerated a third-class carriage, were obliged to use second if they wished to travel daily after around 07.30, since third-class season-tickets were not generally available until the 1900s (not until 1938 in former Great Eastern territory). Suburban ticket revenue was also sustained by refusing to issue season-tickets for a shorter period than three months – weekly seasons were not available until the 1920s.

We now turn to examine a cross section of the railway suburbs which developed during 1850–1914.

Ealing

An anonymous writer of a Victorian municipal guidebook awarded Ealing the title 'Queen of the Western Suburbs', and certainly by 1905 it was a place simply bursting with municipal pride and respectability; no discussion of the present subject could

11   London County Council, *Report on Housing Development and Workmen's Fares, 1912*, and *Housing, 1928*, p. 169.
12   London County Council, op. cit. (9).
13   Public Timetables, London & Brighton Railway, 1845; London, Brighton & South Coast Railway, 1912.

ignore it. Richmond and Kingston were much in the same category but both were important pre-railway settlements. Modern Ealing, like the earlier Surbiton,[14] was very much a creature of the railway.

The railway influence is interesting, since although the Great Western had provided Ealing with a station as early as 1838, that company's indifference to suburban business had not encouraged much residential development. Indeed an ambitious villa-estate project launched in 1863 at Castle Hill, to the west of the ancient settlement, had come to grief, leading its promoter into bankruptcy in 1872. Such development as did occur was mainly to the south of the railway, mostly modest cottages for the families of artisans and workmen. It was not until the last half of the 1890s that the GWR 'conceded a service which at all kept up with the town's rapid growth, or corresponded to its importance.'[15]

Ealing's suburban growth owes much to the Metropolitan District Railway, which on 1 July 1879 opened a terminus adjacent to the GWR station, along with a second station to the south-east at Ealing Common. This provided a good train service right into the heart of the City, playing a strong role in the town's expansion from just over 18 000 in 1871 to 47 500 in 1901, the year in which it became the first of the modern boroughs in Middlesex. By that time there were many substantial middle-class villas, mostly north of the Broadway, conveniently placed for the two stations. Further north, in 1910–14, there appeared the 60-acre Brentham Garden Suburb, sited on the slopes down to the river Brent. A pioneer venture for London, operating on co-partnership principles, its inhabitants needed to be blessed with much physical energy as they faced a walk of up to a mile to reach the District and Great Western stations. With a train service both inconvenient and sparse, the eponymous Halt opened in May 1911 on the Great Western's new Birmingham line did little to alleviate their plight.

To the west, the late Victorian and early Edwardian years saw a considerable expansion of mostly small houses, though some larger residences were erected on the higher ground. This area was served by West Ealing station, belatedly opened in 1871 and at first known as Castle Hill. By 1901 a 3½d workmen's return fare was available from here to Paddington, valid on ten trains between 03.29 and 07.38. South of the Broadway, in an area still being developed in the 1880s, 1890s, and 1900s, the housing was still generally modest in size. This district stretched down to and beyond the District Railway's line to Hounslow and was served by stations called South Ealing (1883) and Northfields (1908).

From its arrival in 1879, the District Railway provided Ealing with two trains an hour all day, a frequency doubled in the peak periods. The centre of the City of London was reached in 48 minutes. Electrification between Ealing and Whitechapel from 1 July 1905 and the opening of an electric tramway through Ealing to Shepherds Bush, where commuters and others could transfer, with through tickets, to the new Central London tube railway, added further to Ealing's attraction as a place for middle-class residence. By 1914, early-rising workers could travel to Paddington for 3d return in any of ten GWR trains leaving between 03.31 and 07.41, but the

---

14   For the development of Surbiton 1838–1914, see R. Statham, *Surbiton Past*, 1996, chapters 4–9.

15   The Revd W.J. Scott, *Railway Magazine*, March 1899.

District Railway did better, offering no less than 17 trains between 04.54 and 07.38 for a 5d return fare to Mansion House, a journey of just under 11 miles each way.[16] The competition from the District Railway jerked the GWR into an immediate introduction of through trains from its suburban stations to the City over the connection to the Metropolitan Railway at Paddington.

With such services, and low fares available, it is hardly surprising that between 1901 and 1914 the town's population continued to expand, nor that many of the houses of this later period were quite small. There was an 85 per cent population increase between the 1901 and 1911 censuses, with a total of 61 200 reached in the latter year. In 1913 the District was offering up to 12 trains an hour at peak periods and eight an hour at other times. Some of these ran semi-fast and one, the so-called *Ealing Express*, passed all nine stations between Ealing and Sloane Square.

East of Ealing Common station much of the housing was built in the 1900s and early 1910s but to the north-west towards Ealing proper, development took place mostly in the 1880s and 1890s. A 1904 commuter journey from this area was meticulously recorded for posterity.[17] From the Ealing house, there was a 4-mph walk to the electric tram stop, consuming 7½ minutes; an 8-mph one-penny electric tram ride followed before a 9½-mile journey on a 3½d ticket to Mansion House by train in 45 minutes – an average speed of 12.6 mph. Finally, another 7½-minute walk to the workplace, so in all, 11½ miles in 67½ minutes (including 5 minutes waiting for tram and train), an average of 10¼ mph at a total cost of 4½ old pennies, or the cost in 1904 of five copies of a newspaper or two small loaves of bread. A year later, electrification was to cut the railway journey from 45 to 35 minutes.

Apart from the northward expansion up to the low lying land of the Brent Valley, Ealing had by 1914 begun to merge with its immediate neighbours: Hanwell in the west, Brentford in the south and Acton in the east. At its centre along the Broadway, there was a handsome Town Hall and lines of busy shops, some of considerable size, also three cinemas and a theatre.

The Cheap Train Suburbs

Away to the north-east of central London, Tottenham, Edmonton, Walthamstow, and Leyton, a group of densely-populated suburbs in the Lea valley, could not show the many fine new villas or the complacent municipal pride that characterized late Victorian and Edwardian Ealing. Their shops, although plentiful, were less prosperous of aspect and often in competition with street stalls. Here artisans and clerks jostled in close proximity to the working class in streets of terraces, built in patterned rows with tiny front gardens and back plots only slightly larger. Unlike Ealing, trees were not much in evidence along these streets, where houses were crammed in at up to 40 to the acre and built to the minimum standards required by the 1875 Public Health Act.

16   London County Council, op. cit. (9).
17   *Royal Commission on London Traffic*, 1905, vol. III table 13, p. 547 (C.S. Meik).

Most of this development had taken place between 1870 and 1900 and the very marked concentration of lower-income groups in these districts was a direct consequence of railway influences. The construction of the large terminus at Liverpool Street had entailed the demolition of much tenement property and as an alternative to rehousing the displaced inhabitants, the Great Eastern Railway had in 1864 accepted a statutory obligation to run one train in the morning from Walthamstow and one from Edmonton, together with matching return trains in the evening, both at a fare of a mere two old pence return. Two things happened. Firstly, over the years, the GER much exceeded its minimum statutory requirement; operational convenience saw the outer terminus of the cheap trains moved on to Enfield rather than Edmonton – a daily return journey of 21½ miles for twopence; and, following the 1894 enlargement of Liverpool Street station, the total of 2d trains was built up to 15 daily, supplemented by a large number of half-fare trains over the company's suburban lines, all arriving at Liverpool Street before 08.00. By 1897 six million passengers a year were being carried on the twopenny trains and four million by the half-fare trains.[18] The second consequence was that the twopenny and half-fare trains were not in the event patronized by the displaced tenement dwellers for whose benefit the whole scheme had been evolved; these unfortunates simply moved to adjacent slums, increasing overcrowding in the inner area. The patrons of the cheap trains were predominantly skilled workmen, market porters and warehousemen, who voluntarily and gladly moved into the Lea Valley districts to rent the thousands of four-room terrace houses erected to meet this specific demand. Many low-paid clerks and shop workers of both sexes also left their beds in the Lea valley suburbs at crack of dawn to catch a workmen's train at the cheap rate, killing time at and around Liverpool Street station until they could gain entry to their place of employment between 08.30 and 09.00.

In Tottenham, on the higher ground, there were also larger terrace houses of 'tunnel back' type with bay windows, a small fourth bedroom for a maid of all work and electric bells to summon her. They even had the unheard-of luxury of a bathroom. This gave the more ambitious something to aspire to, especially those in public or railway service or working for other employers offering prospects of promotion.

The Great Eastern's successful strategy, which it doggedly refused to modify in face of LCC attack, was to limit its cheap train traffic – which only barely if at all paid its way – to a specific corner of its suburban territory. Elsewhere, outward from Leytonstone and Chadwell Heath, no reduced fares were available to commuters other than the discount included in first- or second-class season-tickets. This explains the spacious high-quality Edwardian villas that can still be seen just by the west side of the railway at Woodford.

The Cheap Train Suburbs continued to grow in the final years of our period, showing population increases of between 31 and 44 per cent between 1901 and 1911, by which time almost all available building land was swallowed up. Many of the dwellings, especially in Walthamstow, were of the 'half-house' type, an ingenious design which provided self-contained accommodation for one family on the ground floor and another set on the first floor, at very low rents. Street doors were shared or

---

18   Lord Claud Hamilton (Chairman, Great Eastern Railway), *Railway Magazine*, December 1898.

placed side by side, with separate narrow access passages. A prolific provider of this was Thomas Courtenay Theydon Warner, MP for Lichfield, one-time member of the Essex County Council, who by 1901 owned 300 acres in Walthamstow, inhabited by 2 184 weekly tenants, almost all artisans and clerks travelling daily by train to workplaces in central London. Hindered in his enterprise by the lack of railway accommodation north and south of the Great Eastern's Chingford branch, which bisected the area from south-west to north-east, Warner boldly decided to promote his own railway scheme. After consulting parliamentary solicitors and a civil engineer, he produced a Bill for a railway making a junction with the line from Kentish Town at South Tottenham before running across Walthamstow from north-west to south-east to join the London Tilbury & Southend Railway (LTSR) at East Ham. This secured approval from the LTSR and, much more important, was strongly backed by the Midland Railway, which realized it would be a useful means of reaching Southend and Tilbury Docks as well as providing the company with more suburban coal depots in the London area. For its part, the LTSR was pleased to gain access to a railway system other than the GER. The two railway companies each supplied two directors under Warner's chairmanship and this Tottenham & Forest Gate Railway, authorized in 1890, was completed by the late summer of 1894. Since much of the district was already densely covered with small houses, its construction was costly, with most of the line carried on a brick viaduct to limit the land-take and property demolition. It did not offer anything lower than a 4d workmen's return.[19]

Ilford

Further east at Ilford, 7¼ miles from Liverpool Street on the GE main line, the last decade of the nineteenth century saw a great boom in commuter house construction.[20] There had been a station here since 1839 but the following 50 years saw little growth in the small community it served. Then, in 1894, the train service was much improved following widening of the approaches to an enlarged Liverpool Street station. Speculators who had been busy erecting small houses to the west, bringing the edge of London eastwards as far as Forest Gate and Manor Park now moved into Ilford; its population of 11 000 in under 2 000 houses in 1891 was to grow by 1901 to 41 240 in almost 8 000 houses; by 1911 the 1901 figure had almost doubled.

South of the railway, much of the new housing was of the small-cottage terrace type but north of the station, from the mid-1890s, the houses provided by two enterprising new speculators were larger and generally superior to the majority in the Cheap Train Suburbs. Many had four, some five bedrooms; there were even annexes for servants' accommodation. Archibald Cameron Corbett (later Lord Rowallan) was a prominent developer here between 1898 and 1907, running up hundreds of well-built houses which brought the burgeoning white-collar class into Ilford in strength. Having filled the best sites in Ilford itself, Corbett moved on to build new

---

19 Joint Select Committee on London Underground Railways, 1901, evidence of Warner, 14 May 1901.
20 Alan A. Jackson, *Semi-Detached London ...*, 2nd edition, 1991, pp. 37–42.

eastern appendages, named Seven Kings and Goodmayes. By 1903 he had sold over 3 000 houses. Other developers were also at work, mostly with houses aimed at middle-class types apt to describe their jobs as 'something in the City'.[21] Ilford had no 2d trains but a 5d return was sold until 07.30.[22]

By 1904–5 the Ilford building boom was beginning to fizzle out. There were, however, enough remaining available for those living in the tightly packed terraces of two-up, two-down houses south of the line who nursed a common ambition to move north of the tracks and employ at least one servant. Impressive parades of shops were also soon lining Ilford High Road, and further east, the climate for small shops along the main road was such that front gardens were soon giving way to retail accommodation.

In 1898 Corbett recognized his debt to the railway by contributing towards the cost of a northern entrance to Ilford station. He then went further, entering into agreements with the GER for new stations at Seven Kings (1899) and Goodmayes (1901). For Seven Kings, he guaranteed the railway company a minimum of £10 000 in new season tickets for each of the first five years, depositing £1 000 as security, a pledge more than honoured. A similar arrangement was made for Goodmayes. For its part, the GER provided two extra running lines between Ilford and Seven Kings and on to Goodmayes when that was ready. Ilford station itself was enlarged and rebuilt in 1893–8 and the best journey time into Liverpool Street was reduced to 15 minutes.

The First Electric Railway Suburb

Unique among the railway suburbs of the Victorian and Edwardian era was one which flowered in a few brief years across open country immediately north and north-west of Hampstead Heath. As Britain's first electric-railway suburb, it set a pattern for the great building boom of the late 1920s and the 1930s. Here, between the Midland main line out of St Pancras and the GNR branch from Finsbury Park to Edgware and Barnet across to the east, there was still no railway. Despite the proximity to London, this district was still totally rural in 1901 apart from a thin string of villas along the road towards Hendon.

When a new syndicate led by the buccaneer Chicago financier Charles Tyson Yerkes took over a scheme authorized in 1893 for a tube railway from Charing Cross to Hampstead, the Americans quickly decided to extend the line to the crossroads at rural Golders Green, where they could bring the line into the open air and build the car-depot on land which could be acquired much more cheaply than the site originally proposed. But this was only half the story; the Americans knew from experience that an electric railway brought to the edge of a city in this way and given an intensive service of trains could rapidly prompt increments in land values and, if all went well, stimulate housing and shop construction on a large scale. They had better advice than the luckless GER, correctly judging this could be achieved at Golders Green, where the railway terminus lay almost exactly halfway between the two existing rail lines,

21   Ibid.
22   London County Council, op. cit. (9).

ideally placed to open up what one of their spokesmen described as 'an entirely fresh district'.

Parliamentary approval was secured for the short extension in November 1902; there was also a separate Bill seeking authority for just over five miles of electric street-tramways designed to feed traffic to the tube terminus from nearby Hendon and Finchley. This part of the scheme was sponsored by a separate group under their control, not using the railway's name. Finally, a covert syndicate was set up through agents to develop a new residential area around the railway terminus. This, with prior knowledge, was in place to start buying up land at agricultural prices before the transport schemes came into the public domain. Such manoeuvres, familiar enough in the USA, were virtually unknown in Britain in the early 1900s.

One aspect of these carefully laid plans fell away when the tramway-feeder proposals failed to get through parliament, though a tramway to Finchley was eventually built by another company. The railway extension to Golders Green authorized in 1902 was linked with a scheme for a surface extension to Edgware which in the event would not be achieved until the 1920s. Construction of the deep-level tube railway and its stations proved a lengthy task and the line from Charing Cross through Hampstead to Golders Green was not ready for public traffic until 22 June 1907.[23]

Meanwhile land speculators and estate developers had been very active. Agricultural land formerly sold at £150 to £250 an acre rose in price seven or eight times once railway construction was seen to be assured, often changing hands twice or more at higher and higher prices before the builders arrived and broken up into smaller and smaller parcels in the process. Late in 1904, the Hendon Council passed the first plans. The main road towards Finchley was lit and drained at public expense in 1906 and by then two enterprising firms of estate agents had set up shop close to the site of the new terminus. A year before the railway opened, land across the road from the station was sold at £5 500 an acre for shopping parades.[24]

The first 19 new houses were completed in the last four months of 1905. In 1907, the year of the railway, 73 were built and in 1908, 340. Thereafter in every year up to 1915, completions exceeded 400, rising to a peak of 744 in 1911. The first full year of traffic at Golders Green tube station yielded the Underground Company one-and-a-half million passengers; by 1915 it was handling over ten million.

Half timbered and gabled, with tiled instead of slate roofs, and cottagey in aspect, the Golders Green villas set the trend in suburban housing design for the following three decades. Most were semis with two bedrooms of reasonable size and a third with room for a single bed. Bathrooms and indoor WCs were standard, as was electric light, though all main rooms still had a fireplace. Inflated land values limited plot size and pitched sale prices above the contemporary norm.

No overall plan was laid down for the new suburb, the only public control being over road widths, drainage, and conformance to basic building bye-laws, the latter not

---

23   D.F. Croome and Alan A. Jackson, *Rails Through the Clay*, 2nd edition, 1993, pp. 17, 44, 46, 52–3, 80 et seq.

24   For more detail on the emergent growth of Golders Green, see Jackson, op. cit. (20). pp 42–55.

always adequately policed by an overwhelmed local authority. Street layouts were wayward, responding to pressure to fit in as many houses as possible on the expensive land and also to cope with complicated ownership boundaries. Whilst the overall visual impact was much less dreary than the backward-looking new housing and streetscapes at Ilford, there was little sense of space and virtually no reminders of the countryside so quickly and savagely overwhelmed. No large areas of open land were left for recreation. The mistakes and omissions were to be repeated many times over in the mass of new suburbs that were to appear around London after 1920.

Close by this speculators' paradise and in high contrast, there arose at the same time the misnamed Hampstead Garden Suburb (most of it was in Hendon and Finchley). There was a railway association here too, of a rather different kind.

The tube-line planners had provided for a station on the Hendon/Hampstead boundary a quarter-mile north of the *Bull & Bush* pub. Since it was known that Eton College, the owners of much of the neighbouring land, wished to put it on the market and it was expected to attract building developers, a viable revenue seemed likely. Pending further developments, the station was finished at rail level but nothing was done above.

There were soon second thoughts. Henrietta Barnett (1851–1936), who was wont to relax with her husband in a country cottage nearby, had launched a successful campaign to rescue the 80 acres of meadow attached to Wylde's Farm from the developers' grasp. Having achieved this, she said, somewhat tongue in cheek, that the proposed tube station would be useful to give thousands of inner Londoners access to fresh air and salvaged open spaces at a twopenny fare.

Whilst engaged in her campaign, Henrietta conceived the idea of what was to become the Hampstead Garden Suburb,[25] to be built around the rescued meadowlands, mostly to their north. She envisaged a landscaped, architect-designed settlement, each house or flat with its own garden, where all social classes and old and young could 'live together under the right conditions of beauty or space'. Her emphasis on the classless nature of the development was in part a tactical move, to obtain support for her Wylde's Farm campaign from Hampstead Council, which had a housing problem. Construction of the Garden Suburb began in 1906 and the first two cottages were completed in October 1907. What was to be a very extensive development would have no public houses, few shops and, until 1973, no public transport within its boundaries. In an age when very few had motor cars, those who lived in the original western part, largely completed by 1914, walked or took a tram to Golders Green station.

By the end of 1906, the plans to open the tube station opposite Wylde's in Hampstead Way had been abandoned, its traffic potential judged to be irretrievably damaged by the removal of so much building land from a catchment area which already included the large open spaces of Golders Hill Park and Hampstead Heath proper. Looking back, we should perhaps be grateful that events took the course they did. For had there been no proposals for a station at North End, close to the Barnett's little holiday cottage, there would probably have been no Hampstead Garden Suburb.

---

25   For detail on early growth of Hampstead Garden Suburb, see Jackson, op. cit. (20) pp 46–55.

The new tube railway benefited from the outcome since Henrietta's democratic ideals for the Suburb were not realized in practice. Instead, it quite soon evolved as a predominantly middle-class area. Her planned hostel for 'working lads', with its barns for tools and coster-barrows was soon forgotten. Despite the existence of a 2d return workmen's fare from Golders Green station to Charing Cross, available on any departure between 05.20 and 08.00, costermongers and working lads were never a very substantial element in the Garden Suburb. Instead, in due course, it would attract such residents as Frank Pick and Harold Wilson.

Enough has perhaps been said in this necessarily brief survey, to show how London's first 'railway suburbs' varied both in aspect and in form. Certainly when considering this subject, it is wise to be wary of easy generalizations.

Chapter 14

The Railway and Rural Tradition 1840–1940

Alan Everitt

Railways transformed the economy and society of England in so many ways that it is easy to forget how much they left unaltered or the part they played in awakening a sense of the past. Yet wherever they went they passed through an ancient land, deeply embedded in local custom, rich in the evidence of historic buildings and rural tradition, the creation of generations of country people.[1] Sometimes they obliterated the remains of a medieval castle, as at Northampton; sometimes they cut through the centre of historic villages, as at Weston on Trent; sometimes they strode across market towns on mighty viaducts, as at Mansfield, heedless apparently of historic considerations. In the sixth chapter of *The Victorian Railway* (1991) Jack Simmons has forcefully reminded us of the 'vandalism' of their early years. Eventually, indeed, they were destined to undermine much of the fabric of rural tradition itself through industrial and suburban expansion and other, more insidious, developments.

Yet in his earlier book, *The Railway in Town and Country, 1830–1914* (1986), Jack has also warned us against oversimplifying the impact of the railway on provincial society. Consequences of the kind I refer to were more gradual and less complete than they seem perhaps to us today. They were occasioned, moreover, quite as much by the rise of motor transport and the 'tarmac revolution' of the early-twentieth century.[2] During their first hundred years at least, railways also did much to open up the historic riches of the countryside to a new generation of writers on rural society. Like all historical sources, the works of those writers need to be treated with judgement and good sense: they do not answer all our questions; they rarely supply the detailed statistics we often ask for today. Yet, as Kipling found when he visited Australia, we can sometimes learn more by listening to what people have to say than by asking questions. For us it was fortunate that so much of the fabric of tradition survived into this time of transition, and that so many people were interested in observing and recording it. In their very different ways, writers like W.H. Hudson, Cecil Torr,

---

1   The word 'tradition' is now so often debased that I use it reluctantly and in the absence of an alternative. The following pages will make my meaning clear.
2   'Tarmac' was patented in 1902 but its use spread slowly. Some major and most minor roads were not tarred until the inter-war period. Few developments have done more to conceal the past from us.

J.C. Atkinson, Richard Jefferies, George Borrow, George Sturt, and Flora Thompson – to name but a few of those people – enable us to think ourselves back into a world of the past which has itself now vanished beyond recall.

Railway interests, if not the companies themselves, early came to appreciate the wealth of historic evidence awaiting discovery in the English countryside. Soon after the South Eastern Railway through Croydon, Reigate, and Tonbridge was extended to Tunbridge Wells in 1845, the *Railway Chronicle* issued one of its 'Travelling Charts' for passengers to study on the journey. It was one of a series of 15 published by the paper,[3] and it gives us a fascinating picture of the country on either side of the line. The anonymous author was an ardent ecclesiologist, with a keen eye for the picturesque, a sovereign contempt for churchwardens, and an expert knowledge of the local geology. He expected passengers to be observant people, with an intelligent interest in what they saw from the carriage-window, a wish to be accurately informed about it, and an urge to explore it for themselves – of course on foot. They were encouraged to do so by 50 engravings of places worth visiting, some of which involved a tough walk of seven or eight miles each way through hilly country.

The first 13 or 14 miles of the route down the border of Surrey and Kent, 22 years before the present line through Sevenoaks was completed, are described in the following terms:

> It is a lovely walk, of less than two miles, over Forest Hill, to Dulwich College... Sydenham hills and woods offer many charming walks... Like all old places, [Croydon] has many features of interest. The geologist will find here chalk, gravel, sand, clay and peat; and hence there is no richer spot for the botanist... The walk from Sydenham station to Beckenham, thence to Bromley, on to Chislehurst and Sundridge [in Bromley parish], can be highly recommended for a day's excursion... Of late years the hills have been inclosed, and the pleasant walks over the purple heather contracted... [but] at the proper season the hills, especially Crohamhurst, are luxuriant with the lily of the valley... The pedestrian should alight at the Croydon station, and stroll over and about Crohamhurst to Sanderstead, and then make a circuit back through Addington. The woods in the spring resound with the plaints and the joyous 'jug-jug' of the nightingale. It is a most delicious country.

It must sound strange to modern ears: the whole of this area now lies in Greater London.[4] What interested the writer was plainly its picturesque character: its homely woodland churches, its timber-framed farms and manor houses, then often in some decay, its mingled scenery of woods, heaths, parks, commons, 'bottoms', sunken lanes, winding shaws, irregular little fields, and wealth of wild flowers. Yet other sources confirm his picture and tell us that within this district, or just beyond it,

---

3   J. Simmons, *Victorian Railway*, p. 244. 'Reigate' station was later renamed 'Redhill'.

4   Nowhere south of the Thames became formally part of London until 1855, when 14 parishes in Surrey (22 951 acres) became 'metropolitan Surrey' and eight in Kent (22 591 acres) became 'metropolitan Kent'. They were not fully integrated with London until the creation of the LCC in 1889, based on the area of 1855. The Surrey parishes (upstream from Deptford) were in the diocese of Winchester; those in Kent in the diocese of Rochester: F.A. Youngs, *Guide to the Local Administrative Units of England*, I, 1979, pp. 304–11; *National Gazetteer* (hereafter *NG*), 1868, XII, Appendix, p. 5.

numerous woods, heaths, commons, and half-wild spots remained unenclosed at that time, covering some thousands of acres on either side of the county boundary, and in places reaching nearly to the Thames. They remained not merely as relics of the past, moreover, but with a distinctive life of their own. Over the centuries they had given birth to many green-hamlets and scattered farmsteads, and to an unsuspected range of humble crafts and local industries, drawing on the woods and heaths for their resources. By providing Londoners with some of their more unusual raw materials, such as box-wood for instrument-makers, and with many domestic necessities such as brooms, besoms, mats, baskets, woodware, earthenware, bricks, tiles, charcoal, garden produce, geese and poultry, culinary and medicinal herbs, they afforded them also a glimpse into an unfamiliar world of cottage-farmers, woodlanders, and rural craftsmen.[5]

The extent to which London thus remained embedded in a world of the past until the days of Queen Victoria is a little-known yet fascinating theme in English history. Unique though it was on account of its size, it sheds light on the way past and present were often interwoven in the provinces too. Most of our major cities, such as Birmingham, Bristol, Sheffield, Leicester, Newcastle, and Manchester, have owed something in their history to the woods, moors, commons, heaths, or forests surrounding them. The complex moorland parish of Sheffield covered more than 23 000 acres, reaching far to the west in its outlying townships, and set in an area richly endowed with the mineral resources on which its livelihood was based. The suburbs of Birmingham, though in very different country, are still peppered with names in 'green', 'heath', 'wood', 'end', and 'common', which have something to tell us about the city's origins in a forested district. In industrial regions such as the Black Country, the Potteries, and the Leicestershire-Derbyshire borders, early settlements were often surrounded by tracts of heath, moor, and woodland from which they drew their raw materials, and in which new industrial villages developed in the nineteenth century, like Lye Waste in Old Swinford, and new towns on parish boundaries, such as Coalville and Woodville.[6] They too were embedded in the past.

Forty years after the coming of the railways, the third edition of the Ordnance Survey map makes these points abundantly plain. It is a marvellous document; centuries of historical evidence are inscribed upon it, layer upon layer, much of it still awaiting interpretation. If we take the one-inch sheet which extends from Wednesbury in the north-west to Henley-in-Arden in the south-east, we see one of the most heavily industrialized areas in England. Towards the northern side of the map the city of Birmingham already stretches for some six or seven miles east and west and seven or eight miles north and south. Thirteen railway lines converge upon it from every possible direction. Yet it is not only a major industrial city that is delineated before us but the chief historic market town of north Warwickshire. Despite its extraordinary growth, it was still surrounded in the 1880s by a wholly rural hinterland. Within its orbit more than 80 village-blacksmiths' forges are marked on

---

5   Alan Everitt, 'Common Land', in *The English Rural Landscape*, ed. Joan Thirsk, 2000, pp. 210–35, passim.
6   Ibid., pp. 213–14; David Hey, *The Fiery Blades of Hallamshire: Sheffield and its neighbourhood, 1660–1740*, 1991.

that same sheet, at places like Tanworth and Upper Bentley, Packwood, Feckenham, and Coughton; or in what is now outer Birmingham, but was then still 'deep among the lanes', at places like Oscott and Quinton, King's Norton and South Yardley. During the Victorian era, the city also built up one of the largest retail markets in Britain, with space for 600 stalls every day of the week. Partly as a consequence, it became the centre of one of the most extensive networks of village carriers' routes in the country. Every week, every market day, hundreds of carriers streamed into it from the surrounding district, bringing in the people and the produce from the farms and hamlets of George Eliot's Loamshire, of Shakespeare's Forest of Arden.[7]

If in matters such as these, humble it is true yet essential, town and country still met face to face in the streets of provincial cities, if the rural past still lived on even within the ambit of London, it is not surprising if the curiosity of intelligent people was awakened to make use of the railway to explore further afield. The 'Discovery of Britain', as it has been called, was not a creation of the railways.[8] It had a lengthy ancestry behind it. Its origins went back to the days of William Camden and John Leland. It was a well-established tradition by the time of Celia Fiennes and Daniel Defoe. We find it again among antiquarians such as Robert Plot in Staffordshire, William Woolley in Derbyshire, and John Aubrey in Wiltshire and Surrey. A century later it blossomed afresh and with new insight, in the work of writers like Gilbert White, William Cobbett, George Borrow, and Mary Russell Mitford. But it was never more vigorous and abundant than in the golden age of the railway, between the 1840s and 1940s. By opening up the countryside to a far larger public at a time of rapidly increasing population, rail travel played a significant part both in the rise of a new generation of writers, and in the creation of a market for their publications. In their hands the 'discovery of Britain' took a fresh direction.[9]

The awakening of interest in vernacular tradition at this time – in the lives and customs of the people, and in the landscape, the buildings, and the wildlife of the countryside – was a very widespread movement in English society. It was not confined to the writers on rural life considered here. It also inspired the Vernacular Revival in architecture; it played a notable part in the Arts and Crafts Movement; it influenced the development of the provincial novel; it attracted a multitude of visual recorders – painters, etchers, and engravers; it aroused an interest in English dialects and the study of place-names; it gave a new impetus to the native enthusiasm for gardening; it gave birth to many conservation bodies such as the National Trust; and ultimately it made a far-reaching impression on historiography. These are all matters that also need to be borne in mind, though they lie beyond the scope of the present chapter.[10]

The interests, themes, and viewpoints of the writers we are more particularly

---

7   Alan Everitt, *Transformation and Tradition*, 1984, pp. 6–7, 26–7.
8   Esther Moir, *The Discovery of Britain. The English Tourists, 1540–1840*, 1964.
9   Everitt, 'Common Land', pp. 214, 339–40.
10   As, for example, in the work of George Devey, Norman Shaw, and Sir Thomas Jackson (architects); Ernest Gimson and C.R. Ashbee (Arts and Crafts Movement); Herbert Railton and F.L. Griggs (engravers and etchers); Gertrude Jekyll and Victoria Sackville-West (gardening).

concerned with here varied as widely as the countryside they set out to explore.[11] They cannot easily be summarized. But in one way or another they were all people who were gifted with exceptional powers of observation. Though not historians themselves, they were people who tended to think historically, exploring the past through the lives and customs of country men and women rather than through formal documents. When they went by train, their purpose was to reach a settled destination, explore the surrounding country on foot, and by preference always alone. In Louis Jennings's view, in *Field Paths and Green Lanes* (1878), that was the only way to learn, and the only way to get country people to talk freely.[12] In *Forty Years in a Moorland Parish* (1891), J.C. Atkinson calculated that he had walked more than 70 000 miles in his duties as incumbent of Danby in Cleveland, 'and much more than as many again for exercise, relaxation, or recreation.'[13] The human contrasts thus brought home to writers like Atkinson and Jennings opened their eyes to perceive vestiges of an older civilization in the world around them, just as F.W. Maitland learned to decipher the past in 'that marvellous palimpsest', the Ordnance Survey map.[14] They light up for us as a consequence those human aspects of the past on which official records are largely silent.

In the limits of this chapter, I shall illustrate these themes by comparing two works by one of the earlier writers of the railway age, Richard Jefferies (1848–87). His novels and mystical writings do not concern us; but in his work as a naturalist Jefferies included much about rural society too. For him indeed, as for W.H. Hudson, 'wildlife' and 'human life' were but two parts of one whole. In *Wild Life in a Southern County* (1879) and *Nature near London* (1883), he gives us compelling portraits of two very different stretches of country in north Wiltshire and north-east Surrey. Like most of his books, they were originally written as a series of essays for London periodicals; but they have the edge over his more miscellaneous writings, such as *The Life of the Fields* (1884), in concentrating on a single well-defined district.[15]

Jefferies was born at Coate, a hamlet in the parish of Liddington near Swindon, where his father had a small farm. His ancestors had owned land and farmed in the neighbourhood for some three centuries.[16] In 1877 he moved to Surbiton to be nearer the London publishers, writing prolifically on rural and agricultural subjects for such journals as *The Pall Mall Gazette*, *The St James's Gazette*, *The Standard*, and *The Live Stock Journal*.[17] Though his heart remained in Wiltshire and he often returned

---

11 See Select Bibliography at end of this chapter. For further examples see *The Victorian Countryside*, ed. G.E. Mingay, 1981, II, Bibliography, pp. 639–77.

12 Jennings, op. cit., pp. v–viii.

13 Atkinson, op. cit., p. vii. All the writers mentioned were notable walkers.

14 F.W. Maitland, *Domesday Book and Beyond*, 1960 edn., pp. 38–9. He followed this remark by comparing a district on the Oxfordshire-Berkshire border with one on the border of Devon and Somerset.

15 In *Nature Near London* three essays were added on Brighton and the South Downs. The 1913 edition is that cited in this essay, hereafter as *NNL*.

16 See letter to George Bentley in Richard Jefferies, *Field and Farm*, ed. S.J. Looker, 1957, p. 41.

17 Ibid., pp. 11 seqq.; Richard Jefferies, *The Life of the Fields*, 1900 edition, prefatory Note of acknowledgements, citing ten journals.

there, his time in Surrey came as a revelation to him. All his preconceived ideas, he said, 'were overthrown by the presence of so much that was beautiful and interesting close to London.'[18] The contrast undoubtedly quickened his perception and gave a new depth to his writing on rural society. 'There is a frontier line to civilization in this country yet,' he wrote, in the preface to *Wild Life*,

> and not far outside its great centres we come quickly even now on the borderland of nature. Modern progress, except where it has exterminated them, has scarcely touched the habits of bird or animal; so almost up to the very houses of the metropolis the nightingale yearly returns to her former haunts. If we go a few hours' journey only, and then step just beyond the highway... and glance into the hedgerow, the copse, or stream, there are nature's children as unrestrained in their wild, free life as they were in the veritable backwoods of primitive England. So, too, in some degree with the tillers of the soil: old manners and customs linger, and there seems an echo of the past in the breadth of their pronunciation.[19]

Old manners and customs linger: that was the secret of what appealed to writers like Jefferies. He was not an idealist in what he recorded; he shared the widespread belief of his age in 'progress'; as the son of a small farmer, moreover, he was often harsh on labourers, especially in Wiltshire. But as a naturalist of human life as well as wildlife, the survival of rural custom also exercised an irresistible fascination over him.

Jefferies' Wiltshire

Like other parishes in Jefferies' part of Wiltshire, Liddington straddled the familiar division between 'chalk' and 'cheese'. In his day it covered nearly 3 000 acres, extending for some six miles from the River Cole in the north-west, across the Marlborough Downs, to Liddington Warren in Aldbourne Chase. It comprised dairy farms, water-meadows, cornfields, sheep-walks, scattered remnants of woodland, disused rabbit-warrens, a rough tract of common, and a large 'mere' – actually a reservoir – constructed in 1822 for the Wiltshire & Berkshire Canal. He extended his perambulations along the Downland ridgeway, and then to the densely wooded area of Savernake Forest seven miles to the south. His portrait thus covers a notably diverse countryside, in many ways transformed by parliamentary enclosure in the 1770s, and then in his own lifetime by the railway, yet still rich in wildlife and human tradition.[20] Jefferies gave form to his observations by beginning at the hill fort of

---

18 *NNL*, p. vii, and cf. p. v. Wiltshire was the inspiration behind his autobiography, *The Story of My Heart* (1883); he seems to have had some mystical experience on Liddington Down.

19 Richard Jefferies, *Wild Life in a Southern County* (Nelson edition, n.d. [c. 1910]), p. vii; hereafter cited as *WLSC*.

20 Liddington and its outlying tything of Medbourne were enclosed in 1777: W.E. Tate, *A Domesday of English Enclosure Acts and Awards*, ed. M.E. Turner, 1978, p. 270. That award affected 1 002 acres. Jefferies makes various references to amalgamation of farms in this and other works; one consequence was the degradation of small 'peasant' farmers to the status of labourers. For the impact of railways on milk-production, displacing cheesemaking, see J. Simmons, *Railway in Town and Country*, pp. 50–51, and the references on p. 346.

Liddington Castle, 900 feet above sea level; then descending the downland scarp to the mother-village and church by its springs; proceeding thence to an outlying hamlet of cottage-folk, probably Medbourne; and finally to the medieval farmstead at Liddington Wick.[21]

The dominance of the downs in this stretch of country and their loneliness are strikingly emphasized. The prehistoric ramparts of Liddington Castle were densely overgrown with shrub-like thickets of thyme and furze, untouched by the woodman's tool, and undermined by innumerable flint-pits long since abandoned.[22] For seven long miles the ridgeway linking it with three other hill forts was completely deserted, without so much as a wayside tavern:

> Then the traveller finds a little cottage, with a bench under a shady sycamore and a trough for a thirsty horse, situate where three... modern roads (also lonely enough) cross the old green track. Far apart and far away from its course, hidden among their ricks and trees, a few farmsteads stand, and near them perhaps a shepherd's cottage: otherwise it is an utter solitude, a vast desert of hill and plain; silent, too, save for the tinkle of a sheep-bell or, in the autumn, the moaning hum of a distant threshing machine rising and falling on the wind.[23]

The downland shepherds thus figure prominently in Jefferies' story. In Wiltshire, as in Sussex, their calling was often a matter of family tradition. Sons followed fathers generation after generation, accompanying them to the sheepfolds from their childhood, early becoming inured to the loneliness and silence of the hills.[24] Though without any amusements but those they made up for themselves, these young shepherd-lads were not quite without infantine pleasures. In a little copse by the ridgeway, Jefferies found they had cut their names with their clasp-knives, in the age-old way, on the bark of the beech trees.[25]

> Sometimes in the evening, later on, when the wheat is nearly ripe, such a shepherd lad will sit under the trees there; and as you pass along the track comes the mellow note of his wooden whistle, from which poor instrument he draws a sweet sound. There is no tune – no recognizable melody: he plays from his heart and to himself. In a room doubtless it would seem harsh and discordant; but there, the player unseen, his simple notes harmonize with the open plain, the looming hills, the ruddy sunset, as if striving to express the feelings these call forth.[26]

---

21   These places are not specifically named, but Liddington Wick is recorded as Wick Farm, and described at length in Chapters VII–XII. More than 20 Wiltshire parishes had an outlying 'Wick', or dairyfarm, of medieval origin.

22   *WLSC*, pp. 59–60, and cf. p. 64.

23   Ibid., pp. 62–3. The threshing machine was a comparative newcomer, but widely employed on the large Wiltshire farms, in striking contrast with Surrey. Its use declined during the agricultural depression in the 1880s as fields were put down to grass.

24   For Sussex, see Peter Brandon, *The South Downs*, 1998; for Wiltshire see also W.H. Hudson, *A Shepherd's Life*, 1910; and Alfred Williams, *Villages of the White Horse*, 1913, especially pp. 166–9.

25   *WLSC*, p. 64. Isaac Watts refers to this custom in one of his hymns, c.1700.

26   Ibid., pp. 64–5. Shepherd-boys usually made their own pipes, from elder-wood.

It was in the lambing season that the solitude of the hills seemed most intense. Then the shepherds and their lads lived on the down, lodging in wooden huts for several months by the scattered folds and sheep cotes – a custom which survived well into the twentieth century.[27] At other times they might spend nights there guarding the sheep; but their homes lay in the village or near the outlying steadings. The peasant-life of the common fields had disappeared with enclosure, and great farms, of up to four or five hundred acres in this district, had replaced it. But old habits died hard, and one that Jefferies describes in detail was the annual sheep-washing in the pool below the village:

> At that time the roads are full of sheep day after day, all tending in the same direction; and the little wayside inns, and those of the village which closely adjoins the washpool, find a sudden increase of custom from the shepherds. There is no written law regulating the washing, but custom has fixed it as firmly as an Act of Parliament: each shepherd knows his day, and takes his turn, and no one attempts to interfere with the monopoly of the men who throw the sheep in. The right of wash here is upheld as sternly as if it were a bulwark of the Constitution.

Sometimes a landowner or farmer, thinking perhaps that education and easier travel will have weakened tradition, tries to enclose the approach or utilize the water to irrigate his meadows:

> He finds himself entirely mistaken: the men assemble and throw down the fence, or fill up the new channel that has been dug; and the general sympathy of the parish being with them and the interest of the sheep-farmers behind them to back them up, they always carry the day, and old custom rules supreme.[28]

Towards autumn a second great time of gathering began. The downland tracks were then thronged with sheep from half the southern counties of England and from distant parts of Wales, all wending their way with shepherds and drovers to the great stock-fairs at Tan Hill (6 August), Yarnbury Castle (4 October), and Weyhill (from 10 October). Neither these vast gatherings, where hundreds of thousands of sheep were sometimes sold, nor the local fairs of the Wiltshire countryside, had yet been affected by the railway. From Somerset to Surrey, the annual farming calendar was still governed by them; their dates were familiar to everyone, old men and young girls alike; and they remained notable centres for the exchange of news and rural intelligence. Perhaps the greatest attraction of those in Wiltshire, Jefferies said, 'may be found in the fact that all the countryside is sure to be there. Each labourer or labouring woman will meet acquaintances from distant villages they have not seen or heard of for months. The rural gossip of half a county will be exchanged.'[29]

Throughout the southern counties of England, the downland shepherd held a time-honoured place in the rural economy. For on his knowledge and fidelity the principal

---

27 Ibid., pp. 103–4; Williams, loc. cit., gives a remarkable account of a sheepfold near Aldbourne, c.1913.
28 *WLSC*, pp. 76–7.
29 Ibid., p. 112, and cf. pp. 110–11, 116. There is also much on these fairs, and others in Surrey and Hampshire, in the writings of George Sturt.

profit of a whole season depended on so many farms.[30] By inclination and experience alike he frequently developed singular powers of observation:

> On the hills he has often little to do but ponder deeply, sitting on the turf of the slope, while the sheep graze in the hollow, waiting for hours as they eat their way. Therefore by degrees a habit of observation grows upon him – always in reference to his charge; and if he walks across the parish off duty he still cannot choose but notice how the crops are coming on, and where there is most 'keep'.[31]

So the condition of every farm became familiar, and the fields were like books to him in which the life-history of an entire countryside was written.

In many Wiltshire parishes shepherds thus came to be regarded as natural repositories of local tradition. While labourers, in Jefferies' view, were becoming restless and discontented, there were still

> not a few instances of shepherds whose whole lives have been spent upon one farm. Thus, from the habit of observation and the lapse of years, they often become local authorities; and when a dispute of boundaries or water rights or right of way arises, the question is frequently finally decided by the evidence of such a man.

Sometimes, perhaps, one of the old green lanes or bridle-tracks of the neighbourhood has become a valuable thoroughfare through the increase of population and then the question arises as to who should repair it:

> There is little or no documentary evidence to be found – nothing can be traced except through the memories of men; and so they come to the old shepherd, who has been stationary all his life, and remembers the condition of the lane fifty years since. He always liked to drive his sheep along it – first because it saved the turnpike tolls; secondly, because they could graze on the short herbage and rest under the shade of the thick bushes. Even in the helplessness of his old age he is not without his use at the very last, and his word settles the matter.[32]

The shepherd's enduring place in rural society was one he shared with other characteristic figures described by Jefferies: the blacksmith, the wheelwright, the thatcher, and the older farming families.[33] In Liddington, as elsewhere, a number of people also followed less familiar callings. Among those Jefferies mentions were hurdle-makers, flake-makers, flint-diggers, mop-makers, willow-weavers, basket-makers, wild-fowlers, travelling tinkers, workers in the osier-beds, and furze-cutters for making walking-sticks. Such occupations did not employ large numbers; some

---

30 *WLSC*, p. 103.
31 Ibid., pp. 104–5. The theme is also taken up by W.H. Hudson, op. cit.; his shepherd (on Salisbury Plain) was a deeply religious man, who rarely went out on the hills without his Bible in his pocket.
32 *WLSC*, pp. 104–6. Parallels to these practices can be found as far back as the sixteenth century. Jefferies' views on labourers contrast with those of Alfred Williams, op. cit., in this area, and Flora Thompson in north Oxfordshire in the 1880s, in *Lark Rise*, 1939.
33 *WLSC*, pp. 80–81, 122–30, 132 seqq., 146 seqq.

had begun to decay by Jefferies' time, or to move into local market towns. But like the shepherd and the blacksmith, they belonged to a traditional economy whose products and activities were still a necessity at that time. Like their counterparts in other parts of England, they provided or augmented the livelihood of many humble country men and women.[34]

The impact of the railway upon that economy was gradual and in some ways unexpected. The main line of the Great Western to Bristol passed close to the northern boundary of Liddington. By the 1870s the growth of the London milk-trade had largely displaced cheesemaking in Wiltshire; it had also led to the amalgamation of the smaller dairyfarms.[35] Yet the country people themselves seem as yet to have made little use of the railway as passengers. Their movements were still centred on the sheep-fairs and market towns, and were made on foot or by horse and cart. Early in the spring the local population was augmented by the annual migration of itinerant labourers – south countrymen, Irishmen, and gipsies among them. Sometimes the same families returned to the same farm year after year, lodging in sheds and barns, or in tents by the wayside. They came first for the hoeing on the upland arable farms, then moved down to the water-meadows for the early haymaking, next to the valley-farms for the summer hay harvest, and finally turned east again into Hampshire and Surrey, where the early corn crops were fast ripening. But they, too, came on foot; for that was the way they heard where work was, and where they were most likely to find a welcome from farm to farm. Like George Sturt and Flora Thompson, Jefferies has much to tell us about this little-known migrant army of the southern counties – particularly the Irish – and the bush-telegraph by which they moved and lived.[36] Victorian farmers were heavily dependent upon them in many parts of lowland England.

The survival of local customs in a mingled society of this kind was also in some ways unexpected. Richard Jefferies has much of interest to tell us on that subject too: on the village-feasts at Liddington, on the curious popularity of handbell ringing, on the many superstitions of the hamlet-folk, on lingering beliefs in 'wise women' and in magical herbs, on birds like the swallow still regarded as sacrosanct, on the customs of carters and ploughboys on Oak Apple Day, and of the perambulation of the parish by village children on St Thomas's Day.[37] Such matters still gave a certain ritual pattern to the year. From his intimate acquaintance with the parish and its neighbourhood Jefferies thought that magical beliefs in particular remained a good deal more widespread than a casual enquirer might suspect. But now, through increasing contact by rail with the outside world, at least among farmers, they survived only in outlying places. The people of the hamlet especially, buried as it was among trees and often isolated by flooded lanes or snowstorms in winter, were shy of talking to strangers and kept them increasingly to themselves. But in the church-village, and on the greater farms, any belief in magical cures and even in the

34   Ibid., pp. 60, 71, 73, 78–82, 237–44 passim, 295–6.
35   See note 20, supra; the decline of cheesemaking is often mentioned by Jefferies in *WLSC* and other works.
36   *WLSC*, pp. 320–21; cf. *NNL*, pp. 71 seqq., for a fuller account. Contrary to some modern views the Irish were usually the most welcome.
37   *WLSC*, pp. 106–10 passim, 118–20, 157, 159, 201–2, 227–8.

traditional uses of herbs was discredited. These were conflicting attitudes sometimes found in other parts of England too, as Flora Thompson noted in north Oxfordshire, though not everywhere.

Jefferies' Surrey

The country Jefferies described in *Nature near London* was of course very different from that of Wiltshire, and in more ways than one.[38] It extended southwards for about ten miles from Kingston-upon-Thames and included Surbiton, Long Ditton, Claygate, Esher, Ewell, and the wooded hills beyond. With the coming of the railways suburban development was exceptionally rapid in Surrey, much more so than in Essex or Kent.[39] At the beginning of the century the transpontine suburbs did not extend westwards beyond Lambeth, which was still partly rural.[40] The ancient borough and parish of Kingston, though only 12 miles from London, then had fewer than 4 000 inhabitants, and barely 7 500 when the South Western line to Southampton was authorized in the 1830s. By 1861 the population had shot up to nearly 18 000 and by 1891 to 42 000. Surbiton, where Jefferies lived, was in origin a mere hamlet of Kingston; but with its own station on the new railway, it too began to grow quite quickly. In 1845 it was formed into a separate chapelry; by 1861 its population had risen to nearly 4 700 and by 1891 to more than 12 000. It was a town in its own right. Jack Simmons has aptly described it as 'the oldest suburb in Europe, perhaps in the world, that was called into being by a railway.'[41]

Such rapid expansion is a familiar theme in parishes bordering the Thames. But in this area, as so often, it was only part of the story. For London had never been able to expand over virgin territory, though that impression has sometimes been given by historians. Its new suburbs had to fit themselves everywhere into a long-settled landscape. In Jefferies' district that meant, in the main, a landscape of scattered farms and hamlets, embedded in a wooded, broken, piecemeal countryside: an intricate mosaic of little meadows, pastures, orchards, and cornfields; of heaths, commons, copses, shaws, and 'scrubs'; of winding lanes and holloways, and broad stretches of roadside waste. It was a more varied landscape than that of Wiltshire, and on a smaller scale, where small owners, farmers, and rural craftsmen were still numerous, and local society was more loosely organized. It was not dominated by great farms and landed magnates.

---

38 The sixteen essays on north-east Surrey were written for *The Standard* during periods Jefferies spent at Surbiton between 1875 and 1881.

39 In 1861 'metropolitan Surrey' had 579 748 inhabitants and 'metropolitan Kent' 193 427; the area in each county was similar – nearly 23 000 acres: *NG*, XII, App., p. 5. East of London, the River Lea formed the boundary, so that no parish in Essex was included in 'London' until the creation of the GLC in 1965; by then the Essex suburbs formed a very densely populated area.

40 H.C. Darby, ed., *A New Historical Geography of England*, 1973, map on p. 386.

41 Simmons, *Railway in Town and Country*, pp. 63–4. Its population as a 'hamlet' is not separately recorded in the early censuses. The name ('south barton') is first noted in *The Place-Names of Surrey* (English Place-Name Society, XI, 1982) in 1179.

In one important sense it was also a more ancient countryside, for it had not experienced the same revolutionary impact of parliamentary enclosure. Though open fields had once existed, and those of Kingston did not wholly disappear until after 1808, they were limited in extent and different in kind from those of the Midland counties and Wiltshire. By far the greater part of enclosure in Surrey – parliamentary and otherwise – related to commons, heaths, woods, and uncolonized wastes.[42] Of these Surrey possessed a greater proportion than any other lowland shire: more than 96 000 acres in 1800 – one-fifth of its entire area – 65 000 acres in 1833, at least 43 000 in 1873, and 27 000 even in the late twentieth century.[43] In the south-west of the county, Bagshot Heath alone covered 31 500 acres before it was enclosed, an area larger than Dartmoor Forest.[44] Though there was nowhere on that scale in Jefferies' neighbourhood, there were many lesser tracts of common, heath, scrub, and waste.

Suburban development was often, therefore, itself a piecemeal matter. It was accommodated in the gaps and spaces of a deeply-entrenched society still tenacious of local tradition. Poor though much of the country might be, its diversity provided abundant cover for wildlife and rare species; and an abundance, too, of those homely resources by which many of its people lived. Eventually almost all yielded to metropolitan exigency. Not quite all: for London is still blessed with more open spaces than perhaps any other European capital, and almost all of those spaces – royal parks included – are remnants of ancient rural commons.[45] In Richard Jefferies' time, a good deal of genuinely half-wild land remained, and in no other part of England had he ever observed such an abundance of wild birds. The first spring he resided in Surrey, he said:

> I was fairly astonished and delighted at the bird life which proclaimed itself everywhere. The bevies of chiff-chaffs and willow-wrens which came to the thickets in the furze, the chorus of thrushes and blackbirds, the chaffinches in the elms, the greenfinches in the hedges, wood-pigeons and turtle-doves in the copses, tree-pipits about the oaks in the cornfields; every bush, every tree, almost every clod, for the larks were so many, seemed to have its songster. As for nightingales, I never knew so many in the most secluded country. There are more round about London than in all the woodlands I used to ramble through [in Wiltshire].[46]

Redwings, goldcrests, kingfishers, nighthawks, owls, herons, snipe, corncrakes, cuckoos, goat-suckers, sparrowhawks, kestrels, lapwings, jays, woodpeckers, and shrikes: these were among the many other species he recorded, some of them in quite extraordinary numbers, such as the 2 000 lapwings seen in a single field

---

42   Tate, op. cit., pp. 247–50.
43   Everitt, 'Common Land', in Thirsk, op. cit., p. 223. The 1873 figure is probably an under-estimate. See also *NNL*, chapter on 'Heathlands'.
44   *The Imperial Gazetteer* (hereafter *IG*), 1870 edition, *sub* Bagshot.
45   Everitt, 'Common Land' in Thirsk, op. cit., p. 211. None of them had pertained to London itself, which apparently had no common land.
46   *NNL*, p. 30.

one January.[47] In the comparatively wild or open districts to which he had been accustomed hitherto, he could not recollect ever seeing such vast numbers of birds.[48]

The variety of shrubs and wild flowers was not so remarkable as in Wiltshire.[49] But once again their sheer abundance in Surrey amazed him. Whole tracts of roadside-waste were covered with the common yarrow, for example, a herb still much used by local people. The warm, dry, sandy soils of the commons and heaths encouraged an especially dense growth of shrubs like furze, broom, bracken, bramble, and honeysuckle, and in places rarer species such as dogwood, guelder-rose, wild clematis, and wayfaring-tree. As for heather, he said, 'southern London can boast stretches of heath which, when in full bloom, rival Scottish hillsides.'[50] Points like these, and the lowly wooden cottages, barns, and farmhouses of the locality which equally surprised Jefferies, may seem trivial as matters of history. But they tell us something about the kind of *countryside* upon which the local economy subsisted, and in which habits and customs often lingered which had died out further west. By tiny touches of human detail interspersed with his observations on wildlife, he gradually builds up a picture of that humble world of tradition – telling, if inevitably incomplete.

The economy of the district was not marked by any notable speciality such as the sheep-farming and dairying of Wiltshire. But the people themselves were still very much country people, still manifestly 'of the fields'. Everything distinguished them as a people apart from suburban society, and strangely oblivious to the new world arising round them:

> The stamp of the land is on them. They border on the city, but are as distinctly agricultural and as immediately recognisable as in the heart of the country. This sturdy carter, as he comes round the corner of the straw-rick, cannot be mistaken. He is short and thickly set, a man of some fifty years, but hard and firm of make... He trudges deliberately round the straw-rick: there is something in the style of the man which exactly corresponds to the barn, and the straw, and the stone staddles, and the waggons. Could we look back three hundred years, just such a man would be seen in the midst of the same surroundings, deliberately trudging round the straw-ricks... calm and complacent though the Armada be at hand.[51]

On Saturday afternoons in the towns, the country women too were at once marked out from the newcomers by their speech and dress. 'They have come in on foot from distant farms for a supply of goods, and will return heavily laden. No town-bred woman, however poor, would dress so plainly as these cottage matrons.'[52] Under the burning sun of the summer cornfields, men, women, lads, and girls still harvested together in the traditional manner – whole families of them, side by side – almost

---

47   Ibid., pp. 25 seqq., 137–8, et passim.
48   Ibid., p. 28.
49   But the list of 60 species observed along a single road is surprising today: ibid., pp. 40–41.
50   Ibid., pp. 40–42; cf. pp. 107–8: 'the entire hillside, seen at a glance, is covered with heath, and heath alone. A bunch at the very edge offers a purple cushion fit for a king.' Hence there was ample heather honey in Surrey.
51   Ibid., pp. 74–5.
52   Ibid., p. 78. But their daughters had 'caught the finery of the town'.

unaware of the passing trains in the valley. 'A girl, as she rises from her stooping position, turns a face, brown, as if stained with walnut juice, towards me... but her dark eyes scarcely pause a second on a stranger. She is too busy, her tanned fingers are at work again gathering up the cut wheat.' Gradually, the sheaf grows under her fingers; it is bound about with a girdle of twisted stalks, plucked from the common flowers of the cornfield.[53] Nothing seemed to have changed; the reaping machines of Wiltshire were as yet virtually unknown; the Surrey fields were too small and crooked to accommodate them.

In the woodland parts of the neighbourhood Jefferies remarked on the great pride which the carters took in their horses. Felling and barking were necessarily winter tasks. They had to be done while the sap was rising, and the teams travelled 'with rows of brazen spangles down their necks, some with a wheatsheaf for design, some with a swan.'[54] The last of the great timber carriages

> came by on May-day with ribbons of orange, red, and blue on the horses' heads for honour of the day. Another, which went past in the wintry weeks of the early year, was drawn by a team wearing the ancient harness with bells under high hoods, or belfries, bells well attuned, too, and not far inferior to those rung by handbell men. The beat of the three horses' hoofs sounds like the drum that marks time to the chime upon their backs. Seldom, even in the faraway country, can that pleasant chime be heard.[55]

Though Surrey was not known for its sheep, some of Jefferies' most interesting observations again relate to the local shepherds. It had its own breed, it is true, the 'heath-croppers' of Hindhead and Bagshot, locally esteemed for their delicate flavour – for they fed on heather – though despised by agricultural improvers.[56] But in 1866 only 150 000 sheep were recorded in the early Agricultural Returns, compared with 485 000 in Sussex, 597 000 in Wiltshire, and 731 000 in Kent.[57] Nevertheless, the shepherds here were notably tender over their little flocks and careful not to press them. They led them along the broad roadside verges, where the weary or infirm were left to lie down and rest while their companions fed peacefully on the sward. The older men still wore the traditional white smock-frock, and many of them, as on the Sussex Downs, carried their crooks. 'In districts far from the metropolis you may wander about for days, and with sheep all round you, never see a shepherd with a crook; but near town the pastoral staff is common.'[58] Their lives were so much bound up with these little flocks that they seemed as unaware of the new suburban society around them as 'if they were on the loneliest slope of the Downs... Carriages go past, and neither the sheep nor the shepherd turn to look.' The hollow booming sound of the great guns at Woolwich echoes in the distance,

---

53   Ibid., pp. 80–81.
54   Ibid., p. 21.
55   Ibid., p. 17.
56   They were the poor man's sheep, comparable with the heath-sheep of other areas, such as the Suffolk Breckland: *IG, sub* Bagshot. William Marshall was one of the few agricultural writers who saw their value on poor land.
57   *NG*, XII, App., p. 4.
58   *NNL*, pp. 45, 76.

[but] the shepherd takes no heed, neither he nor his sheep. His ears must acknowledge the sound, but his mind pays no attention. He knows of nothing but his sheep. You may brush by him along the footpath and it is doubtful if he sees you. But stay and speak about the sheep, and instantly he looks you in the face and answers with interest.[59]

Further out among the heaths and commons many other local customs survived, unsuspected by strangers. When the acorns fell thick beside the hedgerows in autumn, the swine were let loose to forage for themselves along with the pheasants, wood pigeons, and red squirrels.[60] Where the bracken grew dense and tall in the hollows, 'the cattle which wander about, grazing at their will, each wear a bell slung round the neck, that their position may be discovered by sound. Otherwise it would be difficult to find them in the fern or among the firs.' Such practices were no longer known, or indeed necessary, in the broad 'leazes' or open pastures of Wiltshire which had become characteristic after enclosure.[61]

In these outlying parts, and among the woods and meadows of the neighbourhood, the Surrey herb-gatherers were also often to be seen. For the poor sandy soils of the commons and the rough meadows by the river were rich in herbal species, and here they were not despised or forgotten as they were further west. The demand for them among druggists and housewives was still substantial in the vicinity of London. In springtime whole sackfuls of dandelions were gathered from the wastes, to be eaten as salad. Yarrow was pulled up in advance of the hay-mowers, and sometimes carried away in cartloads to be boiled down as a remedy for chest-colds.[62] Agrimony was still picked by some local people to make themselves tea, for it was considered 'good for the flesh'. In hot weather borage was collected, and sold even to the London railway stations to flavour the claret-cup ladled out to thirsty travellers. The somewhat dubious characters Jefferies called 'mouchers' came round at all seasons gathering watercress, cowslip mars, and mushrooms, and in winter cutting rushes, for litter perhaps or for thatching sheds and hovels.[63] People from distant parts were 'surprised to find the herbalists flourishing round the great city of progress'.[64] But these were all traditional Surrey customs, in some cases going back to the sixteenth century.[65]

There is much that Jefferies does not tell us about the district. He says nothing of the broom-makers, potters, and basket-makers; of the brick-burners and charcoal-burners in the woods; of the annual harvesting of heather, furze, and bracken among the commons; of the poor people in autumn who came to gather beech-leaves to fill their mattresses; or of the ice-harvest on the ponds in winter, for supplying the

---

59   Ibid., p. 77.
60   Ibid., p. 155. Red squirrels were still common in this wooded area.
61   Ibid., p. 109.
62   Ibid., pp. 46, 153–4; for uses of yarrow see Florence Ranson, *British Herbs*, 1949, p. 113.
63   *NNL*, pp. 79, 165, 46. Rushes were widely used for these purposes.
64   Ibid., p. 154.
65   Elsewhere Jefferies remarks on the loss of interest in herbs among country people in Wiltshire. It was a consequence perhaps of parliamentary enclosure which had uprooted many rural customs. A greater acreage of medicinal herbs was cultivated in Surrey than in any other county: Samuel Lewis, *A Topographical Dictionary of England*, 1833, *sub* Surrey. See also Ranson, op. cit., passim.

ice-wells of London. For matters like these we must turn to other writers, such as Louis Jennings, James Thorne, and James Greenwood, and to other sources.[66] But what Jefferies does tell us relates to customs and characters he had seen for himself. Woven into his observations on wildlife are the outlines of a human society, on the very edge of London, yet living on from the past.

The Enduring Power of Rural Tradition

Far greater contrasts than those between Surrey and Wiltshire can of course be found by turning to an area such as Exmoor, on which Jefferies also wrote in *Red Deer* (1884), or to J.C. Atkinson's moorland parish of Danby in Cleveland. But these examples are enough perhaps to suggest, as Jack Simmons has often reminded us, that the transformation of England in which the railways played so large a part was a more complex process than it sometimes seems. The changes wrought in the economy over the 150 years or so following 1780 were certainly unprecedented. But because they were unprecedented, and because England was also a country in which the roots of provincial society ran deep into the past, other, older traditions survived alongside them too. Until late in the nineteenth century at least, they still had a life of their own.

The enduring power of rural tradition was a matter of the deepest interest to many Victorian people. Not to all, of course. Such was the pressure of business in the City, thought Richard Jefferies, that travellers on the suburban trains in Surrey looked out on the country people in the fields – and saw nothing; they turned back to their newspapers. The nightingales sang on in the copses by the wayside as they hurried to the station in the morning, and they did not hear them.[67] The native English shrubs and trees were banished from their villa gardens; the wild flowers of the neighbourhood were 'pulled up and hurled over the wall to wither as accursed things'. Sown by the birds no doubt, they had taken refuge along the very railway cuttings and embankments, bordering the Brighton line 'like a continuous garden'.[68]

But to many others in that age of contradictions, the imprint of history in the world around them was both a challenge and a source of inspiration. Forty years after coming to this country from South America in 1869, W.H. Hudson remarked in *Afoot in England* (1909), 'I yet have a sense of satisfaction, of security, never felt in a land that had no historic past.' Few writers on rural society had a more varied life than Hudson; few travelled more widely or observed with keener insight. His strange, wild youth on the Argentinian pampas he recorded for posterity in his autobiography, *Far Away and Long Ago* (1918) and his time among the forest-tribes of Guiana in the haunting pages of a novel, *Green Mansions* (1904). But it was the downland

---

66   Louis J. Jennings, *Field Paths and Green Lanes: being Country Walks Chiefly in Surrey and Sussex*, 1878; James Thorne, *Handbook to the Environs of London*, 1876; James Greenwood, *Low-Life Deeps: an Account of the Strange Fish to be Found there*, 1881, chapter on 'The Ice Harvest'.

67   *NNL*, pp. 88–9, et passim.

68   Ibid., pp. 172–3; pp. 181 seqq. give a fascinating description of the abundant wildlife alongside this route out of London.

country of Salisbury Plain, and the homely manners and customs of its people, that he immortalized in perhaps his finest work, *A Shepherd's Life* (1910).[69]

Select Bibliography

Only contemporary works are included. A complete list would run to many pages; see also Note 11, *supra*.

Atkinson, J.C., *Forty Years in a Moorland Parish...* (1891).

Borrow, George, *Lavengro* (1851).

Borrow, George, *Wild Wales: its People, Language, and Scenery* (1862).

Edlin, H.L., *Woodland Crafts in Britain ...* (1949).

Greenwood, James, *Low-Life Deeps: an Account of the Strange Fish to be Found there* (1881).

Grey, Edwin, *Cottage Life in a Hertfordshire Village* (1934).

Harper, Malcolm M.L., *Rambles in Galloway* (1896).

Hudson, W.H., *A Shepherd's Life* (1910).

Jefferies, Richard, *Nature Near London* (1883).

Jefferies, Richard, *Wild Life in a Southern County* (1879).

Jennings, Louis J., *Field Paths and Green Lanes ...* (1878).

Ranson, Florence, *British Herbs* (1949).

Short, Brian, ed., *The Ashdown Forest Dispute, 1876–1882 ...*(Sussex Record Soc., vol. 80, 1997). [Contemporary depositions by the commoners.]

Sturt, George, *The Wheelwright's Shop* (1923).

Sturt, George, *William Smith, Potter and Farmer, 1790–1858* (1919).

Thompson, Flora, *Lark Rise to Candleford* (1945).

Thompson, Flora, *A Country Calendar and Other Writings*, ed. Margaret Lane (1979).

Thorne, James, *Handbook to the Environs of London ...* (1876).

Torr, Cecil, *Small Talk at Wreyland* (three series, 1918, 1921, 1923; combined edition, with Introduction by Jack Simmons, 1970).

Warren, C. Henry, *A Boy in Kent* (1937).

Williams, Alfred, *Villages of the White Horse* (1913).

---

69   Hudson was born in 1841 in Argentina; his parents were American. He came to England in 1869, lived in straitened circumstances in London until granted a Civil List pension of £150 in 1901, and died in 1922. He travelled widely in South America, and later in England, chiefly in the south.

Chapter 15

Tourism and the Railways in Scotland: the Victorian and Edwardian Experience

Alastair J. Durie

One of the most charming and picturesque lines in the Kingdom is the Highland Railway, which with grand energy and perseverance runs through defiles, over torrents and across mountains from Perth to Dunkeld, Blair-Athole, and Inverness, thus placing the tourist in the very heart of the scenery he wishes to explore.
[Murray's *Handbook for Travellers in Scotland*, Fourth edition, 1875, Preface, p. 11.]

Accounts of the economic development of Scotland in the nineteenth century tend to highlight the spectacular growth of Clydeside shipbuilding and the heavy industries of the West of Scotland, or the continuing place of textiles in the Borders and Dundee, or the rise of fishing and the success of agriculture. But another sector which experienced dramatic and substantial growth was that of tourism, which became the backbone of a number of localities and on which their livelihood depended. This chapter seeks to examine in a Scottish context questions around the themes of what did the railways do for tourism, and tourism for the railways. What part did the railways play in the growth of Scottish tourism? Which types of tourist and forms of tourism benefited most – the working-class day tripper and excursionist, the middle-class holidaymaker, or the moneyed sportsman or health seeker? A closely related issue is which areas of Scotland benefited most from the railways' contribution to the development of tourism and what part did the railway companies play in the promotion of tourism? A third is to ask what the railways took from tourism; was it as 'splendidly profitable'[1] as Acworth asserted of the August sporting traffic? And a final issue is what part tourism played in the formation of railway companies in Scotland, a card played to effect in some schemes throughout the century.

It is important to note at the outset that tourism in Scotland was not the creation of the railways from scratch but that they played a very great part in shaping its scale and character, in changing it from an elite to a mass experience. To a lesser extent, the railways influenced the direction of tourism. While they did open up some localities in the north-west Highlands and the Borders not much on the tourist path previously, and the preserve once of the moneyed and culturally motivated few, more often their arrival prompted a marked *expansion* of existing flows, as in the Trossachs or on the

---

1   W. Acworth, *The Railways of Scotland*, London, 1890, p. 63.

Clyde Coast, and encouraged a widening of access to tourists from all levels of society. The railways were part of a nineteenth-century transformation in transport in Scotland, what contemporaries called 'the perfect revolution effected in favour of the tourist'.[2]

In broad terms Scotland could be broken down into three broad categories as far as the tourist industry was concerned. Some areas were substantial beneficiaries of tourist traffic: resorts such as Peebles, Oban, and Rothesay saw their summer populations double or treble. For them and many other places, coastal and inland, tourism was their lifeblood. When asked in 1877 what his community lived on, a Buteman responded cheerfully that while in the winter it was 'tatties and herring', in the summer 'We're all right then: we live on the Glasgow folk'.[3] Then there were the urban, industrial, and mining districts from which a good deal of tourist traffic originated, much of it in the form of excursions organized by works or churches or local societies, a large business heavily concentrated and constricted in its reach by time and cost. And finally there were the places which both provided holiday traffic and attracted it. From June onwards Edinburgh began to empty of its middle- and professional-class residents as they went away for the summer to the seaside. But this temporary loss was balanced by an influx of tourists: most Americans, for example, who came to Scotland, had Edinburgh on their schedule. There was less of a match in Glasgow, where the scale of the outflow 'doun the water' far outstripped the level of incoming tourism. 'There never was a city whence the annual migration to the sea-side is as universal or so protracted as it is from Glasgow', observed one periodical in November 1856.[4] Yet curiously, and this is one great difference from England, there never developed any mass resort in Scotland on the scale of Blackpool or Scarborough, both of which along with Douglas, Isle of Man, attracted Scots in quantity.

The potential at the coast for pleasure traffic as well as for business was firmly established by the paddle-steamers within a very few years of their first operation in the Firth of Clyde in 1812. A network of regular services was quickly put into place from Glasgow to the coastal resorts, and full-day special sailings in the area and on inland waters – 'mere excursions of pleasure'[5] – proved popular and profitable. John Galt's mercantile Mr Thomas Duffle was but one of many who took an 'adventure for health and pleasure' by steamboat down the Clyde to Greenock or one of the other coastal resorts.[6] When the railways came on the scene in the west of Scotland and elsewhere, they were quick to follow the lead given by the steamship companies, and with great success. As Simmons has pointed out,[7] whereas coaches could move a dozen or so passengers at a time, and steamships a few hundred, the railways could move much greater numbers at greater speed, with less discomfort in bad weather and at less cost. It was not just the business of the day-trippers that was tapped, but the

2   Introduction to Murray's *Handbook for Travellers in Scotland*, 4th edition, London, 1875.

3   *Bute and Beauty. A tour to and through the Island*, Glasgow, 1877.

4   *Fraser's Magazine*, November 1856, p. 411.

5   William Daniell, *A Voyage round the North and West Coast of Scotland*, London, 1820.

6   John Galt, *The Steam-Boat*, Edinburgh, 1822, p. 6.

7   J. Simmons, *The Railway in Town and Country, 1830–1914*, Newton Abbot, 1986, p. 237.

growing market for weekend pleasure travel. By 1841, according to Robertson,[8] the Monday morning trains on the Glasgow to Greenock route were packed with *seabathing* folks who had gone down on the Saturday afternoon. A further and related development was the growth of commuting, with families taking holiday houses for the summer at the coast, father returning to his place of business in Glasgow or Edinburgh during the week. This pattern was not unique to Scotland, nor to coastal resorts, but it reached its fullest development in the Clyde region, thanks to the combination of sea and rail services. A contemporary song catches the scene at Dunoon, where Glasgow-bound travellers would catch the early morning steamer to Greenock for the city train. Similar scenes were to be observed at Rothesay, Millport, Lochgilphead, and elsewhere in the Clyde catchment area, or at Aberdour and Largo in Fife, where the passengers were bound for Leith and Edinburgh.

> The steam was up, the wind was high,
> A dark wind scoured across the sky,
> The quarter deck was scarcely dry
> Of the boat that meets the railway.
> Yet thick as sheep in market-pen
> Stood all the Sunday watering men,
> Like growling lions in a den...
> O what a hurry to get to town
> Upon the morning railway.
> [*Chamber's Edinburgh Journal*, 1846]

The potential of the railways to create business out of the growing interest in travel for leisure and pleasure was not confined to the coast. Fair Days – in July – and Fast Days – in October – provided ever-increasing excursion traffic to many parts, fanned by advertising and cheap fares. One of the early Scottish railway companies to take an active interest was the Glasgow & Garnkirk Railway, a rather unlikely enterprise which served the unattractive mining localities of Airdrie and Coatbridge. Though it had been offering afternoon and evening trips during the summer in previous years, the service that it ran on the Saturday of the Glasgow Fair Holiday in July 1834 has been judged by Robertson[9] the first excursion service in Britain – a non-stop journey over the eight miles of the line, with an hour's sightseeing on arrival. Four such outings a day were offered, with accommodation for first-, second-, and third-class passengers. The results encouraged further promotions, and on the General Fast Day in Glasgow, observed more and more as a holiday rather than as a day of reflection and prayer, of 23 October 1834, some 1 250 passengers were carried.[10] The following season the Glasgow & Garnkirk repeated the venture on a daily basis throughout the

---

8   C.J.A. Robertson, *The Origins of the Scottish Railway System, 1722–1844*, Edinburgh, 1983, p. 244.

9   The Company had in fact been running summer evening trips 'for the use of those persons who cannot leave Town until the Evening' during the previous year in June and July, as well as cheap excursions ['ninepence for the closed carriages and sixpence for the open'] lasting two hours during the week. See Don Martin, *The Glasgow & Garnkirk Railway*, Strathkelvin, 1981.

10   Acworth, *The Railways of Scotland*, p. 21.

summer, with one train a day reserved for the better-off: only first-class coaches and no concessionary fares. *Genteel Parties*, it was promised, 'will find the trip an agreeable and healthful mode of spending part of the day'.[11] The pleasure may well have been more in the experience of travelling and the wayside views than the destination: a neighbouring line, the Wishaw & Coltness, drew attention to its 'newly executed and extensive tunnel as well worthy of a visit'.[12] The Glasgow & Garnkirk was enterprising in its efforts to tap the growing demand for non-commercial passenger travel and even attempted in July 1836 to run a Sunday service to allow people from the Monklands area to attend divine service in Glasgow.[13] It is doubtful whether the Sabbatarian lobby was entirely convinced; its leading lights were certainly to be a powerful lobby against the provision of Sunday services of any kind for many decades, galling to the railway companies and a continuing source of irritation and perplexity to visitors from more relaxed cultures.

Though the initial orientation of these early Scottish railway companies had largely been to the movement of coal and other freight, passenger traffic for pleasure became of increasing significance. The volume of such business leapt: the Glasgow, Paisley & Greenock Railway ran out of engines during the August Paisley Holiday in 1840, its first year of operation, but still managed to carry over 8 000 passengers in the one day. The growth of excursion traffic, to be found virtually everywhere in the Scottish railway system from the 1840s onwards, is a familiar story. 'The practice of large pleasure parties making trips by Railway and Steam boat to distant and interesting parts of the country is, we are happy to find, on the increase both in Edinburgh and Glasgow', remarked one paper in September 1843.[14]

Special outings, organized for works or masonic societies or temperance organizations, became commonplace during the spring and summer. Some of these were 'monster' affairs, and the railways did what they could to keep interest up by cheap fares linked to particular attractions: the visit of the Channel Fleet to Queensferry in 1860, the opening of the General Assembly in Edinburgh, royal tours, and the like. Industrialists, philanthropists, church organizations, educationalists, and reformers were sponsors of trips and outings. One such was John Hope, a wealthy Edinburgh lawyer and temperance stalwart. In July 1847 he organized the first of his annual Juvenile Pleasure Outings for Edinburgh and Lothian children in the British Temperance League. Two special trains conveyed the party of 1 255 boys and girls to the Duke of Buccleuch's park at Dalkeith: under the supervision of 47 adults, and no doubt thanks also to beautiful weather, all went well. The next year saw an even bigger trip to the Earl of Wemyss' grounds at Gosforth, the third found them at Hopetoun House, and so on for many years, wherever the railway could take the outing.[15]

---

11   *The Glasgow Herald*, 10 June 1835.
12   Ibid., 10 July 1834.
13   Ibid., 21 July 1836.
14   Ibid., 24 September 1843. The article goes on to describe a day trip by special train to Edinburgh for upwards of 500 Rechabites and teetotallers.
15   David Jamie, *John Hope, Philanthropist and Reformer*, Edinburgh, 1900, p. 185: 'The children sang their temperance melodies and enjoyed the lovely gardens. So well were they kept in hand that not a leaf was touched; even the wild ferns were left unharmed.'

The role of excursion agents, in the business of organizing travel for profit, was important, though the railway companies were to develop their own programmes of tours in combination with the steamship companies, particularly those of Hutcheson and MacBrayne. The best-known of these agents was Thomas Cook – with some justice dubbed 'the King and Father of Excursions'[16] – whose Tartan Tours began in the summer of 1846. Each year thereafter saw Cook himself in Scotland two or three times, personally conducting tours, tightly timed and costed, with seven- or fourteen-day itineraries which took parties to the established tourist 'draws' of Edinburgh and the Trossachs, as well as Iona and Staffa. Linked into his railway-based programme, and to make everything certain for his groups, was a network of arrangements with local hotels, coach and carriage hirers, and the steamship companies.

It would be a mistake, however, to see Cook catering only for a prosperous clientele drawn from the better-off sections of society. Quite a number of shorter trips for people with limited means and little free time were organized by him, particularly from the north-east of England, where he had an exceptional local agent, W.E. Franklin,[17] a railway bookseller and newspaper agent, to drum up business. The *Scotsman* newspaper noted early in July 1861[18] that amongst the tourists then thronging Edinburgh were no less than 1 500 Cook's excursionists from England: a large party of ladies and gentlemen from Leeds, and another from Penzance and the south-west, both set on a minimum of a week's stay in Scotland. There was, however, a third group (400 to 500 strong) arrived by train from Newcastle, 'mostly', the paper noted, 'of working class people and juveniles' who were in the city for one night only. These short sallies could have a temperance basis, as in June 1858 when Cook organized a special train to the 'Land of Scott' for the Ladies Temperance Association of Newcastle; amongst the 118 who signed the visitors' book at Abbotsford on 14 June were day-visitors from Tynemouth, Hexham, Glanton, and elsewhere in the north-east. Even the longer ventures attracted, in Cook's own words, some of the tradesmen, clerks, and 'swart mechanics' who first gained experience of travel to distant parts on one of his short tours. A steamboat proprietor congratulated Cook in 1861 on his part in broadening the tourist flow to the Western Isles: once there had been but few, except the wealthy, 'but now I see large numbers of the middle and humbler class of society coming out this way, and they will constitute the grounds of future success to the proprietors of these boats' [Cook's' *Scottish Tourist Official Directory and Guide*, 1861].

The possibilities of tourism as a source of business and of profit were, therefore, well apparent to the railway companies – and others – in Scotland by the mid-nineteenth century. And railway promoters were not slow to refer to the prospects of tapping the rapidly growing tourist trade. General cultural tourism was attracting growing numbers of English, continental, and even American visitors, a significant proportion of whom were wealthy and therefore liable to travel first-class, as well as

---

16   Thomas Cook Archives; Letter of Endorsement from Lydia Fowler, 1865.
17   See Franklin's Obituary in *The Newcastle Journal*, 10 January 1887: 'He was local agent for Messrs. Thomas Cook and Sons, the Tourist agents, and was in the habit of conducting parties, chiefly through the Highlands... by this means he became very widely known, and was highly regarded as a conductor for his geniality and general affability.'
18   *The Scotsman*, Thursday, 4 July 1861, 'Mr. Cook's Excursion Trains'.

to stay in the best accommodation. Some of these were literary tourists, drawn to the 'Land of Scott' or the 'Haunts of Burns'; others were interested in the scenery and the history, or the natural history and geology. Whatever drew them, the railway companies were interested in their trade. In W.E. Aytoun's splendid squib, *How we got up the Glenmutchkin Railway and how we got out of it* (Edinburgh 1845), two hard-up young men cast around for a railway scheme. England was out of the question, Lowland Scotland spoken for, a Spanish or Sicilian venture too distant. Then they hit on the right, and for them highly profitable, notion: 'Why not try the Highlands? There must be lots of traffic there in the shape of sheep, grouse and Cockney tourists, not to mention salmon and other et-ceteras. Couldn't we tip them a railway somewhere in the west?'[19]

Promoters of railway companies in the Highlands, where traffic was otherwise thin, understandably made much of the potential of tourism for their operations and profits. The prospectus of the Dingwall & Skye Railway forecast in 1864 a large tourist traffic throughout its entire length during the summer and autumn months.[20] One English visitor in 1875, who had greatly enjoyed the scenery –grand trees, glorious skies, and soft lakes – agreed that it was purely a 'tourists' line'. He added that if report were to be believed, the engineers had gone out of their way to secure a route that was 'picturesque'.[21] Sometimes the claims were more than justified by events. The surge of traffic to Oban, when the railway arrived in 1880, exceeded all expectation and caused what even the local paper conceded was a surfeit of custom in a town 'where the tourist is to custom what the pig is to Chicago'.[22] Fort William experienced similar encouragement from the opening of the West Highland line in 1890. Business was steadily going backwards, with much of the trade diverted to Oban, but since the railway arrived, every hotel and shop had certainly benefited, averred one local businessman in 1892 in evidence to the Committee reviewing the bill for the Ballachulish Extension.[23]

But equally there were times when hopes were disappointed: the Portpatrick & Wigtownshire pushed a branch to Garlieston on the Solway, which it intended to develop as an excursion port for traffic to the Isle of Man. But the tidal waters were too chancy, and the scheme foundered with the withdrawal of all passenger services in 1903. Perhaps the most spectacular failure was that of the Invergarry & Fort Augustus, a short and expensive line intended to tap the tourist traffic between the Caledonian Canal and Loch Ness. A complete flop, it closed after only seven years' operation. These were, however, the exceptions. What was a problem, of course, was the relatively short season, perhaps only three or four months over the summer and

---

19   Douglas Gifford, *Scottish Short Stories 1800–1900*, pp. 92–118.
20   *The Railway News and Joint-Stock Journal*, 1, 1864: 'Pleasure traffic is likely to form an important ingredient in the success of this line.'
21   See A.J. Durie, *Scotland for the Holidays*, East Linton, 2000, p. 151. The Dingwall section was advertised heavily, for example in Slater's *Commercial Directory to Scotland* (1873 edition): 'No Railway Route in the Kingdom, and probably not even on the Continent, presents within the same space of time so great a variety of Fine Scenery. The facilities for Driving and Pedestrian Excursions are very great.'
22   *The Dundee Advertiser*, 5 December 1894.
23   Donald MacKintosh, 18 years a hotelier in Fort William.

early autumn. It was no coincidence that Joseph Mitchell, the engineer in charge of the Highland Railway was pushed against his better judgement to open the final section early in September 1863 because his directors 'naturally wished to catch the tourist trade of that season'.[24]

Highland hotels and some steamer services were equally dependent on a short summer trade, but whereas most shut down in October until April, the railways remained open all year, with in some places but lean pickings. The Dingwall & Skye, admittedly an extreme case, found its August revenue four times that of December when trains and traffic alike were at their lowest.[25] An intriguing proposal was made in the early 1880s, strongly supported by local railway and hotel interests, for a line from Callander to Loch Katrine through the prime tourist country of the Trossachs. There was no prospect of any freight or commercial business and it was proposed that the line be open only during the summer, an idea without precedent but which made sense. The scheme fell, however, when one of the largest landowners in the area withdrew his support.[26] The later nineteenth century saw a rush of proposals geared to the tourist sector, including one for a mountain railway up Ben Nevis, and another for a network of fast trams in Argyll, none of which – other than the West Highland[27] – came to anything.

Beyond general travel for pleasure, and traffic to the coast, which, as we have already seen, was well established by the 1840s, there were two types of tourism which keenly interested the railways. These were the search for health, which had led to the growth of a number of spas, and the sporting traveller. Both of these involved a better-off clientele, who were likely to travel first-class – something of great consequence to railway revenues. One first-class passenger on the Dingwall & Skye, or so the figures showed, was worth two in third class, and there was the additional business of their servants and mounds of luggage. When the bill to authorize the Scottish Central Railway was under scrutiny in March 1845, the question was put as to what watering-places and spas the line might serve, perhaps with the growing patronage of Bridge of Allan in mind.[28] Twenty years later the Dingwall & Skye had its eye on commanding what was called the 'lucrative trade with Strathpeffer, so famous for its mineral waters'. Moffat, the oldest and most significant of Scotland's spa resorts, was served for many years by the Caledonian Railway's station at Beattock. Both it and Strathpeffer acquired their own branches, in April 1883 and

24 Joseph Mitchell, *Reminiscences of My Life in the Highlands*, 1884 (1971 edition), vol. 2, p. 197.
25 John Thomas, *The Skye Railway*, Newton Abbot, 1977, p. 44. The contrast could be even more extreme; in 1870 one period of nine days at the end of August saw receipts of £719, and a week in September £520 – the nadir came however in December when one week saw earnings of a mere £89.
26 See John Thomas, *The Callander & Oban Railway*, Newton Abbott, 1990, pp. 53–4.
27 *The Dundee Courier*, 14 April 1894, 'The West Highland Railway. A Journey Over the Line. Its engineering features. Description of the Scenery': 'The railway not only covers large tracts of country untouched by the recognised coach routes, but it provides facilities for an endless number of the new circular tours through some of the most romantic regions of the country.'
28 Peter Marshall, *The Scottish Central Railway*, Oakwood Press, 1998, p. 22.

June 1885 respectively. The directors of the Ben Wyvis Spa Hotel reported to their company's annual general meeting in January 1886 that, as anticipated, the opening of the Strathpeffer railway had added to their business. In 1911 the parent company, the Highland Railway, which in Thomas's words,[29] lavished attention on the resort, built its own hotel there, and provided a special service, the *Strathpeffer Spa Express*, every Tuesday during the summer, with through coaches from London. Strathpeffer never rivalled Harrogate or any of the top English spa resorts, but it did attract a very respectable clientele from the south, and the railway played a key part in this. There was, of course, the unfortunate death of Sir Joseph Dinsdale in August 1912 at the Ben Wyvis Hotel, the result, or so the locals claimed, not of the treatments but a consequence of a chill caught on his journey north. Critics insisted, and with justice, that poor travelling facilities were scarcely conducive to the success of a health resort.

The development of Scotland as a sporting playground for the pursuit of deer, salmon, and grouse was well advanced before the railways were on the scene, and traffic was already heavy in early August on the eve of the Glorious Twelfth and the start of the grouse season. The *Perth Courier* reported in mid-August 1845 that in

> the previous eight days the stream of sportsmen, tourists and travellers had set in strongly to the north, and in the current week a continuous line of private carriages and vehicles of all kinds had passed through Perth, independent of those conveyed by four daily stage coaches to the Highlands. From the east of Inverness there were four steamships in the course of three days – two from London and two from Leith, all freighted with the muniments of a deer-stalking and grousing campaign.[30]

The attraction of this business was that so much of it was likely to travel first-class. Indeed for the many who came as guests, the cost of travel for themselves, their luggage, and dogs, was their main expense: Evan MacKenzie estimated in 1895 that of the 6 000 or so who travelled to Scotland for the grouse, 5 000 incurred no expenditure other than their rail fare. Indeed while they might stay in the one place – house, castle, or lodge – there was quite a circuit of balls in the autumn to attend at Oban or Dalmally or Inverness. 'The population that inhabit the deer forests rush about a good deal... backwards and forwards on the railway'.[31] The sporting season, therefore, brought profitable business for the railway companies, even if it tended to be heavily concentrated in the early part of August, as the station staffs at Perth and Inverness knew all too well. What was also significant is that while the railway companies could trade on the image of Scotland as a place for health and sport, they

29   Thomas, *The Skye Railway*, pp. 89–92. He draws attention, for example, to the remarkable advertising programme of the Highland Railway for Strathpeffer carried on by motor car through the north of England in 1908. The vehicle was brightly painted in Highland Railway livery, and sent out as a 'mobile tourist information office' with publicity leaflets.

30   This is drawn from A.J. Durie, 'Unconscious benefactors. Grouse shooting in Scotland, 1780–1914', *The International Journal of the History of Sport*, Vol. 15, December 1998, pp. 57–73. It was necessary to bring in everything from bedding to pianos to the sparsely furnished shooting lodges: businesses at Perth and Inverness were quick to offer a comprehensive supply service.

31   Cited in John McGregor, The West Highland Railway, PhD thesis, Open University, 1999, p. 183.

had to do almost nothing to stimulate demand for shooting or fishing. They had no part in the letting of grouse moors, deer forests, or salmon rivers: that was in the hands of the estates themselves or specialist agents such as J. Watson Lyall of Perth. But, on the other hand, proprietors who stood to gain from better access through increased rental income from their sporting rights, were more likely to invest in railway development, as were their wealthy tenants who wanted to bring in their guests, some of whom could manage only a short stay.[32]

Where the companies did become involved at a much deeper level was in the provision of golfing hotels. Other sporting activities, including walking and climbing, had their devotees. John Anderson of the Callander & Oban, ever alert to the possibilities for his newly opened line, instructed the hotelkeeper at Crianlarich to arrange for a mountain guide to take excursionists and climbers up nearby Ben More. His many other promotional schemes included a weather station at Strathyre to show how good the climate was there, and advertising literature which billed Glen Ogle, up which the trains toiled, as the '*Khyber Pass of Scotland.*' Not all could shoot, or stalk, or climb, but nearly everybody from the young to the old could golf, and golf became a central part of a holiday in Scotland for many visitors. A long-established game, golf underwent an extraordinary boom in the later nineteenth century, with as many as 200 new courses opened in and around the cities, and at every holiday destination from Aberfeldy to Uist. The railways played their part by providing special services and golfer's fares to get city players to and from their country courses on a Saturday: supplementary trains were arranged for the big competitions such as the autumn medals at St Andrews and Dunbar.

The commercial possibilities became apparent, and while the Scottish railway companies did not become involved in resort development, other than through joint advertising campaigns, they did venture into the provision of luxury golf-hotel and activity centres, as did others. Sir Henry Lunn acquired the Atholl Palace Hydro at Pitlochry for the British Public Schools Alpine Sports Club in May 1914 to turn it into an all-year resort: skiing in the winter, but golf all year round. The Great North of Scotland Railway led the way with Cruden Bay in 1899, and six years later the Glasgow & South-Western followed suit at Turnberry in Ayrshire. Work on the Caledonian's luxury spa and golf complex at Gleneagles, begun in 1913, was brought to a halt during the war, but the facilities were opened in 1924.[33] The justification for such investment was that there was the prospect of profit both on the hotel operations and on the traffic to them. The GNSR had set up an Hotel Committee in 1890,[34] and had acquired the successful Palace and Station Hotels in Aberdeen in 1891; it

---

32 Cf. Lord Abinger of Abinger and Inverlochy's evidence to the proof before the House of Commons 1889 Committee on West Highland Railway Company Bill, (MacKenzie Papers, West Highland Museum): 'Many tourists dislike the steamers and many more cannot afford the long and expensive carriage driving. During the summer and autumn there is also movement of guests and their servants to the different houses in the Country. Very often guests have to decline to come to my home because they can only spare a few days and the time on the journey would consume this.'

33 Jane Nottman, *The Gleneagles Hotel*, London, 1999.

34 See M.J. Mitchell, 'The Cruden Bay Hotel and its tramway', *The Story and Tales of the Buchan Line*, compiled by Alan H. Sangster, Oxford, 1983, pp. 34–44.

looked at the Invercauld Arms at Ballater on Royal Deeside, but decided instead on a new 55-bedroom luxury hotel near Port Errol on the Buchan Coast. Despite two good golf courses and a resident professional, tennis courts, bowling greens, attractive links, and an electric tramway for guests from the station at Cruden Bay, it proved a financial failure. The location was a problem, and so also were the east-coast *haars* (sea-mists) which dogged summer days.

The involvement of Scottish railway companies in tourism was varied and of varying importance to them. Tourist traffic was most significant to the coastal lines and those in the Highlands. Mostly the coming of the railways led to a marked increase in existing flows and to a broadening in the social and economic composition, but in some areas, such as in the north-west, the railways did open up new territory. Those interested in tourism were well aware of their role, and of the need for good railway service. But the relationship could turn against the railway companies. There were years when tourism took a dip and railway revenues suffered. The weather was one continuing source of worry for all concerned in the Scottish tourist trade. The most vulnerable sector was the top of the market; working-class day trippers and excursionists tended to go almost regardless, as did the professional and middle-class summer vacationists. But bad weather in Scotland might well deter the potential visitor from the south who had other options such as a trip to the continent. The summer of 1881, for example, was one of the poorest on record, and its effects were felt. It was, or so the directors of the Ben Wyvis Hotel at Strathpeffer reported, 'unprecedently cold and cheerless',[35] visitor numbers fell by a third, and the season showed a loss, as against a healthy profit in the previous year. In 1873, to bad weather during the summer was added an almost total abandonment of the grouse season on account of a shortage of birds. The half-yearly meeting of the Highland Railway in November was told that 'the passenger traffic usually so good with us in July and August was prejudicially affected by the wet and uncongenial weather; while the shooting traffic suffered a heavy diminution from the failure of the grouse this season in the Highlands.'

If the weather was one problem, another was that of counter-attractions, as in 1893 when the flow of American tourists to Europe was sharply reduced by the Columbian Exposition and World Fair in Chicago, which kept many at home. The number of visitors to Edinburgh was one-tenth the normal, and, while hoteliers suffered the worst, railway business was also slack.[36] If parliament sat late, as in 1909, that also could have serious effects; members who would have come north for the shooting were detained in London, moors and shooting-lodges went unlet, and rail traffic fell.

The fortunes of tourism and the railways – particularly the Highland lines – were, therefore, closely intertwined, and remained so, though the place of the railways came under increasing challenge from the turn of the century. Whereas collusion and cooperation were mostly the order of the day between rail, coach, and steamer services, inland or coastal, the arrival of motor transport from the 1890s brought

---

35   Highland Regional Archives. Uncatalogued papers: Strathpeffer Hotel Company Accounts.
36   *The Scotsman*, 24 August 1893: Great Diminution of the Tourist Trade in Scotland. See also *The Times*, 29 August 1893, where the 'high charges of the hotels and bad management of the Scottish railway companies' were held to be in part to blame.

direct and serious competition which took its toll of railway traffic and profits on a noticeable scale even before 1914. The companies were not passive: the GNSR started its own motor services in 1904 from Ballater, some of which were seasonal and intended to tap tourist business. But the leaching of custom by the motor car, and of general business by the charabanc, was already exacting a toll. Moneyed people, who would once have come north to Scotland by train, increasingly started to motor. And once arrived, they used their cars to get around their locality instead of travelling by train and trap. The Highland Railway felt the impact almost immediately. In March 1902 the company's chairman, William Whitelaw, told the half-yearly shareholders' meeting that in the previous year the Highland had seen a fall in first-class local journeys of over 1 000, which he attributed almost entirely to the rise of motoring.[37] He thought that this traffic was probably lost for ever, but 'as people are beginning to get a little bit sick of the general squalor of long distance motoring', there was still hope for the services from London and the south. But the loss of revenue was serious. Two years later in 1904 the shareholders were reminded of the problem.

> The more extensive use of the motor car by visitors from the south is making serious inroads into the receipts of the Highland Railway from first class fares. A large number of people who used to visit the Highlands by train, and travelled first-class, have in the present year arrived in the Highlands by motor car, and have moved about the country by the same means, instead of taking train and travelling first-class from station to station as they used to do.

The Invergarry & Fort Augustus was the first casualty. Never a viable scheme, its fate was sealed by the new pattern of travel favoured by the rich. Its closure to passenger traffic in January 1911 was greeted by the *Manchester Guardian* with the headline 'A Highland Railway to be Abandoned. Business taken over by Motor Cars.' The reliance of mainland tourism in Scotland on the railways was weakening. Unfortunately for the railways, there was to emerge no replacement traffic in the interwar years to compensate, although they did continue to compete hard for what was still a major sector of the economy. But an analysis undertaken in the 1930s by the management of Crieff Hydro of the forms of transport used by visitors to that popular resort hotel underlines the problem. Only about a third (and the figure was falling) came by rail: 50 years previously nearly all would have arrived by train.[38] In Scotland generally tourism was still significant as a user, but even in high season, the traffic and the revenues did not equal what they had once been The degree of dependence varied from place to place and from season to season, but the role of the railways in Scottish tourism was fading.

---

37 John Thomas, *The West Highland Railway*, Devon, 1965, p. 115.
38 Cited in A.J. Durie, *Water is Best. The Hydropathic Movement in Scotland, 1840–1940*, forthcoming (Tuckwell Publishing, East Linton, 2003), drawing on Crieff Hydropathic Company minutes.

Chapter 16

Railways and the Evolution of Welsh Holiday Resorts

Roy Millward

The North Wales Coast

The first named train in the history of the railways in Britain was the *Irish Mail*. Its first journey was made in the late summer of 1848 with the opening of the Chester & Holyhead Railway, one of the earliest and greatest achievements in the creation of the British railway system, for Parliament had only agreed to its construction in 1844. In the space of four years the most difficult terrain so far encountered by the railway builders had been conquered by a team of as many as 12 000 navvies. The Menai Strait and the tidal mouth of the Conwy River had been crossed by bridges of a highly original design, the track concealed in cast iron tubes, and tunnels driven through bold headlands at Penmaenbach and Penmaenmawr. Today, a journey along the route of the *Irish Mail* as it follows the shores of the Dee leads to a string of holiday resorts. Prestatyn with its golf and caravans is followed by Rhyl, a place of funfairs and day-trippers. Colwyn Bay, retirement bungalows and late Victorian villas, stands isolated from its seaside by the embankments thrown up by those bold engineers long before the speculators from Manchester began the building of a town in the 1860s.

Llandudno evolved aside from the mainline traffic that takes to the darkness of Stephenson's tunnel bridge across the Conwy, an exemplar of the best principles of Victorian town-planning. The grid-iron plan of wide boulevards and the sweeping esplanade locked between the headlands of the Great Orme and Little Orme gave it an air of which the Victorian upper classes were quick to take advantage. Here, so the guidebooks claimed, one might find branches of the most fashionable London shops and the families of Oxford dons and clerics would take up residence in the summer months. This vanished morsel of Llandudno's social life gave rise to the story that Charles Dodgson, author of *Alice in Wonderland*, stayed at Penmorfa, a holiday home built by Dean Liddell, Dean of Christ Church, Oxford. Dodgson, it is claimed, wrote the story of 'Alice's Adventures underground' to be read each evening of a summer holiday to the Dean's daughter, Alice. Recent research by R.L. Green, editor of *The Diaries of Lewis Carroll*, claims that there is no evidence that Dodgson ever spent a holiday in Llandudno.[1] Nevertheless, it has not prevented the discussion of the creation of an Alice in Wonderland theme park at the resort in the 1990s. Even so, Llandudno has not lost completely the Victorian concept of holidays as cultural occasions because, apart from Cardiff, it is the only place in Wales with regular seasons by Welsh National Opera.

The birth of resorts along the North Wales coastline is more complex than the pattern suggested by the map. The railway entrepreneur did not enter on the scene with the intention of creating a tourist industry. A more pressing demand was rooted in national politics. The political union of Britain and Ireland in 1801 brought Irish MPs to Westminster, and with the MPs came the demand for better communications across the Irish Sea. Holyhead, by reason of the shortest sea-crossing, had long commanded official links with Ireland, to which improvements had been made in the 1820s by Telford's work on the Holyhead turnpike and the building of suspension bridges at Conwy and the crossing of the Menai Strait. In 1821 the Post Office established a steam-packet service, worked by the Admiralty and based on Holyhead. By 1839, the spreading railway network in England brought the possibility of a faster service to Dublin through the port of Liverpool. Sea travel was slow; the steam train held out the prospect of a much swifter means of communication. Holyhead, with a sea-crossing half that of the distance from Liverpool, could maintain its position with the construction of a rail link with England.

Access to Ireland provided the main motive for railway schemes in Wales in the middle years of the nineteenth century. Brunel proposed a line through South Wales that would transform Newquay on Cardigan Bay into a primary harbour for the Irish traffic, and Port Dinllaen, on the northern coast of the Lleyn peninsula, was the object of several schemes. But the Chester & Holyhead Bill in the 1844 session of parliament won the day against all other schemes, partly because of Holyhead's established position on the shortest sea-crossing, but also because of the support of two of England's strongest railway companies – the Grand Junction and the London & Birmingham. The latter provided a million-pound investment and half of the board of directors for the Chester & Holyhead, foreshadowing its absorption into the powerful London & North Western by the year 1859. Euston, the headquarters of the LNWR, was in command of a main artery of Britain's railway network.

The shaping of settlements – nascent seaside resorts – depended on individuals and ideas beyond the control of the railway entrepreneur. Much of the land within reach of the new railway belonged to a handful of wealthy and influential families. The Mostyns of Gloddaeth owned the common grazings and salt marsh beneath Great Orme where Llandudno was to appear. At Colwyn Bay, the Pwllycrochan estate had passed, by marriage, into the possession of the Scottish aristocracy, the Erskines of Bute. The two resorts were deeply influenced by the attitudes and actions of their landowners. The Mostyns seized the opportunity for the making of a new planned town in 1847 as the railway was nearing completion. Edward Mostyn employed an Anglesey architect, Owen Williams, to draw up plans for a seaside resort to attract a wealthy Victorian middle class to enjoy the sea-bathing and scenery of a site that the

---

1   J.B. Edwards, 'Llandudno in Wonderland, the enigma of Lewis Carroll', *Transactions of the Caernarvonshire Historical Society*, Vol. 52–3 (1991–2), pp. 111–21.

guidebooks at the end of the century compared with the Bay of Naples. An Enclosure Act for the common land of salt marsh and sand dunes was obtained in 1848, the year when the trains started running. The new town, focused on the sweeping curve of Orme's Bay, was planned to cover 800 acres with a grid-iron of broad tree-lined boulevards and fashionable shops. Strict building regulations were applied by the Mostyn family to the development of Llandudno. A handsome crescent, in the image of Bath and Buxton, faced the sea across a broad esplanade of grass and gravel to satisfy the resort's select English visitors. By 1854 Mostyn Street was complete, a town hall had been built and hotels were in the course of construction. Llandudno's population, scarcely three hundred at the beginning of the century, a cluster of miners' cottages and fishermen's huts, had risen to more than a thousand in the 1851 census. It was not until 1858 that a branch line from the Chester & Holyhead Railway, three miles in length, reached the threshold of the town. When the railway arrived Llandudno already possessed a complete drainage system, gas works, a market hall, and public reading-rooms.

The branch railway's role in the creation of this resort seems to have been negligible. Until 1862 the single track lacked its own steam engines; in summer a tank engine was borrowed from the London & North Western, and winter haulage was by horses. The rapid expansion of the railway's interest in Llandudno, came only towards the end of the century. The LNWR took over the branch line in 1873 when the track was doubled from the mainline junction into a terminus station. In 1885, when Llandudno's population had grown to 5 000, platforms to accommodate day and half-day excursion trains were added. The summer timetable of 1885 shows 28 trains arriving at Llandudno on weekdays to which were added innumerable special excursions that coincided with the industrial holidays, the 'wakes', of South Lancashire, Cheshire, and the Potteries.

For the first half of the twentieth century, railways reigned supreme in the communications of the British Isles. In North Wales the railway provided a fast, reliable, and comfortable link between rapidly developing seaside towns and the commercial cities of Manchester and Liverpool, whose hinterlands contained some of the most noxious industrial towns of Victorian England. The London & North Western served two distinct streams of passenger traffic between the conurbations and the coast. The summer months and especially bank holidays brought thousands of holidaymakers for a week at a time as well as vast numbers of day and half-day trippers. But in addition there was a daily passenger traffic between the north Welsh coastal resorts and the industrial towns of South Lancashire and Staffordshire. Businessmen found that it was possible to travel daily to their mills, clusters of smoking kilns, warehouses, and offices from homes in the healthier, sunnier climate of the Irish Sea coast. The daily business trains could arouse the envy of commuters into the big cities at the present time. Season-ticket holders from Llandudno, Colwyn Bay, Rhyl, and Prestatyn were entitled to places in the 'Club Cars', comfortable saloon coaches where the first business of the day could be conducted over breakfast. Before the amalgamation of the national railway system into four great companies after the First World War, the North Staffordshire, whose lines served only the Five Towns of the Potteries and an outlying rim of market centres, ran a daily Club Train, decked out in the company's distinctive deep-red livery, between Stoke-upon-Trent and Llandudno.

Colwyn Bay illustrates the parts played by landlords and railways in the growth of a seaside resort in the last quarter of the nineteenth century. A traveller in 1857 describes Colwyn Bay as composed of one cottage and a toll bar. By this time trains had been crossing the embankment close to the foreshore for almost a decade; Llandudno, planned into existence by the Mostyns, was already flourishing. At Colwyn, the Pwllycrochan Estate belonged to the Erskine family, who had resisted any development of their property. On the death of Lady Erskine in 1865 the Pwllycrochan lands were sold to Sir John Pender, a Glasgow merchant with business interests in Manchester. Pender appointed John Porter to manage the sale of land. Thus the evolution of Colwyn was set towards its goal as a residential and holiday resort encouraged by the spreading rumour that here one could find the best climate of the whole North Welsh coastline. A further stage in Colwyn's development came in 1875 when Sir John Pender ran into financial difficulties and the property was sold to a syndicate of Manchester businessmen who formed the Pwllycrochan Estate Company to take charge of planning and development. The syndicate included two Manchester architects, Lawrence Booth and Thomas Chadwick. They formed a partnership with a local surveyor, J.M. Porter, the son of the Erskine agent at Pwllycrochan. Within a decade Colwyn was taken out of the hands of a landed aristocracy, opposed to the radical changes of the railway age, and passed over to a group of Manchester's go-getting entrepreneurs. Within the span of a generation, four miles of the coastline and its background of woodland between Old Colwyn and Llandrillo-yn-Rhos had grown into a settlement of almost 30 000. A residential suburb, reached by steeply curving roads, occupied the woods of Pwllycrochan, and substantial villas clearly expressed the Manchester connection in the architects and occupants.

Railways and Resorts in South Wales

The 'Railway Mania', at its height in the middle years of the 1840s, struck South Wales at the same time as the Chester & Holyhead was about to begin a transformation of the life and landscape of the north. The South Wales Railway company obtained an Act in 1845 to build a main line from the Severn to Fishguard. The motive behind the grand scheme was the same as that driving the pioneers along the North Wales coast – the traffic with Ireland. But the facts of geography combined with the vagaries of the capital market to award most of the advantages to the north. The distance of the Pembrokeshire coast from Dublin and the Irish heartland set against the short sea passage from Holyhead was emphasized by the slower speed of ships when compared with that of the railway. And the wide Severn estuary provided a severe obstacle to communication between England and South Wales. The Great Western from Paddington reached South Wales through Gloucester, and it was not until late in the nineteenth century that the Severn Tunnel was opened. The proposed line from London to Fishguard was 25 miles longer than the London & North Western's Euston to Holyhead.

The disadvantages imposed by the passive facts of geography were as nothing compared with the acts and notions of the developers engaged in the Irish route through South Wales. When the railway reached Carmarthen in 1852, the *Irish Mail* had already been plying its overnight journey on the Chester & Holyhead for the previous four years. By that time the first bout of Railway Mania was over and for almost 20 years the capital for railway investment would be scarce. The collapse of trade with Ireland in the wake of the Irish Famine stood in the way of the completion of the South Wales Railway. Brunel, engineer in charge, had reached within ten miles of Fishguard when building had to be abandoned. A revised plan brought the South Wales Railway to the coast at Neyland on Milford Haven, whence a steam packet service opened between Neyland and Waterford. This remained in existence for half a century. The Fishguard–Rosslare ferry began in 1906, a route that curtailed the sea journey from Milford Haven by 40 miles.

The South Wales Railway was absorbed into the Great Western by the last quarter of the nineteenth century. Unlike the LNWR in the north, the connection with the history of coastal settlements is less direct. For most of the distance to the far reaches of Pembrokeshire the track follows an inland path many miles from the sea. The beautiful cliffs and bays of Gower as well as the coast of Glamorgan, westward of Cardiff, lay beyond the economic grasp of the main line. Sea connections played a much more important role in the history of the South Wales seaside towns than those of the north. Traffic across the Bristol Channel reached from the Avon to the harbours of Devon and North Cornwall. Shipping flourished with the trade in coal from Porthcawl, Barry, and Penarth. Passenger ferries, above all for summer tourists, ran daily services between Barry Island, Weston super Mare, and Ilfracombe until the close of the 1930s. The sea trade of the South Wales coastline survived as a result of the inland track of the Great Western combined with the barrier of the Severn that was breached only relatively late with the opening of the Severn Tunnel.

In the story of the relations between railways and the rise of the seaside tourist industry, North Wales reveals the important part played by distant conurbations towards the end of the nineteenth century. South Lancashire, Merseyside, and the Potteries were the source of a great volume of holiday traffic, so much that the railway itself was shaped by its passenger traffic. Palatial stations were built, out of all proportion to the needs of the permanent residential populations of the resorts. For instance, in 1892 the terminus at Llandudno was completely redesigned with five new platforms and before the First World War most of the line between Chester and Llandudno Junction had been quadrupled.

South Wales occupied a different place in the pattern of Victorian geography. To the east, across the Severn, lay the acres of rural England where life and landscape retained most of the features of the eighteenth century. The nearest conurbations where soaring populations and industry provided the right stimulus for the growth of tourism were London and the West Midlands. Both lay too remote to encourage a sufficient volume of the ephemeral summer traffic that underpinned the development of a place such as Rhyl. The holiday resorts of Glamorgan and the Gower coast grew in response to a local, close industrial background where the short, deeply entrenched valleys of the South Wales coalfield were already set on a course of industrial growth before the end of the eighteenth century.

The building of railways, better known at that stage as horse tram-roads, began in the South Wales coalfield long before the Chester & Holyhead presented its case before parliament. As early as 1825 a 17-mile tram-road, the Duffryn Llynvi & Porthcawl, linked the mines of the valleys that converge on Tondu with the coast at Porthcawl. For more than 20 years it depended on horses before conversion to steam in 1847. The Duffryn Llynvi & Porthcawl provided the model for railway development in South Wales. Apart from the east–west axis of the South Wales Railway a network of lines joined the coalfield with the coastline where busy ports came into existence at Port Talbot, Porthcawl, Barry, and Penarth to handle the traffic in coal, iron and the import of pit-props. The railways, largely owned and financed by mining companies, also played a powerful part in the development and management of harbours. At Porthcawl the harbour was completely in the hands of the railway company where the pier, warehouses and dock were all built by the Dyffryn Llynvi & Porthcawl. The coal trade was the dominant interest; the encouragement of holidaymakers and a seaside pleasure industry only appeared late towards the end of the century. North Wales, by contrast, lacked the opportunities for an overwhelming volume of freight traffic. Slate from the quarries of Penrhyn and Llanberis, important as it was in the traffic of the north, could not match the trade that raised Barry to the position of the world's greatest exporter of coal in 1913, with 11 million tons tumbling through the hoists that lined its quays.

Porthcawl, Barry, and Penarth all emerged in the second half of the nineteenth century to enable the export of coal from the valleys, an export trade that reached to the furthest ports and coaling stations of the British Empire. Porthcawl made the earliest entry into the coal trade, with the building of the tram-road from the Llynvi valley before 1830. Penarth entered the scene in 1856 when the Windsor family of Fagan's Castle and manorial lords of Penarth obtained an Act for the construction of a tidal harbour and dock at the mouth of the Ely River. A decade later a new town had arisen on the limestone headland above the Ely estuary – a town carefully planned by the Windsors with broad streets and land allotted to churches and chapels, with open spaces for parks and gardens.

The three great coal-exporting ports of the latter half of the nineteenth century on the South Glamorgan coast share a common pattern of development determined by the railways that led to the coal-mines in the Valleys. The whole cycle of that development was achieved within the compass of scarcely a century between the 1860s and the years after the Second World War, when the traffic in coal faded into extinction. The last cargo of coal was shipped out of Barry in December 1973 bound for the Channel Islands. The rapid rise of the coal trade is reflected in the topography and society of the towns, and the railways played a vital part in shaping their lives and secondary economic activities. The 1861 census records a population of 1 406 at Penarth; within 30 years it had risen to almost 13 000. The men and women concealed in the lump sums of the census returns came from different strands of Victorian life. The foundation of the new town's life, the digging of docks and the construction of wharves and warehouses, was largely owing to a migrant force of Irish labourers. A second stream of more permanent settlers – artisans, craftsmen, shopkeepers, dock-workers – came across the Bristol Channel from rural Gloucestershire, Somerset, and Devon. A third strand of settlers was attracted to Penarth: business families from Cardiff, themselves enriched by the profits of the coal trade. Penarth's qualities as a resort – the sea air, the pleasant tree-lined streets of the planned town and the reputation of its low death rate – provided the incentive. In the 1880s and 90s Penarth's role as a resort was established by commuting businessmen and retired colliery owners and sea captains. The evidence for their shaping of the town is to be seen in the villas of Bridgeman Road and Marine Parade.

Barry entered into the coal trade in the 1880s, revealing the dramatic impact of a railway on the economy and society of a cluster of farming villages – Cadoxton, Merthyn Dyfan, and Barry. For centuries a port, engaged mainly in coastal trade, had existed in the tidal channel that divided Barry Island from the mainland. The Barry Dock & Railway company was founded by a group of Welsh colliery- and ship-owners to overcome the congestion on the tracks of the Taff Vale and Rhymney Railways that served Cardiff docks. Work began in 1884 on the construction of a dock in the Old Harbour of Barry's tidal channel. The speed and scale of the project typifies the high ambition of engineers at the end of the Victorian era. The chief engineer, Thomas Andrew Walker, was also contractor for the Severn Tunnel and the Manchester Ship Canal. By the summer of 1889 the Barry Dock was opened. At the time it was the largest dock in the world and could accommodate the largest ships on the high seas. The Barry Railway and its subsidiary, the Vale of Glamorgan Railway, was able to tap a more extensive hinterland than the rival ports of the South Wales coastline. The Vale of Glamorgan tracks reached westward to Bridgend and brought the collieries that fed Porthcawl into the sphere of the most modern handling facilities at the Barry. Barry's period of high prosperity in the world's coal trade runs from the 1890s until the First World War. From the 1920s the coal trade all along the Glamorgan coastline slowly dwindled into extinction.

The railways that carried coal from the Valleys also provided the means for the growth of tourism. In contrast to the North Welsh holiday resorts dependent on distant conurbations in England, the sources of tourism lay close at hand in the industrial settlements of the Valleys. Consequently the holiday traffic of the summer months was dominated by the day tripper. For a few hours miners and their families could exchange the claustrophobic deep valleys and their heavily polluted atmosphere for the fresh air and sunshine of the coast. The Barry Railway Company set out to satisfy this need. Before the coming of the railway Barry Island consisted of a single farm and a rabbit warren in a wilderness of dunes. On August Bank Holiday 1896 Barry Island station was opened, with the aim of exploiting the tourist trade. Three years later the railway was extended the short distance to Barry Pier so that day trippers could join the paddle steamers that cruised to resorts on the Somerset and Devon coasts. The railway timetables of 1905 reflect the rising interest in the holiday trade. A special *Boat Express* left Cardiff at 9.30 a.m. connecting with the departure of the Ilfracombe steamer from Barry Pier at ten o'clock. The return *Boat Express* to Cardiff left the Pier station at 6.33 p.m. Special cheap day-return fares were issued on this route and, in order to reduce fraud, tickets of six different colours were issued for each day of the week.

The heyday of the port of Barry lasted from 1889 to the outbreak of the First World War in 1914. In 1890 the docks handled three million tons of coal; by 1913 the figures had quadrupled and Barry was the world's greatest coal exporter. From a population of 500 in 1881 the 1921 census registered 38 945. The building of the first dock in 1884 brought a labour force of 3 000 to the site of this cluster of farming hamlets. After the completion of the harbour, labour was still in demand for the rocketing coal trade so that inward migration continued. By 1914 200 streets had been laid down; a new town had come into existence. As one local historian has written: 'A cosmopolitan English-speaking town emerged, sounding the death knell for the Welsh rural communities.'

Against this background of a soaring economy and the creation of a new town, Barry began to acquire the functions of a holiday resort. Already in 1905 the sands of Whitemore Bay, on Barry Island, attracted day visitors, but it was in the interwar years between 1920 and 1939 with the steep decline in the coal trade that the entertainments of the seaside began to play an important part in the life of the town. The early 1920s – a brief period of prosperity before the economic slump of the thirties – saw the making of a seaside resort largely under the direction of the town council. A sea wall and promenade were built at Whitemore Bay in 1922 and two years later a promenade arcade with shops was opened. At the same time, the centre of Barry Island was taken over by Collins' fairground, to become one of the largest fun fairs in the country, and in 1926 a marine boating lake and lido, the largest swimming pool in Wales, were built with financial aid from the government, one of several schemes to relieve the heavy unemployment that afflicted the South Wales coalfield by the mid-twenties.

Barry, like Rhyl on the North Wales coast, was pre-eminently a working-class resort catering for those who could afford only a few hours by the sea on Sundays or Bank Holidays. A history of the Barry Railway records that on Sundays in summer a porter was kept fully occupied chalking up the times and destinations of return excursion trains on the blackboards outside the Barry Island station. Day excursions by railway to seaside resorts rose to a peak in the 1930s, and among the four main companies after the reorganization of the nation's railways in 1921 the Great Western played a prominent part. It had advertised the qualities of the Cornish Riviera from the beginning of the century. After the opening of the Fishguard–Rosslare sea route in 1906 the Great Western, in collaboration with the Great Southern & Western Railway of Ireland, ran cheap excursions to the Lakes of Killarney. In the 1930s the GWR extended the holiday hinterland of Barry with the introduction of half-day excursions from Paddington.

The first half of the twentieth century witnessed the transformation of the coal ports of the Glamorgan coastline into holiday resorts. Porthcawl was the first to display the economic and social trends that accompanied the extinction of the coal trade, and the railway played a vital part in the changes. In 1873 the Great Western took over the management of the network of lines that served Porthcawl and so gained a total monopoly of the coalfield and industries that focused on Maesteg and Tondu. Twenty years later a dangerous rival for Porthcawl's coal traffic appeared with the creation of Port Talbot in 1893. By 1916 the exports of Port Talbot exceeded three million tons. Porthcawl could not compete with the superior coal-handling equipment of the new rival docks at Port Talbot and Barry. The response of the Great Western came swiftly. The railway company closed the docks in 1898; the future of Porthcawl was to depend on its qualities as a holiday resort.

Every stage in the evolution of seaside tourism from the late Victorian years to the present day is evident in the topography of Porthcawl. While the coal port was still active it served above all as a resort for the mining and industrial settlements of the valleys. A spacious convalescent home for miners, *The Rest*, on the edge of the complex of windy sand dunes west of the town, survives from this first phase of Porthcawl's history as a resort. From the beginning of the twentieth century the attractions of the seaside and its climate came to dominate the life and economic activities of the town. The docks were filled in to become the site of the Esplanade, and with the passing of the railway age they provided space for a car park. Holiday yachts are moored inside the tidal harbour, within the shelter of the breakwater that had been built by the railway company in the 1830s. The only visual evidence of the industrial past is Jennings Warehouse, which was used for storing iron brought along the Duffryn Llynvi & Porthcawl Railway. Today, just outside the warehouse is a short stretch of restored railway track, and at the pier head is the harbour light, erected in 1866.

After the closure of the docks the Great Western responded to Porthcawl's new role as a resort. In 1916 a new station was built reflecting the conversion to passenger traffic of a railway that began, almost a century earlier, as a coal-carrying tramway. Already before the end of the nineteenth century a daily express commuter service ran to Cardiff and similar business trains connected Porthcawl with Swansea. The massive railway closures of the 1960s brought an abrupt end to the link between the railway and the evolution of Porthcawl as a resort. In the late 1950s British Rail had put forward plans for a new station at Porthcawl designed, with six platforms, to serve the day-excursion traffic. Later, in 1963, when the Beeching plan reshaped the railway system of the British Isles, all passenger trains were withdrawn from Porthcawl. Today, the phrase 'track of old railway' on our Ordnance Survey maps, fading earthworks in the countryside, and a few metres of rail-bed at the former harbour alone commemorate the long history of the Duffryn Llynvi & Porthcawl. The motor car took over from the railway in the second half of the twentieth century and the face of Porthcawl has changed accordingly. Inland from the sea, across the windy headland, a spreading suburbia of bungalows has swallowed up the once-isolated rural settlements of Nottage and Newton where whitewashed cottages, a mediaeval parish church, and an early seventeenth-century country house, *Nottage Court*, provide a tourist attraction among the streets of seaside suburbia, that least inviting among our modern townscapes. Down by the seashore the motor-car age has stamped its own mark on the landscape in a vast densely-packed caravan park. But that is another story that dominates the recent histories of resorts that have grown up in this century largely beyond the reach of railways.

Holiday Resorts Remote from Railways

Over a long section of the Welsh coastline, railways have played a minor role in the history of seaside resorts and tourism. The wide arc of Cardigan Bay and the southern coast of Dyfed as far east as the Tywi estuary belong to this category. The first decade of railway building, the years of the Railway Mania, brought this radical means of communication only to the north and south of the country. The Chester & Holyhead laid an iron ribbon along the whole of the North Wales coastline and by 1852 the South Wales Railway reached Carmarthen. For the rest of the nineteenth century railway pioneers entered a more chequered history. Innumerable projects failed through lack of sufficient capital or wildly inaccurate forecasts of their economic potential. Both scarcity of funds and little prospect of profitable freight traffic underlie the distinctive railway history of central and south-west Wales. The advantage of the first successful schemes of the 1840s in the north and south rested on links across the border into the Great Western and the London & North Western systems. Only the Cambrian Railway, straddling the moorlands of mid-Wales from Shropshire to Cardigan Bay, aimed to emulate the successful earlier pioneers to north and south. The Cambrian main line from Whitchurch to Machynlleth evolved out of the ambitions of four separate local companies which laid tracks through the largely fruitless terrain between the Severn and Dyfi in the years about 1860. At the same time another locally inspired scheme, the Aberystwyth & Welsh Coast Railway, obtained an Act for a coastal route that connected Aberystwyth and Pwllheli through a string of small ports and nascent holiday resorts – Aberdyfi, Tywyn, Barmouth, and Porthmadog. In 1864 the Aberystwyth & Welsh Coast was incorporated into the Cambrian system, and it became possible to travel to London from Aberystwyth by a through coach that joined the London & North Western at Whitchurch. The Cambrian Railway was the closest that Wales came to a national railway network. For the rest we find a story of undercapitalized projects and single tracks in largely unfruitful territory that frequently ended in bankruptcy even before completion.

The long coastline that extends between Carmarthen Bay and the Menai Strait is richly blessed with features attractive to holidaymaker and tourist – long sandy beaches, sheltered mountain-enfolded estuaries, and remote exposed headlands and off-lying islands rich in wildlife. In the closing years of the eighteenth century tourism became an important feature in the lives of England's upper classes. The Lake District and the Channel coast were the chief centres of attractions – Cumbria for the romantic qualities of its scenery, and the string of growing resorts along the English Channel for a salubrious summer climate and sea-bathing. Cardigan Bay and Pembrokeshire, although less accessible before the Railway Age, saw an early flourishing of tourism at a few favoured places. Tenby, a walled town, straddles a headland between two cliff-lined sand beaches – North and South Bays. By the close of George III's reign Tenby was caught up in the social activities of the English upper classes. A contemporary account describes it as 'a favourite resort in the summer for the fashionable and luxurious'.

The transformation of the port with its ponderous, stone-built merchants' houses, one of which, now the property of the National Trust, survives on Quay Hill, is largely owing to Sir William Paxton. Paxton, a wealthy landowner with estates in the Tywi valley above Carmarthen built the Public Bath House in 1805. Its Greek inscription meaning 'The sea washes away all the ills of mankind' expressed the philosophy of the summer sea-bathers whose patronage brought so many of the resorts of southern Britain into existence. Tenby is the most perfect example of the Georgian resort in Wales. Its pale colour-washed four-storeyed bay-windowed terraces, once the summer-season homes of the rich, are now hotels and bed-and-breakfasting boarding houses.

The railway came late to Tenby after the Pembroke & Tenby Railway obtained an Act to serve this growing Victorian holiday resort. The section to Pembroke opened in 1863, but it was not until the following year that permission was granted for the extension of the railway inland to join the main line of the South Wales Railway at Whitland. For three-quarters of the nineteenth century, railways played no part in the shaping of Tenby. Not until 1897, when the Pembroke & Tenby amalgamated with the Great Western, was this elegant Georgian resort whose past is still pleasingly evident in the town's topography drawn into the main network of Britain's railway communications. The encouragement of tourism was one of the main planks in the economy of the Great Western until after the 1960s when the motor car and the Beeching reconstruction of British Rail virtually brought an end to the role of railways in the tourist industry. But for the first half of this century the railway brought Tenby into contact with distant urban conurbations. 'Saturday Specials' ran from Paddington and Birmingham, and there were through coaches from Manchester on the London Midland & Scottish that travelled via Crewe, Shrewsbury, and the Central Wales Railway to join the Great Western at Llanelli. This long-distance traffic fitted well the established middle-class character of Tenby. It remained too distant from the South Wales coalfield to be deeply influenced by the day-tripper traffic, the life-blood of Barry, Penarth, and Porthcawl.

Railways came late to the shores of Cardigan Bay – not until the 1860s, when the Cambrian Railway emerged out of the projects of several local schemes. Before that time a succession of small harbours, clasped in the wide arc between the Lleyn Peninsula and Strumble Head, were busy with fishing, trade around the Irish Sea, and rare distant journeys to the Bay of Biscay and the Mediterranean. Until the arrival of the railways the main links with England were by sea. The poor state of roads or even their absence made inland communication extremely difficult. Samuel Lewis's *Topographical Dictionary of Wales*, published in 1834, draws a picture of Barmouth and its trade before the coming of the railways. He lists the import of 'corn, flour and meal, coal, limestone, American and Baltic timber, hides and groceries'. It is evident that day-to-day living had come to depend on a busy sea trade. And Lewis's list of exports from Barmouth reflects the poverty of its mountainous hinterland of poor thin soils, excessive rainfall, and uncertain harvests. They comprised timber, especially poles for collieries, copper and lead ore, and slates. Aberdyfi displays a similar pattern of trade, acting as the main outlet for the several lead mines in the hill country on the flanks of Plynlimon. The economy of the years before the impact of the railways is still recognizable in the simple cottage architecture of the narrow lanes close to the former harbours. Occasionally an even more striking relic of the past survives in the form of a shop with the inscription of Liverpool House, the chief English port of that coastal trade in life's luxuries.

In the latter half of the nineteenth century the coming of railways to Cardigan Bay brought the swift extinction of the coastal trade. Tourism came to fill the gap in the local economies, but holidaymakers were attracted to this coastline long before the Cambrian Railway gave easy access to England. Before the end of the eighteenth century the Reverend J. Evans, touring North Wales in 1798, writes, 'The class of tourist in Barmouth was genteel and in season the place was filled with Shropshire and Herefordshire beauties, this being the resort of the indolent and afflicted from the Midland part of the kingdom'.[2]

Aberystwyth, an urban foundation that dates back to Edward I's conquest of Wales, was already a flourishing resort for the leisured middle classes by the beginning of the nineteenth century. In the guidebooks of the time it appears as the Brighton of Wales and the evidence is still there for us to see in the handsome Regency terraces of

---

2   The Revd J. Evans, *A Tour through part of North Wales*, 1798.

Queen's Square and Laura Place. Of all the resorts of Cardigan Bay the coming of the railway had the most profound effect on the life and townscape of Aberystwyth. Ever since the 1830s the small ports of Cardigan Bay had been the object of railway promoters intent on developing traffic with Ireland. Brunel imagined an *Irish Mail* route that would connect Ludlow with Newtown, Barmouth, and Port Dinllaen on the Lleyn peninsula, a railway line that would follow the only existing turnpike road in this part of Wales. Another scheme aimed at trade with southern Ireland was the Manchester & Milford Haven, which planned to pass from Crewe through Oswestry, Newtown, Llanidloes, and Lampeter to the Bristol Channel – a terrain of moorland and deep valleys, thinly populated and of little economic potential. Abandoned earthworks of this enterprise are still visible among the hills above Llanidloes.

The railway promoters of the Cambrian system lacked the economic incentives of the earlier entrepreneurs in North and South Wales. The Irish trade was already firmly focused on Holyhead and the numerous local railways of the Valleys looked to good profits from the growing coal trade. The entrepreneurs of the Cambrian Railway looked towards the development of tourism. Thomas Savin, a railway engineer and contractor for the Aberystwyth & Welsh Coast, gave tourism a primary role in his enterprise, not only for the transport of holidaymakers but also the building and management of hotels in resorts. His most ambitious scheme was a huge neo-gothic hotel on the sea-front at Aberystwyth built in the 1860s. The railway into Aberystwyth opened in the summer of 1864 with two services a day to London. Savin's ideas for the creation of a holiday industry matched those of Thomas Cook, who was building up a business as travel agent in Leicester at the same time. On the opening of the Cambrian line into Aberystwyth, Savin offered a free week's board at his hotel to anyone who bought a return ticket at Euston. Thomas Savin's £80 000 hotel investment at Aberystwyth brought him to bankruptcy; the building survives as the core of the University College founded in 1872.

Thomas Savin's plans for a holiday industry were to be reshaped by other hands. His ambition – vision, perhaps – had taken the whole coastline of Cardigan Bay northwards from Aberystwyth. He planned the building of hotels at Borth and Aberdyfi, and indeed it is true to say that the railway brought a resort into existence among a wilderness of sand dunes at Borth. Until the opening of the line to Machynlleth this scatter of fishermen's cottages was inaccessible across a wide frequently flooded extent of salt-marsh.

The only other railway to reach the coastline of Cardigan Bay, built in the 1860s, branched from the Shrewsbury & Chester Railway at Ruabon to follow the Dee Valley and the shore of Lake Bala to reach Dolgellau and the Mawddach estuary. In 1862 trains were running as far as Llangollen, a place that quickly became a favourite resort for day trippers from Liverpool. By 1865 the Vale of Llangollen Railway opened to Corwen, and four more years were required for another local enterprise, the Corwen & Barmouth Railway, to Dolgellau. Just as the South Wales Railway and the Cambrian system in mid-Wales required links with powerful railway companies in England to function effectively, so the line from Ruabon to Barmouth developed a close tie with the Great Western. Already in the 1870s, the 10 a.m. from Paddington to Birkenhead connected with a daily train to reach Barmouth in the early evening.

The chief influence of the Victorian railways on the growth of holiday resorts on the Welsh coastline lay in the extension of their hinterlands to the rapidly expanding industrial conurbations in England. North Wales, as we have discussed, became closely linked to South Lancashire; the settlements of Cardigan Bay found themselves tied to Birmingham and the Black Country. The original settlement of Barmouth, with its mother church in the medieval parish of Llanaber, two miles to the north, was one of several fishing and trading ports that had long existed on the estuaries of Cardigan Bay. The lower town at Barmouth, between the harbour and the original settlement of scattered fishermen's cottages on 'the Cliff', came into existence in the late 1860s close to the railway. Tall apartment blocks were built by English business families as summer residences; today they are boarding houses for the bed and breakfast trade. By the end of the century the lower town had acquired all the trappings of a Victorian seaside resort. A new church, St John's, built of a warm red sandstone imported by railway from quarries in Cheshire, was financed by a West Midland immigrant, Mrs Dyson Perrins from Worcester. The *Cambrian News* of the 1890s reveals through its advertisements the flourishing tourist industry, brought into existence by the railway. In 1894 the Jerusalem Bazaar was advertised as 'now open at Compton House, High Street, for the sale of olive wood articles, photographs of Palestine, Mother of Pearl carved at Bethlehem' and 'patrons were assured that the stock has been brought direct from Jerusalem'.[3] Here tourism profited from the church- and chapel-going middle classes, who particularly favoured Wales and its resorts with their strict code of Sunday observance. And the activities of the week that appeared in the *Barmouth and County Advertiser* were equally sober. For sea-bathing a section of the beach was set aside for ladies and on the sands the *Royal Magnets*, dressed in straw boaters and blazers, advertised 'a grand coon night at the Assembly Rooms of plantation songs, coon melodies and dances'.

With the passing of the Railway Age in the twentieth century, holidaymaking and the resorts themselves have been deeply changed. The motor car, already of growing importance in the 1930s, came to dominate the tourist trade after the Second World War. In effect, parts of the coastline that had never been reached by the railway speculators – south from Aberystwyth, into Pembrokeshire and the Lleyn peninsula – were opened up for the holidaymaker. Caravan parks and marinas with all kinds of sea-going tourist vessels came to dominate the scene in the second half of the twentieth century. The railway no longer underpins the holiday industry of the Welsh coastline. Oddly enough, it has itself become a tourist attraction and an entertainment, in the form of the restored narrow-gauge railways built at the beginning of the Railway Age for the transport of slate to export from local harbours – the Tal-y-llyn, the Ffestiniog and the Rheidol Valley.

3   *Barmouth and County Advertiser*, 1894, quoted by Ann Rhydderch, *Barmouth*, 1977.

Chapter 17

Sir George Samuel Measom (1818–1901), and his Railway Guides

G.H. Martin

Guidebooks, like railways, are institutions highly characteristic of the nineteenth century.[1] Both had their beginnings in earlier times, but both came to flourish in the age of steam: so much so, indeed, that it might be said of either that they have never been quite the same since. They were not instantly associated with each other, though once they came together, as they notably did in George Measom's railway guides, their conjuncture seemed so natural as to be symbiotic, and even foreordained.

From the first the steam railway was as dramatic in its appearance as it proved to be in its effects. Neither the steam engine nor the dedicated track was new, but their combination was extraordinary. It called for civil engineering, especially in and after the construction of the London to Birmingham line, on a scale which matched the works of Rome, but which was accomplished in a matter of decades rather than centuries and remains impressive still. New lines spread to change the landscape of both town and country. The locomotive engines were recognized from the first as prodigies of strength, and the speeds at which they moved went beyond any previous experience or imagining. There had been nothing like them before, and no change since has matched that first astonishment of travelling faster than horse- or wind-power had ever allowed.

The pioneering steam railways, like their earliest predecessors, were conceived and developed chiefly as industrial devices, extensions of the mine and the mill which mechanized the business of distribution as the processes of manufacture had been mechanized. The amenity that they offered, however, was irresistible, and could not be confined to the transport of mere materials. The Oystermouth railway had been carrying passengers at an amble round Swansea Bay since 1807 without any startling consequences, but the directors of the Stockton & Darlington Railway explicitly took powers to carry passengers on steam trains. The resulting income was less than one-thirtieth of that raised by the carriage of coal, but on the Liverpool & Manchester Railway passenger traffic was heavy and remunerative from its start in 1830.[2]

---

1   Jack Simmons had a lively and informed taste for guidebooks, and for railway guides in particular. Measom's guides appealed to many of his interests, and I often recall his pleasure in their confident style and superabundant contents.

2   J. Simmons, *The Railways of Britain: an historical introduction* (London, 1961), pp. 3–4.

Those who rode on trains for the sake of novelty at opening galas were therefore immediately outnumbered by those who rode for a purpose. It was the scale and spontaneity of the response that was surprising, and significant. There was a remote but interesting portent in what had happened to the steamer piers built at seaside resorts, which were instantly appropriated by strollers as extensions of the promenade.[3] The railways inexorably attracted passengers, and found that they had to cater for them in unimagined ways. By doing so, they changed almost everything.

Railway trains moved people about in great numbers, swiftly, in an unexampled manner, and over increasing distances. From the beginning they were recognized as a formidable economic agent, and they also proved a social solvent, a fact variously painful to Mr Dombey and to William Wordsworth, but a remarkable source of satisfaction to Dr Arnold.[4] Their real power and influence came from their scale and complexity. With the provision and maintenance of track and bridges, of locomotives and rolling stock, of signalling devices, and of stations, they created and sustained a whole new industry in themselves, and contributed strongly to many others. One of them was publishing, which like the railways had recently been energized by the power of steam.

Printing, and with it the publication of books and newspapers, had kept pace with other industries over the previous century in refinements of technique and in mechanization.[5] It was well able to respond to growing demand, and a travelling public offered promising new opportunities to publishers. Guidebooks had a long history by the beginning of the nineteenth century, and by that time were close to their modern form.[6] They had always served both to excite and to satisfy curiosity at large, but their avowed purpose was to inform the traveller, and they became more

---

3   The promenade is an institution probably as old as civil society, but the seaside promenade in its origins was almost a measure of desperation, there being nothing else to do in the neighbourhood of the spa at Scarborough in the later seventeenth century. See *The Journeys of Celia Fiennes*, ed. C. Morris (London, 1949), p. 92, and, on piers as promenades, G.H. Martin, *The Town: a Visual History* (London, 1961), pp. 49–50. On the immediate growth of the railways, see J. Simmons, *The Railways of Britain*, pp. 6–8.

4   'I rejoice to see it, and think that feudality is gone for ever. It is so great a blessing to think that any one evil is really extinct': quoted in J.A.R. Pimlott, *The Englishman's Holiday* (London, 1947), p. 87, where there is a perceptive discussion of the influence of the railways on the uses of leisure. Wordsworth inveighed in a sonnet (10 October 1834: 'Is there no nook of English ground secure') against the intrusion of the railway from Kendal to Windermere, but his *Guide to the Lakes* was widely read, and from 1835 was published by Cornelius Nicholson of Kendal, a promoter of the line. See further notes 6 and 7, below.

5   See, e.g., S.H. Steinberg, *Five Hundred Years of Printing* (Harmondsworth, 1955), pp. 188–98.

6   There are useful surveys and references in E.S. de Beer, 'The development of the guidebook until the early nineteenth century', *Journal of the British Archaeological Association*, 3rd ser. xv (1952); E.A.L. Moir, *The Discovery of England: the English tourist 1540 to 1840* (London, 1964); and J.E. Vaughan, *The English Guide Book, c.1780–1870* (Newton Abbot, 1974). For the development of the picturesque guide in England, see P. Bicknell, *The Picturesque Scenery of the Lake District 1752–1855: a bibliographical study* (St Paul's Bibliographies, Lyminge, Kent, 1990). See also note 7, below.

numerous and more ambitious as travel became easier, or at least more commonplace. A widening readership demanded a wider scope, but all readers had some interests and perceptions in common, and the genre proved highly adaptable.

On the one hand, an inexperienced public was likely to be less exacting than the dedicated traveller, whilst on the other, particular needs and tastes could be satisfied by specialism. The most authoritative and comprehensive guides of the age were the celebrated *Handbooks* produced by the house of Murray, which had their origins in the travels of John Murray III in 1829–35, and which inspired amongst others Karl Baedeker, both by their content and their form.[7] Maintaining a high standard of accuracy and consistency, and commanding a relatively high price, they effortlessly and long lived up to the promise of their scarlet binding and elegant gilt titling.

Both at home and abroad, Murray's handbooks came naturally to base their itineraries on the railway, but there were no railways overseas when the first volumes were published, and down to the First World War and its demise the series catered primarily for those who wished to travel widely, and to see what was notable by whatsoever means were available to them. The railway companies themselves, having other things to do, were not quick to produce their own guides, and their intensive use of publicity came in the first decades of the twentieth century, when they produced some notable pictorial art in their promotional posters and at least one work of scholarship.[8] They were nevertheless soon provided with guides by others.

From the outset the impulse to inform and instruct the railway passenger was primarily and necessarily commercial, but like much else in the century it often had some missionary fervour about it. One striking example, immensely and ingenuously artful, is *A Season at Harwich* (1851). It presents an overwhelming review of the setting, history, topography, and prospects of the packet port and its neighbouring resort of Dovercourt, then under development, in the guise of a physician's address to his long-suffering patient and her family as they set out from the Shoreditch terminus of the Eastern Counties Railway in search of rest and recuperation.[9]

The first railway guides were literally descriptive, seeking to explain what and where the railway in question might be.[10] The economic and social impact of the innovation came so soon and so suddenly that it is difficult now to appreciate that for a brief time in its very earliest phase it was a curiosity, more striking than some, but

---

7   See particularly Jack Simmons's introduction to his reprint of *Murray's Handbook for Travellers in Switzerland, 1838* (Leicester, 1970).

8   M.R. James's *Abbeys on the Great Western Railway* (London, 1926). On publicity at large, see further the article on Sir Sam Fay (1856–1953), of the Great Central Railway, in the *DNB*, and 'posters' in *Oxf. Comp.*

9   W. H. Lindsey, *A Season at Harwich, with excursions by land and water, to which is added research: historical, natural, and miscellaneous* (London, 1851). Though remorselessly intent on promoting Harwich and Dovercourt, Lindsey does mention that Miss Benson and her kin made their way successfully to the railway at Shoreditch by consulting 'that very useful little book, Bolton's omnibus guide', so he had some wider interests: *A Season at Harwich*, p. 3.

10   See the examples cited under 'guidebooks' in *Oxf. Comp.* and the references there to George Ottley's *Bibliography of British Railway History* (2nd edition, London, HMSO 1983).

to those without experience of it not necessarily more portentous. It was the rapid growth of passenger traffic and the general accessibility of the train that accounted for the difference between J.S. Walker's *Accurate Description of the Liverpool and Manchester Railway* (1830), which relates the history and the works and functions of the enterprise, and the *Railway Companion. By a Tourist* (1833) and its many successors, which describe the actual amenities of railway travel. One of the earliest pictorial guides, G. Dodgson's *Illustrations of the Scenery on the Line of the Whitby and Pickering Railway in the North Eastern part of Yorkshire* (London, 1836), was perhaps the first to bring the traditions of the picturesque into play, where they long remained.

Some ambiguities lingered. The most famous and durable of railway advisers was not a topographical guide, but a set of timetables, first published under that name (*Bradshaw's Railway Time Tables and Assistant to Railway Travelling*) in 1839 and continued from 1841 as *Bradshaw's Railway Guide*. A wide variety of railway itineraries continued the established tradition of the road book, and *Fowler's Railway Traveller's Guide* (1840) offered countrywide train times, a table of distances, and 'a skeleton map of all the railways of Great Britain', one of the first of a whole gallery. The thirteen *Railway Chronicle Travelling Charts; or, Iron Road Books, for perusal of the journey... with numerous illustrations* (1846) offered similar material. *The Intelligible Railway Guide for Great Britain and Ireland*, published in parts in 1858–9, was addressed to the bemused but game traveller. So was, much later, in the heyday of the developed system, W.J. Scott's *The Best Way There: a way-book for the rail-faring fool: being a summary of competitive services for the chief towns* (1892), which appealed simultaneously to self-deprecation and acquisitiveness.

Those were all practical aids. The railway was also, has always been, a talking point which might be recruited to any end. The term 'Railway Mania' has come to be associated with those periods of the nineteenth century when it was believed that investment in such companies' stock was a high road to riches, but it could as well be applied to the practice of attaching the word 'railway' to anything that might be supposed to sell the better for it.[11] Railway song-books, containing quite ordinary songs, and such primers as *Cousin Chatterbox's Railway Alphabet* (1845), took their place in a long series of books, clothes, sticks, bags, unspillable flasks, and other commodities which on the one hand might enhance or mitigate, and on the other even merely evoke, a journey by rail. The association was in fact neither idle nor fanciful, for the railway was both the industrial base for the first great age of commercial expansion, and then something more.

The railway itself, travel by rail, and the industry and commerce which the railways fed came decisively together within the covers (paper, one shilling; stamped cloth and gilding for two shillings to three shillings and sixpence; see Plates 17.1 and 17.2) of Measom's *Railway Guides*. George Samuel Measom was born in Blackheath in 1818 and educated at a preparatory school there. He became an accomplished wood-engraver, presumably by apprenticeship, and in the 1840s was established as a designer and engraver at 29 Upper Seymour Street, Euston Square, London NW1, a

---

11   See 'manias' in *Oxf. Comp.*, with its sobering cross-references to 'speculation' and 'abandonment'.

Plate 17.1   The multi-coloured paper cover of a one-shilling Measom guide (1856)

locality since obliterated. In 1849 he published *The Bible: its elevating influence on man*, a pious album of six wood engravings depicting a drunkard reclaimed from the gin house by exhortation, his regeneration by regular study of the Bible, the benefits to his family of a settled life and attendance at church, his rescue of a former companion in sin, and finally his happy death, attended by his family and his penitent friend. The covering leaves carry a number of advertisements, one of them Measom's own. He promises to 'continue to execute all commissions in the first style of art, combined with moderation in charges and punctuality in delivery'. No less might be expected of the deviser of the album, and his subsequent works suggest that his claims were well considered.

About the time that he published *The Bible*, Measom turned to what was in a worldly sense a much more ambitious undertaking, a series of topographical and commercial railway guides. He continued to work for some time as an engraver, however, and in 1856 produced an illustrated volume of stories, *Light from the East: tales moral and instructive of oriental origin and character.* The tales are moral, instructive, and adequately exotic, but their 194 pages of text are followed by 32 pages of *The Light from the East Advertiser*, on the sound principle of never letting an opportunity go by. A year earlier Measom had published *The Crystal Palace*

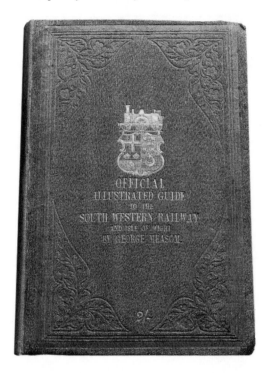

Plate 17.2   The maroon and gilt glazed cloth cover of a Measom guide (1856)

Alphabet: a guide for good children. Those were now secondary interests, though he approved very strongly of the Crystal Palace (Figure 17.1).

In 1865 he speaks of having worked for the past 15 years as 'a railway topographer', and he evidently worked purposefully. A set of 34 illustrations to Thomas Hughes's *The Stranger's Hand-book to Chester and its Environs* (1856) was probably a by-product of his travels. The earliest direct result appeared in 1852 as *The Illustrated Guide to the Great Western Railway*. The guide to the Great Western established the general nature of Measom's new project, but not its full style. That followed in 1853 with *The Official Illustrated Guide to the South-Eastern Railway and all its branches*.

'Official' was an inspired refinement: all the subsequent guides, to the Great Northern and Great Eastern Railways (1857); to the London & North Western (1859) and Lancaster & Carlisle, Edinburgh & Glasgow, and Caledonian Railways (1859); and on through the Midland Great Western and Dublin & Drogheda Railways (1866) and the Great Southern & Western Railway in Ireland (1866) were 'Official' and 'Illustrated'. Their status was established by a respectful dedication, by permission, to the chairman and directors of the company, and their nature by a repletion of illustrations. There could be no doubt about either quality.

Figure 17.1 **The Crystal Palace: the transept and lily pond** (*The Official Illustrated Guide to the London and South-Western, North and South Devon, Cornwall, and West Cornwall Railways*, Griffin & Co., London, 1864)

The series continued, with revisions and couplings of volumes, into the 1870s, when the system was approaching completion and Measom gradually turned to other concerns. From the beginning it was characterized by zest, a high degree of self-confidence, and a manifest deal of hard work and attention to detail. The title page of each volume carried the arms of the company, which were also printed or embossed on the cover, sub-titles with the names of any subsidiary lines, and under the author's name a lengthening list of his other guides. The dedication was followed by a list of the directors and principal officers of the company, sometimes including the stationmasters of the largest stations, and then by a preface. The preface included a portrait of the author, described when first making his way as the proprietor. In 1859, from 74 Charrington Street, Oakley Square, he observed that he had long held his prominent position as an authority 'notwithstanding the feeble attempts of similar rival publications to compete with his own', and attributed his success to his personal supervision of the enterprise. His portrait, latterly from a photograph by Mayall, was kept scrupulously up to date, and shows a handsome and dignified figure, self-assured, and maturing over the decades (Figure 17.2).

Measom's dignity was well enough founded; he set himself high standards and manifestly lived up to them. The serious business of the guides starts immediately after the preface, and goes on rigorously to the end. The earlier guides, besides general information on the railway system, included practical travelling hints, introduced with rustic lettering as a 'Gossip Introductory', but informed by stern common sense. The most important considerations in starting a railway journey are to decide whither one is going, by what train and when, and whether the journey will entail a change of carriages, and if so, where. In subsequent volumes and editions, as the novelty of the experience and with it some of the apprehension wore off, those admonitions gave place to further information about the companies and the system (which might not be entirely out of place today). In the 1860s there was usually, besides other statistical material, prefatory information from the census and local sources. The guide proper then began with an account of the terminus, the incidents of the route, and descriptions of the places through which the line ran or to which, like the Lakeland fells, it gave access.

That much was commonplace, though there is evidence enough of the grounds of Measom's claim that he had travelled over the entire system. The descriptions of the towns, however, go closely and consistently into their manufactures and trade, in the text as well as in the advertisements which the guides carried as a matter of course. There are also descriptions of public undertakings, factories, and shops as places to be visited and admired as susceptible enthusiasts might admire the picturesque and the sublime. The industrial and commercial encomia were plainly contrived as a source of revenue, but that would only have made them more tolerable and not any easier to compose and assemble. There is a good deal of thought and action behind the enterprise, though as the series takes shape they become less apparent simply through familiarity. The fact remains that every town had to be scrutinized, its principal institutions and traders identified and approached, and some thousands of bargains struck. There were no readily accessible and consistent sources of local information available; public libraries were rare and chambers of commerce still emergent. The results gathered had then to be presented in print. The typography is undistinguished

THE AUTHOR.
(*From a Photograph by Mayall.*)

Figure 17.2   Mayall's portrait of Measom (*Guide to the London and South-Western*, 1864)

but serviceable, and the information closely packed, but the text is always legible and the standard of proof-reading is high.

The apparatus of research is quite often displayed, and the display is itself an advertisement for the system. Many of the illustrations of Great Yarmouth in the guide to the Great Eastern Railway (1865) are acknowledged as engraved from photographs by G. Nall, and the first entry under 'The commercial aspect of Yarmouth', an appendix which is a regular feature of the descriptions of the more substantial towns, is 'The premises of our agent, Mr George Nall, bookseller, publisher, stationer and newsagent'. The usefulness of Nall's guide to Yarmouth has already been acknowledged in the text, but visitors might well wish to borrow books as well as to buy them, and Nall's lending library is said to be regularly stocked from London by Mudie's. The accompanying illustration shows Nall's neat shopfront. Its windows are packed with books displayed behind rolled plate glass, itself a mark of technical and commercial progress, and beside it there is a view of Nall's steam-printing works. The two engravings are set under a rustic double arch, and may have been copied from a pair of stereoscopic views and mounted together. There is a slightly different picture of the premises in the general gazetteer of advertisements at the end of the guide.

As Measom's agent Nall would readily secure his place in the forefront of Yarmouth's commercial undertakings. The interests of the publisher and the bookseller were evenly balanced and, as Measom observed, beyond the herring and mackerel fisheries local trade was chiefly concerned with the services offered to summer visitors, amongst which guides to the use of leisure would be prominent. Nall's shop could sell Measom's books, and Measom's pages advertised the whole range of Nall's wares, including scenic photographs, and cartes-de-visite with albums in which to mount them, the picture postcard lying still some decades in the future. The only other substantial business particularly noticed in the town was a wine-and-spirits merchant's. Of 137 pages of print devoted to Yarmouth, Nall has two pages to himself, frequent references to his guidebooks, and credits for half-a-dozen photographs used to illustrate the text.

In other places and with other businesses, however, the relationship between the advertiser and the publisher would be different. Advertising is an ancient craft, and advertising in newspapers, the most widespread and influential kind of the day, was well developed. Many of the intricacies of the sport had been worked out in the provincial press of the eighteenth century, but intensive industrial and commercial development had brought new opportunities and pressures to bear.[12] Measom needed to fill space in a convincing and interesting fashion. He had to subsidize his printing and other costs, but he also wished to sustain and enhance his reputation both with his readers and his clients. His clients wanted beneficial publicity, and to maintain their ground against their competitors with some kind of return in sales. The question for them would be how the quality of Measom's readership complemented its size.

The contents of the guides show that Measom was as persuasive as his appearance and the general tone of his publicity might suggest. His generous use of illustrations

---

12   See, for example, C.Y. Ferdinand, 'Serial conversation: the dialogue of the eighteenth-century country newspaper', in *Compendious Conversations: the method of dialogue in the early Enlightenment* (Frankfurt am Main, 1992), pp. 116–28.

was probably a strong selling point. He seems to have had a general rule not to have more than two openings of continuous print without an illustration or an ornament of some kind, or both.

His care in advertising himself may also have been infectious, for he consistently practised what he preached. The only connection that *Light from the East* has with railways is that Measom identifies himself on the title page as the author of his official and illustrated guides. The stories themselves are set in the age of the magic carpet, but its readers were likely to take the train. *The Light from the East Advertiser* contains 'A note by the author of the Official Illustrated Railway Guides: Smoking is not allowed (see rule iv of the by-laws) – not even the cigars, the fine foreign cigars, sold by Benson of 133 Oxford Street. It was not to be supposed that an exception could be made even in favour of Mr Benson'. He might almost have added 'or of the author of the Official Illustrated Railway Guides', but did not, perhaps to show that he knew when to stop. The spread of shot is commendable, and there are many other examples. Their zest is evident, and manufacturers and merchants seem to have responded to it.

The commercial panegyrics are variously fitted into the guides' itineraries, in and out of towns, or into appendices (Figure 17.3). In either instance there is much enthusiastic attention to detail. The survey in the *Guide to the London and South-Western Railway* of Price's Candle Company's works at Belmont by Vauxhall and Battersea includes packaging as well as the chemical processes involved. 'The manufacture of these boxes is not the least interesting part of the manufactory': it combined broad wood-shavings with a straw paper to avoid the duties which made ordinary paper-board expensive. Firms like Price's would have their own technical literature and illustrations to offer, as would some public utilities.

There is a plan of the works of the metropolitan sewer's great southern outfall at Crossness, by Lesnes, which was included in the guide to the Great Eastern Railway in 1865. The text notes that the best view of Woolwich and its dockyard, which are on the south bank of the Thames, is to be had from North Woolwich, which is in Essex. It goes on to describe the Arsenal, Lesnes, and the outfall, which are all in Kent. The reason for the excursion is that work on the northern outfall was still in progress, and its low level sewer, which required more elaborate engineering, was some years from completion. In the meantime the works on the south bank had recently been opened by the Prince of Wales, and it would have been a pity to confine a great engineering and sanitary enterprise to the pages of the guide to the South Eastern Railway when it could at least be seen from North Woolwich. The fact that the plan was available may have played its part, but public works were public works.

The editor of *Black's Guide to the County of Surrey* (1887) was not convinced that the Woking Necropolis would be attractive to tourists, but Measom had no such doubts, and allowed it six pages in the guide to the London & South Western Railway. It was not only that he saw it as 'the final terminus... at the close of life's swift railway journey', nor that the plantations and the surrounding views were beautiful. It

**Figure 17.3   Horne the house-painter, at home and away. London was the
universal starting-point, and London firms were regularly
advertised in the Guides** (*Guide to the London and South-Western*,
1864)

commended itself as a remarkable enterprise, distinguished amongst other things by the efficiency and delicacy of the arrangements for its funeral trains.[13]

In more mundane business catalogues were an obvious source of copy and illustrations. Croydon in 1858 was still a country market town, but its population had risen from 6 000 to 60 000 over the half-century, and the railway had recently tied it firmly to London. It already had a commuting population, and the great wave of suburban building that eventually overwhelmed north-eastern Surrey was just beginning. The *Guide to the South-Eastern Railway* makes much of the parish church of St John, Croydon, the largest in Surrey. It was gutted by fire in 1867, and the guide would have done a further service to knowledge if it had included views of the interior, with its archbishops' tombs. St John's was still a prominent feature of the town, but the centre of Croydon was changing, and additional churches were already rising. There were some light industries locally, but Croydon was becoming a place where money was spent rather than made. Though it had been a posting town, and kept livery stables, the carriage works which were there served local residents rather than transients. One of the firms, C. Lenny & Co., evidently met Measom's needs with a catalogue. The visitor to Croydon was accordingly invited to inspect Lenny's depôt, and consider the various advantages of the Holford coach, the Paratempo, which provided for all weathers with a patent folding roof, and the ingenuity of the Cruphabena, a patented step which unfolded when the door opened. The Sardinian, named after the king of Sardinia, was a popular and elegant vehicle, and prices started as low as £18. It is still difficult to leave the page without buying something.

It would have been impossible to cover the enormously wider range of Birmingham's manufactures in that fashion, but those taken up in the guide to the North-Western Railway (1859) are presented with the same – or rather with similar and comparable – flourishes. In the premises of J. Hardman & Co, metal workers, jewellers, and glass stainers, the reader would readily recall the splendours of the firm's medieval court in the Crystal Palace in 1851, and its subsequent displays in the exhibitions in Dublin in 1853 and Paris in 1855. The opulence and elaboration of the goods described, of the descriptions themselves, and of the illustrations might overwhelm all but the most frenziedly acquisitive, but there were still other objects of wonder in Birmingham. Besides small arms, engines, clocks, and watches there were pen nibs. That 'the steel pen trade is one of the most interesting of those practised in the town' is a claim substantiated by an account of the organization of the work, with its minute division of labour, the huge scale of production, using many tons of steel every week, and the cheapness of the individual nib. All well considered, the trade could take its place amongst the most imposing of the town's enterprises. Its presence in the text is also a witness to Measom's belief that industry and commerce were interesting and valuable things in themselves. Pen nibs were in universal use, but the demand for them was comparatively inelastic. A description of their manufacture was not likely to produce a surge in sales, but they were a marvel of Birmingham's skills and ingenuity, and deserved notice.

---

13 See *Black's Guide*, p. 122; Measom, pp. 187–92. *Black's* was robust enough on other occasions: compare (p. 36), at Virginia Water, the 'lunatic asylum for middle-class patients... built and started on its career of usefulness by Mr Holloway of Oxford Street. The building itself is stately, though its appearance is marred by the number of its small windows'.

Birmingham did offer a remarkable concentration of business and industries, and in that sense was only comparable to London. In general the further the guide moves from London the more it attends to the ordinary concerns of the sightseeing traveller, though it always responds to impressive undertakings and willing advertisers. Antiquities and scenery were a valuable standby (Figures 17.4 and 17.5). All the guides offer historical sketches and anecdotes, and both ancient and modern buildings are noted. Measom approved and commended medieval churches, though he had little use for monks, whose 'mistaken piety led its deluded followers to waste their lives in passive indolence, shut up from the opportunities of practising that active virtue which is most pleasing to the eyes of a Deity who formed man for social happiness'. Their houses nevertheless offered 'the finest remains of antiquity our country boasts'.[14]

The fells and waters of the Lake District provided copy and views for the guide to the Lancaster & Carlisle Railway (1859), as did the scenery of Scotland to its successors. However, the full tide of tourism in Scotland came rather later, and the guides to the Irish railways, which are the latest in the series, are together more substantial than the Scottish guide. They are not exclusively touristic, and in the guide to the Midland Great Western and Dublin & Drogheda Railways (1867) the 50 pages devoted to the city of Dublin are followed by 86 on the Commercial Aspect of Dublin. The encomium of Kennan & Son's lawnmowing machines includes the reassuring observation that the tilt-action delivery, 'strongly recommended for every mower above 18" wide', has entirely overcome the prejudice with which some had at first received it. At the end of the volume, however, there is an appendix of entries on London firms, one of which, a glass and pottery showroom, introduces an account and map of the Staffordshire Potteries. The Potteries could be related to the line to Holyhead, and the information on London shops is offered for the convenience of customers everywhere, but it may also be that its copy was more readily come by, and would have a guaranteed appeal.

From the first Measom saw the railways as aids to travel abroad, and the guide to the South Eastern Railway (1859) had as a supplement a guide (an official illustrated guide) to the Northern Railway of France 'with six days in Paris'. It recommended creosote for seasickness, and after notes on the amenities of Boulogne, where Mr Seal kept an English library, and Amiens, set out a practical and quite well balanced scheme for a week in Paris. Besides recommending the usual cynosures, it observes almost inevitably that 'the water works in the Bois de Boulogne are highly interesting'. The guide to the Great Eastern Railway (1865) carries a map showing connections through Harwich and Lowestoft to Naples, Vienna, and Danzig. It also has an excursus on the principal towns of the Netherlands and on Antwerp, managing the transition from Amsterdam to the Scheldt by supposing 'the traveller to have started once more from Harwich, and to have landed at Antwerp'. There are also notes on the Rhine railways and steamer services.

Measom continued to publish engravings into the 1870s, but he turned to a variety of other, and especially charitable works in the last decades of his life. He was twice married, the second time in 1867, to Charlotte Simpson of Richmond. By that time

---

14   *Guide to the London & South-Western Railway* (1864), p. 378, in an appreciative discussion of the ruins of Netley Abbey.

NORTH TRANSEPT, WELLS CATHEDRAL.

Figure 17.4   An interior view of Wells Cathedral (*Guide to the London and South-Western*, 1864)

DAWLISH AND SEA WALL.

Figure 17.5   Man and nature: the sea wall and railway at Dawlish (*Guide to the London and South-Western*, 1864)

he had moved from central London to St Margaret's Lodge, Twickenham, which Kelly's *Middlesex* described in 1874 as 'a handsome modern mansion with a picture gallery and large grounds', but his chief interests were still in London. They were not narrowly focused. He was a pertinacious champion of the Cancer Hospital, now the Atkinson Morley, which first made its way against strong prejudices. He also promoted the foundation for lost and stray dogs, which subsequently became the Battersea Dogs' Home, and he served as president and treasurer of the Royal Society for the Prevention of Cruelty to Animals. He became a Justice of the Peace, and was knighted for his public services in 1890. He lived into the twentieth century, dying at Twickenham on 1 March 1901.[15] In his entry in *Who's Who* he gave a summary account of his more recent works, and under Recreations remarked 'Never had time'. He was probably not exaggerating.

---

15   Since this volume was submitted for press Mr J.D. Bennett has kindly shown me a paper of his own on George Measom which has now appeared in *BackTrack*, xv, no. 7, July 2001, pp. 379–81.

SECTION IV
HERITAGE AND HISTORY

The North Eastern Railway Museum, York – 'the germ of a truly National Railway Museum'[1]

Dieter Hopkin

It is remarkable that the history of Britain's first railway museum is so poorly documented in historical museological and specialist railway-heritage publications. This is despite the facts that the origins of the National Railway Museum (NRM) can be traced in the North Eastern Railway (NER) Museum at York and that the latter's national (and an international) reputation made it Britain's *de facto* national railway museum from the 1930s until the 1960s. Jack Simmons, in his introduction to the NRM book *Dandy-Cart to Diesel*, gives a brief overview of the museum's predecessor institutions, but little else has been written on the subject.[2] This chapter draws on the previously unstudied records held by the NRM – principally the minutes of the various committees successively responsible for the museum – and partial files from the British Transport Commission's museums. Various NER and London & North Eastern Railway (LNER) papers held at the Public Record Office, Kew, have also been examined.

The origins of the York Railway Museum can be traced to the strong sense of regional pride in the north-east of England that recognized its role as the birthplace of the railway (identified with the opening of the Stockton & Darlington Railway [S&DR] in 1825). From the middle of the nineteenth century the railway companies began celebrating the early history of their industry. The most notable was the NER, which saw itself as successor to the traditions of the S&DR and was steeped in the memories of pioneers like Stephenson and Hackworth. The company's consciousness of its heritage is shown by its commitment to celebrating the fiftieth anniversary of the opening of the S&DR in the Railway Jubilee of 1875.[3] The centrepiece of the

---

1   Sir Ronald Matthews, Chairman of the London & North Eastern Railway, on the occasion of the reopening of the Railway Museum, York, 18 July 1947: 'The Railway Museum', *LNER Magazine*, 37/9, September 1947, pp. 202, 204–5.

2   J. Simmons, *Dandy-Cart to Diesel, The National Railway Museum* (HMSO, London, 1981).

3   See Dieter W. Hopkin, 'Railways, Museums and Enthusiasts', unpublished MA dissertation, University of Leicester, 1987.

celebration was an exhibition of 27 locomotives, ranging from the S&DR's first engine, Stephenson's *Locomotion No.1*, to contemporary examples of NER rolling stock. While the NER showed an acute awareness of its heritage, it is clear that the company was also conscious of the publicity value of the event.

Before the Railway Jubilee, *Locomotion No. 1* had been mounted on a pedestal outside Darlington North Road Station since 1857. It was therefore the first locomotive to be preserved and publicly exhibited. Other examples of this sort of historical presentation by the NER are also documented. These machines were monuments to the engineers who created them and memorials to the pioneering days of the railways with which they were associated. They were, however, retained through local initiatives rather than any overall policy. By 1907 the NER had 'several old locomotives and vehicles at Darlington, Newcastle and the York shops' which were viewed as potential exhibits for a railway museum.[4] Suggestions for forming a railway museum for Britain had been raised in the professional and railway-enthusiast press in the 1890s and the late 1900s, but there was little official interest either from government or from the multitude of railway companies.[5] As a result, the retention of historical material was left to local initiatives and the interest of local managers with antiquarian inclinations.

The NER had on its staff two individuals who saw it as part of their work for the company to gather together historically interesting items. H.J. Rudgard, Assistant Engineer, and J.B. Harper, Assistant Superintendent, had duties that took 'them into every hole and corner of the old North Eastern Railway'.[6] Rudgard was responsible for a collection of railway civil engineering and trackwork components that was 'over a long period of years' gathered together in 'a spare shed and as the collection grew one heard it spoken of, but it was not on public view'. Harper's interests lay in smaller items, ranging from tickets to staves, that were shown off to a few visitors in his office at York. According to one account, he had been collecting material[7] as far back as 1880 'in the hope of a museum being formed'.[8] Harper and Rudgard were to play a more formal role in developing the preservation of historical material with the formation of the NER Museum Committee.

On 29 March 1922, Robert Bell, Assistant General Manager of the NER, convened a meeting in the company's offices at York. In opening the meeting he referred to the 'efforts that had been made departmentally in the past to preserve interesting relics and said that there was a feeling that the time had arrived when some definite all-line scheme should be embarked upon'. Furthermore, he said that it was

---

4   *Newcastle Daily Journal*, 16 October 1907, in National Railway Museum, British Transport Commission Museum cuttings file XX70.

5   Charles Rous-Marten, 'British Locomotive Practice and Performance', *Railway Magazine*, 22, 1908, pp. 186–9 (p. 188).

6   NRM, BTC Museum file XX72, E.M. Bywell, 'The Railway Museum, York', unpublished and undated typescript (1928?).

7   This may even have included material previously gathered together for the 1875 exhibition.

8   'York Railway Museum', *LNER Magazine*, 17/7, July 1927, pp. 279–80 (p. 280).

especially desirable that this should be done while the N.E.R. still preserved its individuality. The Railway Centenary would take place in about three years and would no doubt be celebrated in the neighbourhood of Stockton and Darlington, and those responsible for the celebrations would look to us for relics of the pioneer line which was now part of the N.E.R. system.[9]

Bell made suggestions for immediate actions to get a museum project under way: firstly, that a basement room in the NER Headquarters Building in York 'be allocated for the preservation of plans books and other small articles'; and secondly, the committee was to seek a suitable building for the storage of 'such things as signal frames, turn-tables, and the larger exhibits connected with the permanent way'. The locomotives preserved at Newcastle and Darlington were to remain there for the time being, but a complete catalogue was also to be prepared of objects of historical interest on the line. He indicated that he thought that the General Manager would recommend 'that the Company should face any reasonable expenses that might be incurred both in preparing housing accommodation and renovating exhibits'. While it is clear that this initiative had official sanction at the highest level of officers of the NER, there are apparently no references to the setting up of this committee in the surviving minutes of the company's main board or principal management committees.

The Museum Committee was chaired by J.B. Harper and its six members were senior managers from the principal operating departments of the company, including Rudgard. E.M. Bywell, from the General Manager's office and Editor of the NER's house magazine, was appointed as Secretary and Curator.[10] As one of its first actions the new committee published a notice in the *North Eastern Railway Magazine* announcing that 'A COMMITTEE has been formed for cataloguing and bringing together where practicable, books, plans and objects of historical interest connected with the North Eastern Railway' and going on to request that 'members of staff will forward to the Secretary or make known to the member of the Committee representing their department, particulars of any relic which they consider worthy of preservation'.[11] The committee also arranged an inspection of the Old Plumbers' Shop behind York passenger station and the former machine-shops of the Queen Street Works at York to see whether these might provide suitable accommodation for the larger items.

By the date of the second meeting of the North Eastern Railway Museum Committee, 11 April 1922, arrangements were being made for fitting out 'Room B'

---

9   NRM, North Eastern Railway Museum Committee Minutes (NERMC), 29 March 1922. The NER was to be absorbed into the new London & North Eastern Railway under the terms of the Railways Act, 1921.

10   The involvement of Bywell is significant in documenting the early development of the museum collection as he ensured that brief articles on new acquisitions regularly appeared in the LNER Magazine. It is interesting to note that the acquisition of some items that escaped formal notice in the committee minutes is recorded in Bywell's short pieces. Bywell's involvement with the railway museum, which continued well after his retirement from the railway, was to last over a quarter of a century and he is undoubtedly one of the most important characters throughout its history.

11   *NER Magazine*, 12/136, 1922, p. 100.

in the basement of the Headquarters Offices in York to display the collection. Similarly, the old machine-shops at the former York & North Midland Railway's Queen Street Works were identified as suitable for the large exhibits and the company engineer was asked to prepare estimates for the necessary improvement works.[12] The minutes also list as an appendix the exhibits that had been accepted for the museum by the committee, including an early Edmundson dating-press, which formally became Exhibit No. 1, and several international exhibition medals. These items had previously been kept in the General Manager's Office. Bywell ensured that these first acquisitions were also recorded in the staff magazine.[13] During the remainder of 1922 the committee made arrangements for the conversion of the allocated accommodation. In 'Room B' cases, cupboards, and picture-rails were fitted for the smaller items, and at Queen Street the building was to be divided into three sections to display signalling, permanent way, engines and rolling stock, and so on. It required refurbishing for museum use, and the estimated costs of the basic works at Queen Street were £360.[14] The work was, however, delayed, as it was of low operational priority and was considered a 'wet day job' for the company's tradesmen.[15] The building was not ready to receive exhibits until early 1923.[16] In order to differentiate between the two buildings that housed the collection, the committee agreed to refer to the Queen Street building as 'Section A' and 'Room B' as 'Section B'.[17]

At the same time the committee received and recorded various 'finds' of items from the company. It is notable that many groups of items listed in the appendices to the minutes are from NER senior managers – including J.B. Harper – who appear to have had their own small local collections of historical items that they transferred to the museum. These included everything from old share-certificate printing-plates through tickets and other printed ephemera to general railway books. All old items associated with the former NER were considered, and there was no clear division between archives or muniments and other sorts of relics, as is shown by the acquisition of the archive from Shildon Works in 1923.

Having secured suitable accommodation, the committee also considered in 1923 large items like the former S&DR Weatherhill stationary winding-engine and what it might cost to install them in the museum. Similarly, the committee planned to acquire examples from two NER locomotive classes.[18] Searches for vehicles were not restricted to within the company, and approaches were made to the Forcett Limestone Company to acquire an early S&DR locomotive-tender and coach in exchange for suitable replacement items.[19] The first two locomotives to be acquired were NER examples from the 901 (Fletcher) and 1001 classes, 2-4-0 No. 910 of 1875 and 0-6-0 No. 1275 of 1874. The minutes show that the locomotives had scrap values to the company of £252 and £245 respectively; these were presumably written off.

12   NRM, NERMC, 11 April 1922, Minute 5.
13   'NER Museum', *NER Magazine*, 12/136, 1922, p. 215.
14   NRM, NERMC, 19 May 1922, Minute 9.
15   Ibid., 14 September 1922, Minute 22.
16   Ibid., 24 January 1923, Minute 31.
17   Ibid., 7 March 1923, Minute 42.
18   Ibid., 7 May 1923, Minute 53.
19   Ibid., 21 March 1923, Minute 49; 21 February 1924, Minute 87; 5 June 1924, Minute 90.

It was clear that the locomotives had to be prepared for presentation to the public, and 'the Committee ... suggested that the engines should be restored to their original appearance, even though partly restored in dummy'.[20] Restoration was to be undertaken with the aid of the original engineering drawings used when the locomotives were built, and then the locomotives were to be stored at Queen Street 'until at any rate, the Centenary of 1925'.[21] However, local managers did not always cooperate with the Museum Committee by transferring to it historic items in their care. In 1923 there seems to have been considerable friction between the committee and the Chief Mechanical Engineer's Department over the S&DR carriage (No. 31) housed in York carriage works. As a result, the carriage remained at the works rather than being transferred to the museum.[22] Despite such local difficulties, arrangements for developing the NER museum at its two sites in York were generally going well.

It seemed logical that the future museum would be near the LNER's headquarters. There was, however, a rival claim. Darlington town council put forward a proposal to set up a new railway museum in association with the proposed Railway Centenary. The proposal was considered by a special committee of LNER directors formed in 1924 to oversee the company's arrangements for the Railway Centenary celebrations on the route of the Stockton & Darlington Railway.[23] Darlington's proposal was that the council would provide half the cost (estimated at £20–25 000) of a building and the running costs on condition that the company would 'be willing to make a contribution and to present exhibits'. Senior officers in the LNER clearly gave serious consideration to this alternative to their embryonic York museum.[24] The scheme appears to have been favoured by the Chief General Manager, and a heads of agreement between the LNER and the council was drawn up.[25] The NER Museum Committee, however, vigorously opposed the proposal to hand over the collection to a third party. It minuted that it felt 'very strongly that the relics should not pass out of the Railway Company's possession permanently' and further felt that 'it would be a breach of faith with the donors of many of the Museum exhibits'.[26] However, in the face of apparent support from senior management within the railway, the committee felt forced to back down at its meeting on 11 August 1924 and to accept terms for a possible transfer to Darlington.[27] The committee was not at this time apparently aware that the directors had, for their own reasons, already ended any negotiations with Darlington several days before and decided 'that the company is not prepared to take part in a scheme for providing a permanent Museum'.[28] The Darlington museum project appears to have died in 1925; it was not to be resurrected until the 1970s.

20 Ibid., 30 October 1923, Minute 65.
21 Ibid., 15 January 1924, Minute 81.
22 Ibid., 23 November 1923, Minute 73.
23 PRO, LNER Special Committee of Directors to consider the Celebration of the Stockton & Darlington Railway Centenary, 1925, RAIL 390/14.
24 Ibid., 2 May 1924, Minute 1.
25 NRM, NERMC, 11 August 1924. Copy appended.
26 Ibid., 28 July 1924, Minute 96.
27 Ibid., 11 August 1924, Minute 98.
28 PRO RAIL 390/14, 1 August 1924, Minute 7.

From August 1924 to February 1925 the North Eastern Museum Committee did not meet, since several of its members served on the local centenary committees that reported to the special directors' committee.[29] The museum collection featured in the Railway Centenary exhibitions at Faverdale, and arrangements had to be made for items to be transported to and from York. The public interest in railway history, in part stimulated by the Railway Centenary, led to other loans from the museum collection to the British Empire Exhibition, Wembley; the Science Museum, South Kensington; and Norwich Castle Museum. Some of the material went direct from the centenary exhibition to the loan exhibition and returned to York only later.[30] The Railway Centenary also brought to the Museum Committee's attention further important historic railway material preserved by other railway companies and elsewhere. Its members examined the centenary exhibition catalogue in detail as it was seen as a useful tool for identifying further items for the museum collection.[31]

When NER locomotives Nos. 910 and 1275 returned to Queen Street following the Railway Centenary, they were joined by the Great Northern Railway (GNR) 4-2-2 No. 1, which had also been exhibited. A further arrival was the Stephenson 0-4-0 Hetton Colliery locomotive of 1822 which had also appeared in the centenary exhibition. It was transferred from the Lambton, Hetton, and Joicey Colliery.[32] Other offers of exhibits in 1926 came from outside the north-east, including, from the Southern Railway, the Bodmin & Wadebridge Railway third-class open coach from the 1840s, and from the Stephenson Locomotive Society, Britain's first independently preserved locomotive, the former London Brighton & South Coast Railway 0-4-2 *Gladstone*, which the Society had purchased and restored.[33]

The inclusion of these vehicles is a clear indication of the changing role and increasing importance of the embryonic York museum. Debate amongst managers and Museum Committee members in 1925 over what the museum should be called led to the adoption of the name 'The Railway Museum, York'.[34] The committee saw that it had a role to play in offering a home to historic railway material under threat, even from companies outside the LNER. This is evidenced by their interest in the City & South London Railway coach of 1890 which had been exhibited at the Railway Centenary and which was apparently threatened with destruction in 1925.[35] The collection of smaller items also attracted material from throughout the country, including from private collectors. The acquisition of the large collection of early written and printed material from the private collector Isaac Briggs not only significantly increased the scope of the collection but also, because of its size, meant

---

29   NRM, NERMC, 26 February 1926, Minute 99.
30   NRM, The Railway Museum, York Committee (RMYC), 1 March 1926, Minute 118.
31   Ibid., 29 October 1925, Minute 112.
32   Ibid., 29 October 1925, Minute 116.
33   For further background see Dieter W. Hopkin, 'Railway Preservation in the 1920s and 1930s', in *Perspectives on Railway History and Interpretation*, ed. Neil Cossons (NRM, York, 1992), pp. 88–99.
34   NRM, RMYC, 29 October 1925, Minute 110.
35   NRM, NERMC, 30 September 1925, Minute 108. (This acquisition was pursued over almost a decade and the coach [No. 10 of 1890] was finally accepted into the collection in 1936.)

**Plate 18.1   The Small Exhibits Section of the Museum, with E.M. Bywell
(Curator), J. Dixon (Museum Sub-Committee) and Mr Lister
(Museum Custodian), 18 July 1947** (National Railway Museum –
37/00)

that an alternative to 'Room B' in the Headquarters Offices had to be found. An
option for putting all the collection in the Queen Street building was explored, but this
was seen as too costly.[36] Alternative rooms were, however, identified in the west side
of the Old Station Building, and these were refurbished to display this important
collection by June 1928.[37]

With the accumulation of what was becoming a significant collection and the
considerable effort and expense of preparing and housing it, the committee members
addressed the issue of opening this company collection to the public in April 1927.
There is, unfortunately, no indication in the committee minutes of when the two
sections of the museum were generally opened to the public. The first formal visit to
the museum collection was by a party of LNER directors on 24 February 1923; they
'expressed pleasure and surprise at the excellence of the collection'.[38] The Museum

---

36   NRM, RMYC, 21 February 1928, Minute 139.
37   Ibid., 5 June 1928, Minute 148.
38   NRM, NERMC, 7 March 1923, Minute 41.

Committee also visited from time to time as part of its work. The first visitors' book entry was made on 28 April 1927 by a visitor from the Hungarian State Railways. There are similar entries from overseas visitors for the rest of 1927 and early 1928 which indicate that the museum was part of the itinerary of their official visits to the LNER in York.[39]

What appears to be the first significant public opening of the Railway Museum displays was during York Civic Week, 3–9 June 1928, when it was proposed to open 'to the public within the hours of 10.0am and 5.0pm'.[40] Later in the month, on Sunday 17 June 1928, a special railway exhibition was held in York in aid of charities. This event included not only opening the two museum buildings but also a display of modern locomotives and rolling stock at York station. The day was apparently a remarkable success, and a note, by E.M. Bywell, in the museum visitors' book records that the visitors to the Queen Street Museum totalled 17 648 and to the Old Station Museum 10 691. The note further records that the maximum number of visitors per hour was 3 489 and that 'the attendances at both Museums may be regarded as about the maximum attendance in a single day as there were very large queues during the whole of the time: both Museums had to be kept open until long after the official closing hour'.[41]

Surprisingly, the Museum Committee minutes are silent on this popular event. Indeed, the committee does not appear to have met between June 1928 and September 1929. However, the Railway Museum was clearly becoming popular, and arrangements were made for staffing it. The first appointment of 'a suitable man to assist in keeping the Museum in order and to act as a Guide to visitors' was apparently made on a temporary basis from June to August 1928.[42] In the following year, because of the increasing popularity of the museum, especially for school parties, arrangements were made for appointing a Museum Attendant as a 'Light Work Post' on a more regular basis.[43] Opening seems to have been kept on a seasonal basis, with an attendant employed from the spring to the autumn, and opening on Wednesdays and Saturdays only for the rest of the year. In a minute referring to the employment of former LNER Police Sergeant Horn in 1930 the committee noted that his appointment had 'allowed the Museum doors being kept open each day' and that 'the number of visitors increased enormously'.[44] Horn appears to have been the mainstay of the museum staff during the 1930s. Entry to the museum was free and by 1931 a guidebook and postcards of key exhibits were available for visitors to purchase.

During the 1930s the Museum Committee made progress in building up the Railway Museum's collections. The minutes record the acquisition of small donations from the LNER's operating area and beyond. Major acquisitions put pressure on the available space at Queen Street and the layout of the locomotives and other large objects was regularly changed to accommodate new items, especially locomotives. The Great Western Railway offered the record holder *City of Truro* in 1930, the NER

39   NRM, Railway Museum, York, Visitors' Book 1.
40   NRM, RMYC, 21 February 1928, Minute 142.
41   NRM, Railway Museum, York, Visitors' Book 1.
42   NRM, RMYC, 21 February 1928, Minute 141.
43   Ibid., 20 September 1929, Minute 155.
44   Ibid., 22 September 1930, Minute 165.

locomotive *Aerolite* came from the LNER, the London Midland & Scottish Railway offered *Columbine* in June 1934, *Agenoria* came on loan from the Science Museum in 1936, and the pioneer GNR Atlantic, *Henry Oakley*, in 1938.[45] Not all offered vehicles were accepted: the NER locomotive No. 991 was declined in 1926, as was No. 1622 in 1934.[46] A Jersey Eastern Railway 2-4-0T *St Brelade* was accepted by the committee in 1937 but never arrived.[47] The committee was also active in the acquisition of other large objects, often with the cooperation of LNER senior managers like the Chief Mechanical Engineer, Nigel Gresley, who, in 1931, was keen to save a large stationary steam engine from King's Cross shed, London.[48] The museum also re-erected the world's first iron railway bridge, from Gaunless on the S&DR. This had lain disassembled at Darlington for many years.[49]

While the acquisition of large objects like vehicles is better documented, the committee and its sub-committee minutes record a regular stream of donations from individuals and the continued accumulation of local collections such as that of the District Engineer at Norwich.[50] The minutes regularly have long appendices of new additions to the museum collection. It was clear that the museum was attracting material from all over the country and was becoming a *de facto* national railway museum. To rationalize this, the committee decided 'that the Museum Committee should consider the acceptance of any relic of exceptional railway interest from whatever source it originates, but that their policy should be passive rather than of an active character'.[51]

With the growth of the large exhibits collection and the scarcity of space in the Queen Street building the Museum Committee had to consider taking hard decisions about what might have to be disposed of. As early as 1936 thought was given to the possible destruction of the Forcett coach, the S&DR tender, the four-wheeled coach-body, and one of the railway fire engines.[52] It is clear that at various periods some items had to be left outside due to shortage of accommodation. Similarly, lack of space in the Small Exhibits section was a consideration when any offers of large collections were made. This was a recurrent issue for the Museum Committee members throughout the museum's history. The displays became increasingly cramped, but they were not entirely static. Major re-arrangements were made to allow large acquisitions to be displayed or when locomotives were moved out for special publicity events. GNR No. 1 was away from the museum for much of 1938 and used to haul special commemorative trains associated with the introduction of new

---

45 E.M. Bywell, 'The Railway Museum, York: Important Additions', *LNER Magazine*, 24/7, July 1934, pp. 364–5; NRM, RMYC, 22 September 1930, Minute 163, Minute 179; Railway Museum Sub-Committee (RMYSC), 19 February 1936 , Minute 10; 20 October 1936, Minute 3.
46 NRM, RMYC, 1 March 1926, Minute 114; 9 May 1934, Minute 179.
47 NRM, RMYSC, 29 July 1937, Minute 2.
48 NRM, RMYC, 2 February 1931, Minute 169.
49 Ibid., 20 September 1929, Minute 161.
50 NRM, RMYSC, 15 June 1938, Minute 15.
51 NRM, RMYC, 19 November 1936, Minute 207.
52 NRM, RMYSC, 7 April 1937, Minute 3.

high-speed *Flying Scotsman* services. The Hetton Colliery locomotive also left briefly in 1936 when it was lent to Gaumont Films for an educational production.[53]

The immediate impact of the outbreak of the Second World War on the Railway Museum is not documented in the minutes. Regular opening probably ceased, but there seem to have been special openings, since the minutes record that 'during the period of the War a fair number of people, including members of the Overseas Forces have asked for admission to the Railway Museum'. Indeed the visitors' book for this period has many entries by British and Commonwealth troops.[54] In October 1940 the Museum Committee turned its attention to 'Protection against War Risks' and set out a list of the rarest items to be packed up and put in secure storage. It also decided to take action to protect the museums by bricking up windows and fitting shutters.[55] By April 1941, however, the plans were revised following the intervention of the General Manager and arrangements were put in hand to evacuate most of the locomotives and coaches from Queen Street. By August 1941 eight locomotives and two coaches had been stored in railway sheds at Reedsmouth (Northumberland), Sprouston (Scottish Borders), and Ferryhill (County Durham), where they were placed under the care of local staff.[56] Concern about the safety of the smaller items was also expressed following the damage done to the nearby York station in an enemy air raid on 29 April 1942. Mr W. Naylor, chairman of the Museum Sub-Committee, recommended their removal to 'some area less likely to be exposed to enemy action than York'.[57] The rural station at Rillington, Yorkshire, was considered, but the more valuable of the small exhibits were eventually transferred to the Bowes Museum, Barnard Castle, where a small display was set up in late 1942 or early 1943. While the Railway Museum was busy moving its collections to secure accommodation, other material from the LNER considered to be at risk was sent to York, as was the collection of relics from the Great Eastern Railway boardroom at Liverpool Street, London.[58] Despite these precautionary arrangements, collecting more material for the Railway Museum continued throughout the war years and major items like the former London & South Western Railway royal saloon of 1848, then on the Shropshire & Montgomeryshire Railway, were given serious consideration.[59]

As early as November 1943 the Museum Committee began considering the possible post-war development of the Railway Museum, the key element of which was the need to unite the two sections under one roof.[60] Although the committee did not meet again until July 1945, it is clear that the members were keen not simply to return the collection to its previous places but to develop the museum further. The preferred option was for a new museum based in the York Old Station terminus, using the remaining parts of the old train-shed for the large vehicles. Plans were

---

53   NRM, RMYC, 19 November 1936, Minute 206.
54   NRM, Railway Museum Visitors' Book 4.
55   NRM, RMYSC, 10 October 1940, Minute 39.
56   Ibid., 19 August 1941, Minute 46.
57   Ken Appleby, *Britain's Rail Super Centres – York* (Ian Allan, Shepperton, 1993), p. 23; NRM, RMYSC, 2 September 1942, Minute 55.
58   NRM, RMYSC, 19 August 1941, Minute 48.
59   Ibid., 28 November 1941, Minute 51.
60   Ibid., 11 November 1943, Minute 69.

prepared, and the scheme was roughly costed at £31 000.[61] However, the LNER's Traffic Department insisted that for operational reasons it could not give up the siding space that would have been incorporated into the museum scheme, and it blocked the proposal.[62] The committee did not, however, give up its plans for the old station, and they were still being considered as an option in 1947.[63]

It was, then, with some reluctance that the committee made arrangements for the basic refurbishment of the Queen Street building, its displays, and the Small Exhibits section. The exhibits returned from the Bowes Museum in early 1946 and the locomotives and coaches in June 1947.[64] The displays were apparently put back largely as they had been before the war with the major addition of NER class M 4-4-0 No. 1621, the acquisition of which had been championed by the LNER's Chief Mechanical Engineer. The arrival of a further locomotive meant that the already cramped Queen Street building was over-full and major exhibits like NER No. 1463 and the Port Carlisle Dandy were moved out to poorer accommodation in the adjacent former Lancashire & Yorkshire Railway engine shed.[65]

The Railway Museum re-opened to the public on 18 July 1947 with a degree of ceremony. The formalities were led by Sir Ronald Matthews, chairman of the LNER, who was accompanied by a party of directors and principal officers of the company.[66] In his speech he remarked that 'from an early stage the Museum was recognised as containing the germ of a truly National Railway Museum' and alluded to the fact that 'the importance of the museum... would be greatly enhanced by the changes in the ownership and organisation that loomed ahead'.[67] From this point the Railway Museum was regularly open to the public from Monday to Saturday between 10a.m. and 4p.m. (extended to 5p.m. in the summer months from 1949) with a museum custodian in attendance. By September 1947 the daily attendance averaged 130.[68] Admission continued to be free, but visitors were able to purchase a guidebook prepared by the honorary curator E.M. Bywell and (from 1949) postcards produced by the Locomotive Publishing Company depicting the principal locomotives in the museum.[69] Money donations continued to be given to railway charities; these were split between the NER Cottage Homes & Benefit Fund and the Railway Benevolent Institution.[70]

Sir Ronald Matthews alluded to the forthcoming nationalization when he spoke of the major changes that would affect the Railway Museum. One of the early acts of the new chairman of the new nationalized British Transport Commission (BTC), Sir Cyril Hurcomb, was to address the issue of the preservation of historical railway material. Public concern over the government's possible intentions with regard to

61 Ibid., 30 November 1945, Minute 78.
62 Ibid., 10 October 1946, Minute 87.
63 Ibid., 4 June 1947, Minute 119; 5 September 1947, Minute 137.
64 Ibid., 24 June 1946, Minute 80; 4 June 1947, Minute 115.
65 Ibid., 14 February 1947, Minute 96; 23 April 1947, Minute 111.
66 'The Railway Museum', *LNER Magazine*, 37/9, September 1947, pp. 202, 204–5.
67 *The Yorkshire Post*, 19 July 1947, cutting in NRM, BTC file XX.
68 NRM, RMYSC, 5 September 1947, Minute 129.
69 Ibid., 31 January 1949, Minute 141.
70 Ibid., 31 January 1949, Minute 144.

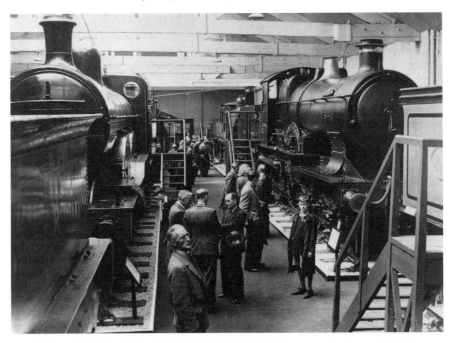

Plate 18.2   The first public visitors to the Queen Street Large Exhibits Section following reopening after the Second World War, 18 July 1947
(National Railway Museum – 36/00)

material transferred from the four private companies had been expressed in the correspondence pages of *The Times* on 31 December 1947 and at subsequent meetings with the BTC.[71] One particular issue was the lobbying to preserve former Midland Railway 2-4-0 of 1866 which the LMS had offered to the Railway Museum in 1946 but which had had to be declined on account of shortage of space.[72] As a result, the BTC commissioned a report on the current situation relating to its inherited relics and records, and the York Museum Sub-Committee was asked to provide assistance to G.R. Smith, who had been given the task of drawing together information from throughout the newly nationalized railway network. However, Smith's report of March 1949 appears to have been unsatisfactory and further investigation was undertaken.[73]

What resulted was the 1951 report *The Preservation of Relics and Records*, which set out the BTC's future policy as a national strategy. It recognized that the BTC had

---

71   See Hopkin, *Railway Preservation*, 1987, pp. 27 ff.
72   NRM, RMYSC, 10 October 1946, Minute 89.
73   PRO AN 104/1, British Transport Commission (BTC) Relics and Records Committee: correspondence and reports, papers relating to the *Relics and Records* report.

acquired 'a considerable collection of relics... accumulated by the four main-line railway companies'.[74] It acknowledged the major contribution made by the former LNER and the York Railway Museum, but also noted that both the museum's buildings were full. Furthermore, it noted that none of the former companies had felt justified in appointing full-time curators and that E.M. Bywell worked on an entirely voluntary basis.[75] The national strategy had a short-term aim of securing the material already held and setting up a regional framework of museums, the first of which was to be the existing York Railway Museum.[76] The main proposal was for the bulk of the existing collections to form the basis for a new Museum of British Transport in London and all material to 'be brought under the Commission's Museum organisation and properly catalogued'.[77] To undertake this work the Commission appointed John Scholes, formerly of the Castle Museum, York, as Curator of Historical Relics.[78]

Although the report commended the work that had been done by the York Railway Museum, there appears to have been some local hostility to the perceived hidden threat that the collection would be withdrawn to the proposed new museum in the capital. There were fears that 'Londonisation' – the popular contemporary term used against the 'Powers that Sit in London' – would be detrimental to the north, and at least one member of staff feared that York would lose its treasures.[79] The Museum Committee was assured by the Railway Executive of the BTC 'that no drastic changes at York are contemplated, at least in the foreseeable future'.[80] While its members were no doubt wondering what the future held for the York Railway Museum, they were occupied with arranging the special railway exhibition held in York in association with the Festival of Britain on 4–16 June 1951. The Railway Museum was given a general tidying up and some of the museum collection was exhibited in the special display of historic and modern rolling stock held in the carriage sidings adjacent to the Old Station. This exhibition no doubt demonstrated to some members of the committee the potential of their original ideas for the conversion of the site to form a new railway museum.

The details of the changes to arrangements in the early 1950s under the BTC at the York Railway Museum are unfortunately not documented in the Museum's minutes. What appears to be the last Sub-Committee meeting was held on 21 March 1952, in the course of which the chairman reported that the BTC's officers responsible for historic material, John Scholes and archivist Leonard Johnson, had visited the museum and that Scholes had expressed a wish to meet the committee.[81] Several sets of minutes remain on the file, apparently uncirculated. What probably happened

---

74   BTC, *The Preservation of Relics and Records* (HMSO, 1951), p. 9.

75   Ibid., p. 10.

76   Ibid., p. 15.

77   Ibid., p. 16.

78   Scholes, John Hornby, b. 14 June 1914, Southport, responsible for Inauguration of Corporation Museum, Southport, 1938, Director of the Castle Museum, York, 1947 to 1951 (*International Who's Who in Art and Antiques*, ed. E. Kay, Cambridge, 1972, p. 11).

79   *News Chronicle, Manchester Edition*, 26 January 1951, in NRM, BTC cuttings files XX.

80   NRM, RMYSC, 7 May 1951, Minute 180.

81   Ibid., 21 March 1952, Minute 188.

**Plate 18.3   The entrance to the Queen Street Museum showing Stephenson's
Hetton Colliery locomotive and admission prices (Adults 6d,
children 3d) c.1960** (National Railway Museum — 894/89)

was that Scholes took the Railway Museum at York directly under his control. The
surviving collections records at the National Railway Museum indicate that the
BTC's museum collection was managed as a single entity overseen by Scholes and
his London staff. Changes at the York museum must have been low on the list of
the curator's priorities, and Scholes's main effort was put into building the collection
for a planned Museum of British Transport and displaying material in temporary
exhibitions in London and elsewhere. There were, however, further fears for the York
Railway Museum when it was reported in late 1957 that the Queen Street building
was under threat from redevelopment at a time when the Railway Executive was also
being publicly criticized for the operation of its preservation policy.[82]

Despite these rumours, the Museum remained very much as it always had been.
The material in the collections was of high quality but the displays had become
cluttered and had little interpretation for visitors. There were temporary changes, like
the removal of *City of Truro* in 1957 for a return to operation on special trains and a

---

82   *The Times*, 3 December 1957, cutting in NRM, BTC file XX27.

tidying and re-display of the Small Exhibits Section in 1958.[83] An admission charge was introduced for the first time, probably in the 1950s. It is perhaps ironic that a museum that had been free to visitors when company-owned introduced charging when under public ownership.[84] The museums, however, were popular with visitors, attracting well over 100 000 a year and achieving a record number of 152 800 in 1959.[85]

By the early 1960s British Railways' interest in maintaining a large historic collection and operating museums open to the public was waning. Although the Museum of British Transport opened in 1963, it was in a converted bus-garage in Clapham, London, and was not the great purpose-built museum that Scholes and others had hoped for. British Railways came to view museums full of antiquated steam locomotives as conflicting with the modernizing image it wished to project. It increasingly starved its museum operation, including York, of resource and allocated a budget that would scarcely permit even essential maintenance.[86] By 1966 the government recognized British Railways' desire to shed its responsibilities for its transport museums and set up a working party to consider future plans.[87] The preferred option was to transfer the collections to the Department of Education & Science and to establish a National Railway Museum on a single site, to operate as an outstation of the Science Museum, London.

At the Railway Museum in York the impact of these political changes was significant. Little or no development occurred, and the Small Exhibits Section was forced to close in 1966.[88] The Queen Street Museum did not close to the public until 31 December 1973, in preparation for the opening of the new National Railway Museum, which, after much political wrangling, was to be located in York. Over a two-year period most of the collection was transferred to the new NRM, which opened on 27 September 1975. Not all the material from the old museum was passed to the NRM, for under the terms of the Transport Act 1968 the collections were reviewed and some of the long-standing collection material like early archives and much of the Briggs collection was transferred to the Public Record Office. Other material that had suffered long years of neglect and was considered beyond economic restoration or conservation was destroyed.

Today the collections from the Railway Museum, York, form an important part of the historic core of the National Railway Collection cared for by the National Railway Museum, and they owe much to the work of all those involved in building Britain's

---

83 Ken Hoole, 'The Railway Museum at York', *Journal of the Stephenson Locomotive Society*, 51, 1975, pp. 275–8.

84 The admission fee introduced for the Large Exhibits Section was Adults 6d, Children 3d. The Small Exhibits Section remained free.

85 BTC Press release, 'Historical Rail Treasures Attract More Visitors', 18 February 1960, in NRM, BTC file XX17.

86 William O. Skeat, 'The Consultative Panel and the Transport Museums', *Journal of the Stephenson Locomotive Society*, 42, 1966, pp. 263–73 (p. 273).

87 See Hopkin, *Preservation* (1987), pp. 41 ff.

88 Ken Hoole, 'The Railway Museum, York', *Journal of the Stephenson Locomotive Society*, 51, 1975, pp. 275–8 (p. 277).

first railway museum. Perhaps, too, the inheritance of this 'germ of a truly National Railway Museum' remains today within the National Railway Museum, which continues to celebrate, in York, the unique contribution that railways make to our lives.

Chapter 19

Transport Museums and the Public Appreciation of the Past

Colin Divall and Andrew Scott

When in 1970 Jack Simmons published his pioneering study *Transport Museums in Britain and Western Europe*, the dedicated transport museum was already well over 70 years old – the first, a modest railway museum at Hamar, Norway, opened in 1897. But, as Simmons noted, the most spectacular growth came after the Second World War, and not only in those parts of the world he had visited.[1] Reliable figures are hard to come by, but even a few national statistics give some idea of the size and geographical extent of the sector today, and its rate of growth over the last three decades. Thus Britain, which in the late 1960s already had over 80 museums falling under the categories of 'Transport' and 'Shipping', now has well in excess of three times that number. There are nearly 250 sites dedicated to maritime heritage in the United States, and over 200 railroad museums. Australia has 26 railway museums; there are at least four in Latin America, and five in each of the continents of Africa and Asia. When one includes heritage transport – the operation of archaic vehicles for public display or carriage – the number of museums, broadly conceived, becomes much larger. The number of heritage railroads in the United States, for example, all of which have opened in their present form since 1945, adds more than half again to that of conventional railroad museums.[2]

Levels of visitation are equally impressive. Several transport museums in Europe and the United States of America can reckon on around half a million visitors a year

---

1   Jack Simmons, *Transport Museums in Britain and Western Europe*, George Allen & Unwin, London, 1970, pp. 7, 19, 23, 265–72, 276.

2   Simmons, *Transport Museums*, p. 19; *Museums and Galleries 1997*, Hobsons, London, n.d. (c. 1996), pp. 130–41; J.D. Storer, 'Conservation of industrial collections', *Transport Museums*, vol. 15/16, 1988/89, pp. 57–65, at p. 59; Kenneth Hudson and Ann Nicholls, *The Directory of Museums and Living Displays*, Macmillan, Basingstoke, 3rd edn, 1985, pp. viii, 814; 'Trains 2000 Guide to Recreational Railroading', supplement to *Trains*, May 2000; Glenn Godinier, 'Mystic Seaport', Maritime Heritage Conference: Carried Along by the Currents?, University of Portsmouth, April 1999; Robert F. McKillop, *Guide to Australian Heritage Railways and Museums*, Australian Railway Historical Society, Redfern (NSW), 6th edn, 1997; June Taboroff and John Tillman, 'The Bank and cultural heritage: transport sector perspectives', mimeographed internal discussion document for the World Bank, July 1999, annex 1.

– the leading example in Britain is the National Railway Museum (which owes its existence partly to Simmons's tireless lobbying in the late 1960s and early 1970s).[3] Many more museums exceed the hundred thousand mark, a figure which in the 1960s was reckoned possible by only a tiny elite. Yet for all this growth the challenges facing transport museums in the twenty-first century do not differ in kind from those identified by Simmons. Chief among these is the problem of maintaining adequate numbers of visitors. A hostile observer in the 1960s, he commented, could point to a decline as evidence of the passing nature of public interest in transport museums. Although in absolute terms the situation today is far healthier, few museums can afford to be complacent. Most welcome far fewer visitors than they did 20 years ago.[4]

Museums, as Simmons averred, 'have a triple task to perform: to collect and preserve, to display their exhibits, and to explain them, telling us why each has been thought worthy of occupying the space allotted to it'.[5] Succeeding in the first of these functions, a precondition for the second and third, presents severe difficulties, now as then. The costs of storage and of conservation over the long term are still a considerable charge to budgets that remain chronically inadequate. While progress in the provision of accommodation has certainly been made, too many transport museums still have to put up with unsuitable and overcrowded premises in which to house their collections, or parts of them. And even those museums fortunate enough to be able to rely on large amounts of volunteer labour are having to face the fact that many of the specialist skills needed to conserve and restore objects, particularly the vehicles that lie at the core of many collections, are increasingly hard to come by.

What of display and explanation? One can hardly quarrel with Simmons's remark that the 'main purpose of museums is unquestionably educational'. He congratulated the 34 leading museums he surveyed for resisting the temptation to neglect exhibitions in the face of difficulties of finance and staffing; 'in each of them', he said, 'one can see that things are on the move, somewhere if not all along the line'.[6] It was a typically perspicacious remark. In the three decades since his book, exhibits crowded together in unsuitable, badly lit, and poorly heated buildings with little or no explanation beyond traditional object-labels have given way, at least in the better places, to visually attractive and well interpreted displays set in comfortable and attractive environments. In this respect the best transport exhibitions are now on a par with those in other kinds of history museums.

Yet even this elite could be better still. Exhibitors – public historians, curators, and others with a responsibility for mounting exhibitions – need to engage more with wider debates about the theory and practice of museum exhibiting, although the fault is certainly not all one way. Commentators on cultural politics and policies still seem largely unaware of the sheer size and importance of the transport sector, while Simmons's volume remained the only book-length critical study, at least in English, until the publication of our own over 30 years later.[7] This chapter then explores some

---

3   Simmons, *Transport Museums*, pp. 283–6.
4   Ibid., pp. 282–3.
5   Ibid., pp. 18, 274–6, quotation at p. 18.
6   Ibid., pp. 276, 281–2.
7   Colin Divall and Andrew Scott, *Moving On: Making Histories in Transport Museums*, Leicester University Press, London, 2001.

of the ways in which the public learns about the past in transport museums, and, more particularly, what can be suggested by conventional exhibitions of static vehicles and other objects. Heritage transport remains outside our purview for reasons of space – its rapid growth from the 1970s and subsequent importance is one of the very few trends in the public history of transport that Simmons failed to anticipate.

Visitors and Representing the Past

Does a visit to a transport museum change the way people understand the past? There are few studies of this important issue, and still fewer clear answers. Research on visitors is however quite well established in many other museum sectors, particularly in the USA. Some scholars are deeply sceptical that people learn anything in museums, suggesting that exhibitions merely reflect back visitors' preconceptions. However, most studies suggest that visitors can learn. Much turns on what is meant by terms such as 'learning' and 'education'. Learning in a museum, at least through the medium of an exhibition, is clearly an informal process; it is voluntary and, largely, self-directed. These points are crucial. Visitors are in a real sense in the driving-seat; their personal agenda play a part – probably quite a large part – in what they will attend to during a visit.[8] As Simmons noted, for visitors to transport museums these interests may plausibly embrace a concern for the political and other ramifications of contemporary transport policy, nostalgia for the once-familiar forms of transport of their earlier years, or curiosity about vehicles that are wholly alien to their own experience.[9]

It is hardly surprising that visitors' sense of the past is strongest when it relates to personal experience. This is an important lesson for anybody with a responsibility for transport exhibitions. What was very familiar to an earlier generation of exhibitors and visitors might not be so to modern visitors – Simmons could refer to trams as 'old friends', but they are hardly so now for most people in Britain. Of course, some of the current generation will be naturally curious about these 'lumbering, plebeian, utilitarian vehicles', but how to make them interesting to a wide audience is not easy without systematic studies of visitors and would-be visitors.[10]

Personal memories are not all there is to museums. Many writers understand them as places that help to shape a sense of identity through the collective memories they trigger and sustain. These memories constitute a shared popular consciousness of the past, helping to define us all as social beings.[11] Sometimes, however, museum exhibitions trade on a popular consciousness of the past which tends to exclude or denigrate the experience of particular social groups.

8   Lois H. Silverman, 'Personalizing the past: a review of literature with implications for historical interpretation', *Journal of Interpretation Research*, 2/1, 1997, pp. 1–12; George E. Hein, *Learning in the Museum*, Routledge, London, 1998, passim.
9   Simmons, *Transport Museums*, pp. 276–7.
10   Ibid., p. 277.
11   John Urry, 'How societies remember the past', in Sharon Macdonald and Gordon Fyfe, eds, *Theorizing Museums: Representing Identity and Diversity in a Changing World*, Blackwell, Oxford and Cambridge, Mass., 1996, pp. 45–65.

These linked senses of belonging and exclusion are intimately bound up with the ways in which exhibitions represent the past. Through their arrangement of objects, labels, graphic panels, and so on, exhibitions tell stories or narratives about the past. One of the most important kinds of story is myth – or 'heritage' in one sense of that much abused term. Myth/heritage is often contrasted with 'history'. The distinction is a subtle and important one, and it is worth spending a moment understanding its implications.[12] In David Lowenthal's formulation the essential difference is between a representation of the past orientated primarily towards the many and often conflicting purposes of the present ('heritage') and that which tries to understand the past in its own terms ('history'). In its application to the public or collective realm, 'heritage now mainly denotes what belongs to and certifies us as communal members', it 'passes on exclusive myths of origin and continuance, endowing a select group of people with prestige and common purpose'. In other words, it is a crucial element in the way we define our particular identities as social beings, functioning in an analogous way to memory in the formation of personal identity.[13] For this reason the truth or falsity of popular or collective memories is, as Patrick Wright remarks, 'often peripheral to their practical appropriation in everyday life'.[14] Heritage, in Lowenthal's sense, wins out over 'history'.[15]

Lowenthal is more positive in his evaluation of myth/heritage than those writers who understand heritage simply as 'bad history'; a biased, ideological representation of the past sustaining the interests of dominant groups.[16] As Lowenthal puts it:

> The crux of most aspersions against heritage is that it undermines 'real' history, defiling the pristine record that is our rightful legacy. Critics who idealize this unsullied past view history as true because innate, heritage as false because contrived. History is the past that actually happened, heritage a partisan perversion. Substituting an image of the past for its reality, in the typical plaint, heritage effaces history's intricate coherence with piecemeal and mendacious celebration, tendering comatose tourists a castrated past.[17]

By contrast, Lowenthal argues that heritage's engagement with the past in terms of the present is in a sense inescapable. 'Heritage', he writes, 'no less than history,

---

12   David Lowenthal, *The Past is a Foreign Country*, Cambridge University Press, Cambridge, 1985, and *The Heritage Crusade and the Spoils of History*, Viking, London, 1997. See also, Raphael Samuel, *Theatres of Memory: Past and Present in Contemporary Culture*, Verso, London, 1994; David Brett, *The Construction of Heritage*, Cork University Press, Cork, 1996; Kevin Walsh, *The Representation of the Past: Museums and Heritage in the Post-Modern World*, Routledge, London and New York, 1992.

13   Lowenthal, *Heritage*, pp. 67, 128.

14   Patrick Wright, *On Living in an Old Country: The National Past in Contemporary Britain*, Verso, London, 1985, p. 188.

15   See also Hein, *Learning* , chapters 6, 7; Silverman, 'Personalizing the past', pp. 1–12.

16   See, for example, Robert Hewison, *The Heritage Industry: Britain in a Climate of Decline*, Methuen, London, 1987, pp. 83–105, 131–46; Walsh, *Representation of the Past*, pp. 130–32.

17   Lowenthal, *Heritage*, p. 102.

is essential to knowing and acting. Its many faults are inseparable from heritage's essential role in husbanding community, identity, continuity'.[18]

Transport museums peddle myths, at least some of the time. Insisting that we were the first, or the best, with regard to some particular facet of transport is one means by which museums can shape social identities in all sorts of ways. It was, for example, the English nationalist historian G.M. Trevelyan who said that 'Railways were England's gift to the world', and, as we now argue, for better or worse displays of transport often consist of little more than trophies and icons symbolizing such historically selective sentiments.[19]

Displaying Transport – Traditional Approaches

One of the most striking and prescient aspects of *Transport Museums* was Simmons's insistence that museums should do more to explain transport's historical context. His delineation of this wider set of purposes remains a model of clarity and conciseness, and one that anticipated many of the criticisms levelled more recently at transport museums:

> 'transport' is a very wide-ranging subject. It comprehends the movement of men and women and their goods by all possible means from place to place; and properly considered, it ought to take account of the changing demands that affected it and the consequences it produced. It is a mere device for serving human needs, and unless those needs are kept in mind the study of it loses all touch with reality. It may become a romantic pursuit of curious or beautiful machines. But they were not created as such. They were intended to meet needs that arose, in precise and specific forms, for conveyance that should be quicker or safer or cheaper or more commodious, or that they should combine several of these advantages together. We shall not understand them or appreciate their true importance unless we understand the jobs they were designed to do.[20]

Simmons was worried by the failure of many transport exhibitions to achieve such comprehension. As he remarked, the transport museum 'tends to concentrate on the object itself, on its arrangement and display and on the proper description of it as a piece of equipment or machinery' and hence such museums 'often fail to make clear the relationship between technological progress and economic and social change'.[21]

Matters are not quite so straightforward, however. At one level many transport exhibitions, particularly those mounted before about 1970, do seem to have little to say about the past beyond the strictly technical. On the other hand, the display of transport objects has always been informed by a set of conventions, dating back at least to the Great Exhibition of 1851, that understands them as symbols of

18　Lowenthal, *Heritage*, p. xi. See also for instance, Brett, *Construction of Heritage*, pp. 1–37.

19　G.M. Trevelyan, *English Social History*, Longmans Green, London, New York and Toronto, 2nd edn, 1946, p. 531.

20　Simmons, *Transport Museums*, p. 18.

21　Ibid., p. 278.

wider social processes and particularly of 'progress'. Both historically and, more problematically, in the present day, a belief in the myth of progress often informs a sense of collective identity. Thus, as the brief quotation from Trevelyan suggests, one construction of what it means to be English turns partly on a sense of the country's long history of achievement in transport.

Even the cluttered jumbles of objects typical of some transport museums can then say more to visitors that one might at first suspect. Beautifully restored vehicles, for example, may, both individually and *en masse*, suggest that society has developed as a technological cornucopia based on more power and speed.[22] This idea of a close and positive connection between technological and social progress is suggested particularly strongly by another, very common type of arrangement in transport museums – namely, serried ranks of vehicles. Car museums are particularly keen on this kind of approach, ordering vehicles in rows according to one, or more, of all sorts of criteria – chronology, usage, make, model, and so on. At the most obvious level this kind of arrangement is a visual catalogue, reflecting curators' semi-formal categorizations of the objects that have come into their care. But at another level the common time-line of vehicles may suggest a story of progress – most immediately of technical improvement, but at another level, of social betterment.

Technological and social progress are not however one and the same, and there is a problem if visitors (or for that matter, exhibitors) are not encouraged to make the distinction and then end up confusing the two. As we have remarked, Simmons anticipated aspects of this criticism. It is plausible to suggest that at the time he was writing most people assumed a positive correlation between technological progress and social and economic changes. It seems unlikely that the same is true today, not least because the public – or, more accurately, a certain segment of the public – is more doubtful about the balance of social benefits and costs. The apparent convenience and freedoms of personal motoring, for instance, do not look as convincing in an era of chronic congestion and high levels of pollution as they did in the 1960s and early 1970s. On the other hand this kind of debate has scarcely started in the realm of air transport – we all want our cheap holidays in the sun, it seems. Indeed it still takes an exceptionally critical and self-aware exhibitor to face up to the challenge of how to interpret the past in ways that do not tend to assume that more transport technologies were, and will continue to be, automatically beneficial. And it certainly remains the case that much more work needs to be done if we are to understand the way in which the public takes a sense of the past from even the most object-centred of transport exhibitions.

---

22 George Basalla, 'Museums and technological utopianism', in Ian M.G. Quimby and Polly Anne Earl, eds, *Technological Innovation and the Decorative Arts*, Winterthur Conference Report, Charlottesville (Virginia) 1973, pp. 355–73.

Displaying the Social History of Transport

In 1970 very few transport museums did more than 'glance aside from the machine to consider the end it has served'.[23] But 30 years later no museum that appeals to a general audience can afford to ignore the wider social context of transport in the past, for the general public seems to have a large appetite for exhibitions that deal with the history of everyday life. Mounting such displays should encourage exhibitors to think more deeply about the narratives they try to communicate. Yet in practice, reflection is often not very deep. Few transport museums provide their visitors with opportunities to reflect critically on the myths that inform contemporary identities.

One striking and popular exhibition, 'Royal Palaces on Wheels', at the National Railway Museum, must serve to make the general case. Visitors are encouraged to enter the display at one end, pass by a set of graphic panels setting out a historical background to royalty's use of railways and the difficulties this caused the operating authorities, and then inspect the oldest of several royal carriages. A video sketches the evolution of royal travel by train from the nineteenth century. Visitors then contemplate the rest of the royal coaches, gaining, one hopes, some sense of historical change. (This narrative is subverted by those visitors – a good proportion – who enter at the 'wrong' end!) Attention to visual and aural detail very effectively enhances the exhibition's appeal. A red carpet running throughout the exhibition, a mock triumphal arch, and a sound loop of cheering crowds suggest the spirit of a royal visit. Thoughtfully positioned lighting highlights the superb workmanship and luxurious appointments of the carriages, and sound cones, activated when visitors step within their circumference, give potted accounts of each vehicle's history and use. These are reinforced by graphic labels associated with the coaches.

It is easy to comprehend 'Royal Palaces' as an exhibition defining Britain as a stable nation with a long history of being at peace with itself. More specifically, the royal trains are displayed chiefly – both by commission and omission – as symbols of a monarchy that was always, and still is, permanent and uncontested. Yet in fact the monarchy as an institution was deeply unpopular for much of the earlier part of the nineteenth century. Winning public support for the monarchy, and with it a very particular vision of the nation state, was a matter of considerable importance for the country's social and political elite. The railways played an important part in this. By enabling Victoria to tour the country with great pomp and ceremony, they assisted in creating 'traditions' that helped to popularize the monarchy. But the museum has missed the opportunity to present this fuller, more scholarly, and more critical view of the past. Instead it largely trades on and reproduces the myths of longevity, harmony, and stability that remain central to the defence of monarchism in the present.

Not all visitors will read the exhibition in this way. While many no doubt wallow in what they take to be a romantic and nostalgic presentation – the Museum's marketing department saw those of late-middle-age and older as an important target-audience – other visitors probably do not take it very seriously at all, particularly given that accoutrements such as the triumphal arch have, quite deliberately, something of the look of a cardboard cut-out. There are hints that the exhibitors knew there was more

---

23   Simmons, *Transport Museums*, p. 18.

to the royal trains than mere ceremony and appearance: a careful reading of the introductory graphic panel, with its detail of the huge workload and operating problems caused by royal trains, might well incline some visitors to a critical appraisal of royal travel. Indeed, given the modern monarchy's uncertain fortunes and, arguably, dwindling popularity, especially among the young, it seems possible that public perceptions of the display are more critical than some academic critics perhaps allow.[24] Nonetheless 'Royal Palaces' is essentially descriptive 'history', presenting as given fact a view of events which was contested at the time and is so today.

This kind of criticism can be levelled in its fundamentals at almost every transport museum that displays the social context of transport. The objection is not that exhibitions are selective in their representation of the past, for any exhibition, however scholarly, always involves choices – of objects, themes or subjects, and their interpretation. Our concern is that usually these 'choices' – they may be made almost subconsciously – almost without exception place transport, its operating institutions, and uses in the most favourable light. Such perspectives tend to reflect ideologically conservative points of view, although this is not always the case – the Railroaders Memorial Museum in Altoona, Pennsylvania, for example, depicts the development of the company town of Altoona largely from the point of view of the Pennsylvania Railroad's workers and their families. But for the most part transport museums do end up being more concerned with myth than history, presenting what is necessarily a selective account of the past as something that may not be questioned.

A Future for Transport Exhibitions

What then is the future of exhibitions as far as their role in enhancing the public's appreciation of the past is concerned? The thrust of this essay is that transport museums should take their visitors, and their potential visitors, more seriously. There are then two extremes to avoid. One is the argument that since most visitors do not have the same deep and disciplined interest in the past as historians there is no need to worry overly about the historical content of exhibitions – in other words, the public will be happy enough with myth. The other extreme is to say that if museums are truly to become places of learning about the past, then simply mounting more comprehensive and historically accurate displays is enough.

Both these positions can be understood as granting the visiting public certain rights. The first might be interpreted as saying that visitors have a right to be entertained, or perhaps a right to have their prejudices confirmed; the second as offering a right to be instructed.[25]

In either case, exhibitors have, in a manner of speaking, an easy task. Granting visitors a right to be entertained means that no one has to question the habitual modes

24   See particularly Richard Sykes, Alastair Austin, Mark Fuller, Taki Kinoshita, and Andrew Shrimpton, 'Steam attraction in Britain's national heritage', *Journal of Transport History*, 3rd ser. vol. 18/2, 1997, pp. 156–75, especially pp. 160–61.

25   The idea of visitor rights we take from Tony Bennett, *The Birth of the Museum: History, Theory, Politics*, Routledge, London and New York, 1995, pp. 89–105.

of thinking about the past that we all acquire simply by living in a particular culture. No one has to break through the myths about the past that lurk in our collective consciousness, inform our sense of collective being, and are difficult to detect and discard. Yet saying that exhibitions should be about the historical 'truth' can also be an abdication of responsibility. Leaving to one side the very difficult issue of what exactly is meant by 'truth' when it comes to writing about and displaying the past, this approach can make it too easy for exhibitors to ignore the needs and wants of the public that the museum is supposed to serve. If visitors do not engage with or understand an exhibition, then – the argument might run – so much the worse for them. Going down this road runs the risk of turning transport museums into something like the more reactionary of art galleries, making them less, not more, socially inclusive. Although we exaggerate to make the point, curators in these institutions are rarely interested in addressing the interests of that great majority of the public which, for good reasons, is ignorant of the aesthetic codes that allow meaning to be taken from visual art.[26]

What is needed instead is an approach to exhibitions that recognizes that people have a right to make use of the museum as they see fit while not denying them opportunities to develop and stretch their appreciation of the past. Transport museums should be places that help people to understand more about the process of making sense of the past, that help them make up their own minds about history and how this affects them in the present and in the future.

Is this not still to advocate an easy life for exhibitors, dressed up now in a rhetoric of rights of intellectual or cognitive access? No, for two reasons. First of all, recognizing this right implies that at the very least museums have to find out much more about visitors' concerns, interests, and existing state of knowledge; only thus can exhibitors work with visitors to ensure that displays are of interest, that they are pitched at the right level and will not be ignored. This becomes even more important if transport museums take the next logical step and try to develop exhibitions that will appeal to those segments of the public that at present do not visit.

Simply appealing to the public's taste is not enough however. As others have argued, the past is rather like a foreign country: a cultural resource that in being repatriated to the present is domesticated as myths as it inevitably becomes bound up with the making and remaking of identities of all kinds.[27] Finding out more about visitors is not something to be done just so that their cultural preconceptions can be reflected back more effectively. Indeed it would be a dereliction of exhibitors' professionalism if this were all that happened, for as Eric Hobsbawm has remarked the 'deconstruction of political or social myths dressed up as history has long been part of the historian's professional duties'.[28] People can hardly be blamed for failing to break free of the myths that help to make us all what we are if museums offer no

---

26   Bennett, *Birth of the Museum*, pp. 163–73; William H. Treuttner, 'For museum audiences: the *Morning of a New Day?*', in Amy Henderson and Adrienne L. Kaeppler, eds, *Exhibiting Dilemmas: Issues of Representation at the Smithsonian*, Smithsonian Institution Press, Washington DC and London, 1997, pp. 28–46.

27   Lowenthal, *Foreign Country*, passim; Eric Hobsbawm, *On History*, Weidenfeld & Nicolson, London, 1997, passim.

28   Hobsbawm, *On History*, p. 273.

other ways of understanding the past. Under the right circumstances 'the shock of non-recognition' can help visitors to reinterpret the past, bringing myth more closely into alignment with historical 'reality'.[29] But the circumstances will only be right when people feel engaged by an exhibition's themes and confident enough to allow themselves to be challenged. Once again, we find ourselves back with the need to find out more about museum visitors and might-be visitors.

Potentially, then, people could be attracted to transport museums to make and explore more critical popular histories of transport. How to reconcile scholarly approaches to history with the personalized ways through which most people connect with the past is the main challenge facing transport museums in their role as educational institutions.[30]

---

29 Ivan Karp, 'Culture and representation', in S.D. Lavine and I. Karp, eds, *Exhibiting Cultures: The Poetics and Politics of Museum Display*, Smithsonian Institution Press, Washington DC and London, 1991, pp. 11–24.

30 Steven Lubar, 'Exhibiting memories', in Henderson and Kaeppler, eds, *Exhibiting Dilemmas* , pp. 15–27.

Chapter 20

Writing the History of British Railways

Terry Gourvish

When I first started writing professionally about the railways I was acutely aware of the contribution that Jack Simmons was making to the subject and the need to position my researches in areas not already covered at length by one of the masters of the art. Based in Leicester, my home town, Jack had already provided signposts for most of the modern research agenda. His book on *The Railways of Britain* in 1961 provided particular inspiration, encouraging me to choose a railway topic – the career of Captain Mark Huish – for my PhD.[1] Later on, when I thought I had some claims to being a professional railway historian, Jack showed me how it should be done, notably with the first book in his series on *The Railway in England and Wales 1830–1914* (1978), and with *The Oxford Companion to British Railway History* (1997), a magisterial and deservedly successful volume which he edited with Gordon Biddle. Above all, his personal encouragement to me to write, and continue writing, about nationalized railways from 1947, was much appreciated. In 1986 I published a commissioned history covering the first 25 years of nationalized railways, 1948–73.[2] Hardly had it been published than some critics were suggesting that a follow-up volume was required. It took several years to convince the British Railways Board that a sequel was worth promoting, and I know that Jack played no small part in lobbying for such an eventuality. I therefore owe him a considerable debt. The aim of this piece is to explain the salient features of my work as British Railways Board's commissioned historian, with particular reference to my experiences in preparing 'Volume II'.[3]

Writing history while history is being made is the inevitable lot of the contemporary historian. When in 1979 the chairman of the British Railways Board Sir

1   Jack Simmons, *The Railways of Britain: an historical introduction* (1961, 2nd edition, 1968); T.R. Gourvish, *Mark Huish and the London & North Western Railway: a study of management* (Leicester University Press, 1972).
2   T.R. Gourvish, *British Railways 1948–73: A Business History* (Cambridge, 1986); Terry Gourvish, 'Writing British Rail's History', *Business Archives*, 62 (November 1991), pp. 1–9.
3   This chapter began life as a paper delivered to rail lobbyists and historians in Worcester as part of the events to mark the 150th anniversary of the opening of the railway to Worcester. My thanks to all concerned for their comments, and especially to Mike Anson, Dil Porter, Julian Palfrey and Gerry Douds.

Peter Parker asked me to write about the early years of nationalized railways, there was a new spirit of optimism in the railway industry. Parker's infectious enthusiasm had quickly revived a morale that had been worn down by some of his predecessors, by Dr Richard Beeching's rationalization plans in the 1960s and Richard Marsh's battles with government over railway investment in the 1970s.[4] Parker inspired several initiatives to engage the railway passenger, or customer, as the passenger came to be known. There was the free day of travel for holders of Senior Citizen Railcards in 1978, for example, and the Commuter's Charter of 1979–80, which began the dialogue between the railways, the government, and passenger users' groups to establish quality measures for passenger travel. All this was over ten years before John Major's much publicized Citizen's Charter. Parker also offered railway staff the new (and somewhat unfamiliar) notion of worker participation: this he did by means of a newly-constituted British Rail Council, designed to bring the trade unions into major strategic debates about the future of the railways. And he argued passionately that a dialogue with government would produce meaningful and supportive objectives for the industry, including substantial investment for re-equipment, electrification, the APT (Advanced Passenger Train), and the Channel Tunnel.[5]

However, as I began to write about the 1940s, 1950s and 1960s – the book took me six years to complete – the climate swiftly changed. A new Conservative government led by Margaret Thatcher gathered confidence and acquired a new mission: rolling back the public sector, or privatization. Her antipathy against the public sector and its managers was well known. Furthermore, the recession of 1980–82 put paid to any hopes railway managers had of additional investment, while their ambitious railway electrification plan of 1981 was shelved after the intervention of the Tories' economics guru Alan Walters. Relationships with the trade unions quickly deteriorated, notably when Parker took on the drivers' union, ASLEF (Associated Society of Locomotive Engineers and Firemen), over flexible rostering in 1982. Privatization, at first a tentative rallying-cry which promised public–private participation as well as outright sale, soon became the latter. Railway managers' hopes of new opportunities were transformed into a more ruthless search for profit; downsizing, evident in the shedding of subsidiary businesses; competitive tendering, which decimated the railway engineering industry; and the first whispers of privatising the railways themselves. So in 1986, when I completed my book on the period 1948–73, the railways were a very different industry from that of 1979.[6]

I should not have really been surprised by the pace of change. After all, the first 25 years of nationalized railways had been a rather turbulent period too. First, there were the well-publicized squabbles between the British Transport Commission at the top and the professional managers of the Railway Executive lower down, which dominated the running of the railways in the period 1948–53. Then the evident failures of the Commission under General Sir Brian Robertson had culminated in the

---

4   Sir Richard Marsh, *Off the Rails: an autobiography* (1978); Gourvish, *British Railways*, passim; R.H.N. Hardy, *Beeching: Champion of the Railway?* (Shepperton, 1989).

5   Sir Peter Parker, *For Starters: The Business of Life* (1989).

6   See T.R. Gourvish, 'British Rail's "Business-Led" Organization, 1977–1990: Government-Industry Relations in Britain's Public Sector', *Business History Review*, 64 (Spring 1990), pp. 109–49.

Stedeford Committee's secret report of 1960. Commissioned by Transport Minister Ernest Marples, it led to the end of the Commission and the famous – or infamous – Beeching report of 1963 with its controversial prescriptions for a more efficient railway.[7]

From 1979 to 1986 equally important changes had occurred, though the time-period was more compressed. Business-led sector management had been introduced in 1982, and gathered strength under Sir Peter Parker's successor as British Rail chairman from 1983, Sir Bob Reid: Sir Robert *Basil* Reid, known as Bob Reid I, to distinguish him from *his* successor, Sir Robert *Paul* Reid, chairman from 1990, known as Bob Reid II.[8] In 1983, a year after the introduction of the new approach, the Serpell Report on *Railway Finances* was published. Although discredited by an astute public relations campaign orchestrated by British Rail, the report was as influential as the Beeching report had been in encouraging cost-cutting within the railways, particularly in engineering.[9] In addition, the 1980s saw large chunks of the British Railways Board's assets sold off – family silver such as surplus property, British Transport Hotels, Sealink, and the engineering workshops. The first sales were of British Rail's laundries and wines businesses. Both were in bad shape; the wines business, for example, had built up huge stocks of Chablis and Burgundy, which had to be disposed of at a loss! And so, by 1986 the railway world had changed a great deal; in short, the balance between the railways' historical goals of satisfying financial expectations and providing a public service had most definitely tilted in favour of the former.

Thus, as I began work in 1997 on the sequel to my first book – the second volume deals with the period 1974–97 – I was acutely aware that events around me were once again transforming the way the railways were run and governed and the relationship they had with both the government and the public. However, on this occasion, the book was written with the outcome – a peculiarly complicated form of quasi-privatization – established, even if the long-term implications are far from certain.[10] Clearly no one in 1993–7 listened to the business historians who invoked the work of the scholars Ronald Coase, Oliver Williamson, and Alfred Chandler in warning about the perils of excessive transaction costs![11] I have also learned to be rather diffident

7 See, for example, Gourvish, *British Railways*, passim; British Railways Board, *The Reshaping of British Railways* (1963); Sir John Elliot, *On and Off the Rails* (1982); Michael R. Bonavia, *The Nationalisation of British Transport: The Early History of the British Transport Commission, 1948–53* (1987).

8 Inside the business they were also known as 'half-price Reid', and 'full-price Reid', since their salaries were very different. Reid I's salary on appointment in 1983 was £63 600, and his final salary was £91 000; Reid II's starting salary in 1990 was £200 000 plus performance-related bonus.

9 *Railway Finances*. Report of a Committee chaired by Sir David Serpell KCB KMG CBE (HMSO, 2 vols., 1983).

10 Terry Gourvish, *British Rail 1974–97: From Integration to Privatisation* (Oxford, 2002). See also, W.P. Bradshaw, 'The Rail Industry', in Dieter Helm and Tim Jenkinson (eds), *Competition in Regulated Industries* (Oxford, 1998), pp. 175–92, and Roger Freeman and Jon Shaw (eds.), *All Change: British Railway Privatisation* (Maidenhead, 2000).

11 See Oliver Williamson, *Markets and Hierarchies* (1975); Alfred D. Chandler, Jr., *Scale and Scope* Cambridge, Mass., 1990.

about my role as the British Railways Board's commissioned historian. One of the things you quickly learn about railways is how much everyone else knows about them. Sir Richard Marsh, chairman of British Rail from 1971 to 1976, was not entirely joking when he observed (in 1974): 'Together with labour relations and singing in the bath, knowledge of how the railways should be run is provided by the Almighty at the moment of birth. It is a well-known fact that the nation is divided between 27 million railway experts and 190 000 of us who earn our living on the railways.'[12]

My intention in this short piece is to draw upon past and current experiences in order to explore some of the salient features affecting the railways over the last 40 years, and in so doing to reflect both upon the national picture and the implications for local rail operations and local communities of rail users – something which Jack Simmons has always emphasized. I take the main developments of the period to be as follows:

- the move to a more business-led management with the creation of business sectors from 1982;
- the government pressure on British Rail to reduce the subsidy paid under the 1974 Railways Act;
- the Serpell Report of 1983, which contended that rail costs, particularly engineering costs, were unnecessarily high;
- the shock to the industry at the end of the late 1980s boom that the Clapham accident of December 1988 presented (further emphasized by more recent accidents such as Southall and Ladbroke Grove); and
- the drive to privatization from 1991.

All have had their impact on the nature of Britain's railways, on the historically difficult relationship with government, on the way the railways are managed, and on the operation of local lines and services.

Let me start by examining the changes that have taken place under the banner of sector management and their impact on the price and quality of passenger services. Sector management, to recapitulate, was an organizational change introduced from 1982 on a modest basis which gathered pace in the mid-1980s. It was part of the government's drive to reduce the cost of subsidizing the railways through the PSO (Public Service Obligation – itself an innovation: the railways were the first public corporation to have social obligations identified and priced separately). One way to reduce the PSO was to discourage the cross-subsidization of unprofitable elements of the railway businesses and to encourage the development of improved formulae for attributing joint costs to particular activities, whether passenger or freight, or within the passenger business, between, say, express trains and local services, trains running between, say, King's Cross and Edinburgh (InterCity), Dartford and Charing Cross

---

12   Richard Marsh, STC Communications Lecture on 'The Economics of Indecision', 17 May 1974, cited in *Modern Railways*, July 1974, p. 253. Marsh unwisely excluded women from the group of experts.

(Network SouthEast), and Stourbridge Town and Stourbridge Junction (Regional Railways). A 'business-led' focus was all the rage in business schools; there was much management-speak in justification of the division of the railways into five businesses in 1982: InterCity; London & South East commuting (later Network SouthEast); a collection of trains once known as 'Other Provincial Services', later renamed simply 'Provincial', then 'Regional Railways'; Freight; and Parcels. The idea was to give railway managers a measure of control over the assets that they used; and in the process, to break the domination of the regional general managers. Known as the 'barons' of the railway, these railwaymen had followed in the footsteps of the gods of the Great Western – Sir Felix Pole, Sir James Milne, Keith Grand, and Gerry Fiennes – and the equally impressive managers of the Eastern Region (once the London & North Eastern) – such as Sir Ralph Wedgwood, Sir Charles Newton, and Sir Henry Johnson. The regional structure had encouraged 'production-led' management – railways which seemed to be run for the benefit of engineers and operators, rather than for customers.

With the break-up of British Rail into separate sectors, the aim in the 1980s was to create 'business-led' management, to ensure these businesses made a profit, or if this was not possible for social reasons, to reduce their financial deficits and the amount of subsidy provided by central government. The change was not introduced overnight and was not without numerous battles internally; indeed, sector management did not reach its full flowering until the eve of privatization in 1992. Having acquired responsibility for dedicated operating assets, the sectors were asked, in 1990, to incorporate engineering assets and functions, and thus to move, in effect, to separate company status, with the British Railways Board becoming a holding company. This last organizational change in a long list in the history of nationalized railways was undertaken under the slogan 'Organizing for Quality', or OfQ for short.[13] Worse still, no sooner had the troops dutifully marched up the hill to implement it (and it was not without its traumas), they had to march down again, because the government decided instead to privatize the railways under a quite different model with the creation of Railtrack and the separation of the track and engineering from rail operations.

What impact did sector management have in the period 1982–92? I think the short answer is that it sharpened up management attitudes to the railway businesses, improved cost-control and marketing, and encouraged customer-oriented investment. Over time, regions were downgraded, divisions eliminated, and more power was delegated to Area Managers: this was important for the revival of lines such as the Cotswold and Malvern (which were in a bad way in 1982). However, the efforts to improve the railway, business by business, cannot be divorced from the government's use of the sector structure to fix tough financial targets for each sector. Thus, tentative targets in the early 1980s gave way to much more precise ones. In October 1983 British Rail was required to cut the PSO by a quarter in real terms in three years, a requirement which was met. Freight was to earn an operating profit of five per cent, and InterCity was expected to achieve a full commercial objective, then, from August 1984, to make an operating profit of five per cent by 1988/9. New targets set in

---

13   The acronym may be pronounced in different ways. Supporters tended to refer to 'OforQ', cynics to 'OferQ'.

October 1986 required a further 25 per cent reduction in the PSO by 1989/90, with the commercial businesses (all except NSE and Provincial) required to earn 2.7 per cent on their net assets by 1989/90. Again, this was achieved. The December 1989 targets for 1992/3 were even tougher: a 7.5 per cent profit for InterCity, Network SouthEast to eliminate its subsidy, and Regional Railways to cut its subsidy substantially.[14] Unfortunately, the long recession of 1989–93 put paid to all this.[15] Of course, in the process of managing the railways under such a regime some good and bad things resulted for passengers and customers, especially as government support for investment was linked to the attainment of performance targets. Some hard decisions had to be taken about particular lines and services. Some services were moved from InterCity to Provincial or vice versa to suit financial rather than operational criteria. The treatment of London–Bournemouth services in the mid-1980s is a good example of the manoeuvres which took place.

On the other hand, some localities clearly benefited. They were not candidates for rail closure or a study advocating bus substitution: the latter a popular pastime in the late 1980s. Electrification and new rolling stock enhanced the East Coast Main Line, though the West Coast remained neglected. The Advanced Passenger Train was a disaster, but the High Speed Trains, slower but much more reliable, improved InterCity services, while a new generation of diesel multiple units – 'Sprinters', 'Super Sprinters', and 'Turbos' – enhanced commuting and encouraged Provincial to change from being obsessed with cost-cutting and rationalization to a business attracting new customers through new services, for example Norwich–Liverpool. Marketing and pricing became more imaginative, if periodically the proliferation of fares got out of hand (notably before the simplification of 1985, while in the late 1990s there were signs of excessive complication again). Marketing is currently more about APEX fares booked in advance than walk-on fares, which have declined. Operators routinely seek premium fares for higher quality services. Fares on average have increased in real terms, despite extensive discounting; longer-distance commuters have suffered as InterCity became more 'commercial'.

Rail travel in Britain has remained expensive by European standards. It has been suggested that particular localities benefited from the sharper attitudes produced by sector management. Worcester/Hereford is one example, though activists from the Cotswold Line Promotion Group (formed in 1978) would argue that their efforts were just as important in the process of reviving the railway's fortunes, and they would also note that there remains much still to be done. The Oxford–Hereford line, consciously neglected by British Rail (which banned track renewal work in the period 1979–83) and probably scheduled for closure or severe rationalization in the early 1980s, in common with the Leeds–Settle–Carlisle line, was revived thanks to strong lobbying; but we should not forget the work of dedicated railway managers given greater freedom to run the line by sector management principles.

The remaining issues in the list are best considered under the heading of 'the balance sheet of change'. It is difficult to pass judgement on very recent events, but

---

14   BRB, *Report and Accounts, 1983–89/90.*
15   Ibid. 1990/1–93/4.

we can highlight some key elements since 1979 by asking a series of pertinent questions.

1   Were the railways better managed in 1994, on the eve of privatization, than they were in 1979?

I think the answer is a qualified 'yes'. The railways were certainly better managed financially, with improved systems for allocating costs, controlling these costs, and making managers aware of what assets were under their control. The Serpell Report of 1983 was an important, though certainly not the only, influence in this process. On the other hand, the operating function, and elements of engineering too, did not necessarily improve under the cosh of rationalization and frequent organizational change. Peter Rayner, a vocal critic of current railway operations, is a good example of the charismatic, enthusiastic (and even cavalier) operating manager, whose presence can do so much to improve operating at grass-roots level. There was no room for him under the post-OfQ organization. After his criticisms of the new regime were given some considerable publicity, he was sacked, or, more exactly, he left his job as Regional Operations Manager, London Midland Region, to become Project Director (Psychometric Testing)![16]

2   To what extent has improved railway performance been a factor of improved economic conditions?

Naturally, when there is a boom railway traffic goes up, and when there is a slump it does down – something current operators would do well to remember. Bob Reid I was lucky to work during a sustained boom; Bob Reid II was unlucky to work during one of the most severe recessions – and certainly the longest – of the twentieth century. The private operators are currently enjoying Bob Reid I's good fortune today. This said, we should not be sanguine about the decimation of the railways' freight traffic, where the market share has fallen from 23 per cent in 1963 to only four per cent in 1993–4, and it remains to be seen whether the current boom in passenger traffic will survive the next recession.[17]

3   The railway industry became more profitable, or shall we say, less unprofitable, in the 1980s. Why was this?

One element was the acceptance, popularized by Peter Parker, that the central government's subsidy was really a payment made for services deemed socially necessary. The subsidy was then treated as part of turnover, and not as part of any profit and loss. Second, the upturn in railway profits in the late 1980s owed much to the boom but more to the sale of assets. From 1980 to 1990 the sale of much of the railways' extensive property portfolio, plus the privatization of the subsidiary, non-rail businesses, produced no less than £1 400 million, 87 per cent of it from property

16   Peter Rayner, *On and Off the Rails* (Stratford-upon-Avon, 1997).
17   Department of the Environment, Transport and the Regions, *Transport Statistics 1999*.

sales. This generated most of the surplus cash at the end of the decade. Cost-cutting contributed also, but for all its pain, was less significant than the elements I have just described.

4   Did the drive to greater efficiency compromise safety?

This is an extremely difficult question to answer, and anyone who has examined the causes of railway accidents will recognize that events are often highly complex.[18] Nevertheless, the accident at Clapham in December 1988, in which 35 people died, and which was quickly followed by accidents at Purley and Bellgrove in March 1989, was a traumatic one. It followed a period of relatively safe operations, but one in which operating costs had been subjected to considerable scrutiny, and financial performance had improved. Clapham, together with the fire at King's Cross underground station in November 1987, in which 31 died, caused much soul-searching in the British Railways Board, and had a profound effect upon the then chairman, Bob Reid I. The anxieties which followed were naturally linked with earlier warnings by railway engineers about the dangers of excessive cost-cutting and inadequate investment, expressed by Parker when he referred to the 'crumbling edge of quality'.

   In the first half of the 1990s British Railways Board embarked on an extensive review of safety procedures, and many positive elements emerged in day-to-day operations. However, improvements to the safe running of trains required extensive investment, and the money promised by the Conservative government in the wake of Clapham was not forthcoming. Nevertheless, the last two years of the Board's independent responsibility for Britain's rail network, 1992–4, were among the safest in British Railway history. Since then, the complexities of privatization have broken up the integrated safety approach introduced by the Board in the early 1990s. The unfortunate accidents at Southall (1997) Ladbroke Grove (1999), Hatfield (2000) and Potters Bar (2002) have all served to highlight the need for a re-examination of safety management under the new regime. Some commentators, and notably Christian Wolmar, have argued that the current fragmented system has seriously compromised safety standards.[19]

5   Did the drive to greater efficiency remove some of the positive elements of the railway culture?

Certainly, 'Mother Railway', as Cliff Rose (ex-Western Region manager and British Railways Board Member until his premature death in 1983) used to call it, has been battered a fair bit over the last 40 years, and some once-positive elements have been lost. The huge amount of loyalty the industry once inspired, in the 1950s and 1960s, has dwindled, along with the reduction of the workforce (256 000 in 1974, only 121 000 twenty years later); today's managers have very little corporate memory for the railways of old, as was the case before. And although the break-up of the industry

---

18   Stanley Hall, *Hidden Dangers: Railway Safety in the Era of Privatisation* (Shepperton, 1999).
19   Christian Wolmar, *Broken Rails: How Privatisation Wrecked Britain's Railways* (2001).

after 1994 has had a lot to do with this, privatization is not the only explanation. Before this, the notion of railway community was attacked: by the removal of the regional loyalties of managers, which lost the industry the services of people such as Leslie Lloyd and Frank Paterson (General Managers, respectively, of the Western and Eastern Regions); by the rationalization and sale of the railway workshops in Swindon, York, Derby, Crewe, and Doncaster; and by the determined taking-on of the unions by railway management. The drivers' union, ASLEF, was challenged by the flexible rostering battles in 1982/3; the larger, more generic National Union of Railwaymen (now the RMT – the National Union of Rail, Maritime, and Transport Workers) was affected by the move to Driver-Only Operation from the mid-1980s and the promotional simplifications of the traincrew concept from the late-1980s. Unfortunately, these intense battles never seemed to deliver the savings promised when the arguments started. One cannot help but feel that in many ways the baby was thrown out with the bathwater. As a result, some pride appears to have deserted the grass-roots management. It is harder today to find the spirit shown in earlier years, for example when coping with operational difficulties, particularly those caused by 'leaves on the line' and the 'wrong kind of snow', or when heralding new technologies such as TOPS in the 1970s or APTIS in the 1980s.[20]

6   Did government intervention disappear, after the introduction of more precise targets?

The answer is a firm 'no'. There was still regular intervention on the extent of fare increases, and investment controls and financial controls remained firmly in place via cash limits and the External Financing Limit (EFL). Intervention was particularly severe during the Labour government of the late 1970s, when Sir Peter Parker complained of a 'perpetual audit'. However, intervention was also evident during the Conservative administration in the early 1980s recession, and again during the early 1990s recession. During the period of privatization, it was of course intense. In addition, railway managers have had to put up with a high level of public audit, and in some quarters, an anti-rail bias. Inherently a comparatively safe form of travel, the railway was always subjected to intense scrutiny when accidents occurred.

In more general terms, and contrary to the popular belief stoked up by politicians such as Margaret Thatcher, and media references to Thomas the Tank Engine and Fat Controllers, railway managers have not been incompetent: indeed, some of the best British managers were attracted to what was a large industry in the 1960s – for example Bob Reid I, John Welsby, John Edmonds, and John Prideaux. Ministers and civil servants, on the other hand, have often displayed much greater enthusiasm for road transport. Indeed, one Transport Minister, Dr John Gilbert, confessed to Richard Marsh in 1976: 'I am a motorway man myself'.

---

20   TOPS = Total Operations Processing System, computerized monitoring of wagon-load freight operating, but as a result of the large number of Newcastle managers involved was also known inside the industry as Tynesiders' Own Promotional System. APTIS = All-Purpose Ticket Issuing System, computerized ticketing and fare calculation.

The Treasury has never really liked the railways and, after the failure of the Modernization Plan of 1955–65, never warmed to ambitious investment projects if put forward by public sector managers. Note, for example, the recent charades over the choice of and support for the high-speed rail link to the channel tunnel in the period 1989–96. For much of the period of nationalized railways, government and railway managers have been engaged in what the last BR chairman, John Welsby, has called a 'dialogue of the deaf'. Civil servants would tell railway managers, 'We've got so much money, what can you do for it?' while railway managers would tell the civil servants, 'Tell us what sort of railway you want, and we'll tell you how much it will cost'. The setting of precise targets, and then the complexities of privatization itself, have not resolved this fundamental divide.[21]

---

21   John Welsby and Alan Nichols, 'The Privatisation of Britain's Railways', *Journal of Transport Economics and Policy*, 33 Pt. 1 (January 1999), pp. 58 ff.

Chapter 21

'Bibbling' the Railways

George Ottley

Jack Simmons and I first met in the Reading Room of the British Museum Library in 1954.

Two years earlier an enquiry desk had been set up to deal with readers' problems. Previously, newcomers had to try to find out for themselves how to use the catalogue and how to apply for books. If other readers were not sure that they could advise adequately, the enquirer would need to go to 'the centre', as it was called by the staff. This was the administrative hub of this round room, arranged on a raised circular dais, where sat the Superintendent and four clerks. It was to any one of these four that a reader would go for advice. The main responsibility of these four, however, was to control the issue and return of books, and with close on 400 reading places in the 1950s they could be fully occupied in this primary duty. Rarely could more than summary advice be given. It was not only new readers who sought guidance, and requests calling for a *studied* response would be referred to the Superintendent.

The establishment of 'Reading Room Enquiries' brought order to the handling of readers' problems. Three of the existing staff were appointed to run it, and as front-line reference librarians our days were busy, but happy. I think we all felt privileged to have been chosen for this work.

It sometimes fell to me to deal with transport-related enquiries and I soon found that there was a pressing need for a reference guide to books on this subject. There were, of course, general reference works with contents arranged by subject, notably the BM Library's own *Subject Index* for books published since 1880 ('Fortescue') and since 1950 the familiar *British National Bibliography*; but there was no one-volume guide to books on transport, a facility much needed in the 1950s when books on that subject, especially on railways, were appearing in growing numbers. This, I felt, was a challenge which I could respond to with enthusiasm, but it would have to be a spare-time venture and for practical reasons railed transport would be quite enough for me to take on.

With nationalization on 1 January 1948 our railways were suddenly no longer a group of families, each with its own locomotive classes, livery, and traditions. They had all become 'British Railways' overnight, and a veritable boom in railway-book publication was the result. Much of it was for armchair reading and among works of this kind were some outstandingly attractive ones published by George Allen & Unwin, written and illustrated with colour plates, by Cuthbert Hamilton Ellis, the pioneer being *The Trains we Loved*, which appeared on the eve of nationalization, in 1947. It was evident that this firm, with Phillip Unwin's encouragement, was setting itself on course for a spirited venture into the nascent readership of enjoyable, well-illustrated histories.

From the 1950s on there has been a good number of railway histories for readers who value fact and informed comment more than nostalgia, and Hamilton Ellis's *British Railway History* (Allen & Unwin vol. 1, 1954, vol. 2, 1959) was an important work which introduced the subject rather more seriously and objectively while still retaining this author's happy avuncular style of presentation.

A revival of interest in railway history at academic level was pioneered by Jack Simmons's *The Railways of Britain: an introduction* (1961) and Michael Robbins's *The Railway Age* (1962, 3rd edition, 1998). It could be said that Hamilton Ellis's two volumes led the way from the armchair into the study where Simmons and Robbins were more appropriately to be found.

How remarkable it is that in spite of the fading from living memory of steam locomotion the magic of 'steam' persists, and *grows*, it would seem, for since the early 1950s there have been no fewer than 300 popular railway books with 'steam' in their titles (up to 50 between 1950 and 1963, over 100 from 1964 to 1985 and another 150 from 1986 to 1995) and not one of them is about the water vapour exhausted by locomotives. Modern popular 'railway books' consist in the main of reprints or second, and so on, editions, sometimes with altered titles, in paperback more often than in hard covers. Original works are often about relatively minor topics such as a branch line previously not the subject of a monograph, memoirs of individual railwaymen, and occasionally a work of promise such as a biography of a prominent figure in our railway past. Albums of illustrations vary in usefulness. The best of these, products of no more than two or three firms, are of hitherto unpublished photographs, fully annotated, and are important contributions to our picturing of the past.

Writings of substance and originality continue to break through into publication, and the dichotomy, which since the 1920s has characterized railway literature has widened through the years. Today we have on the one hand commercial publishers trying to fill whatever gaps are still to be found and to re-issue works now out of print – endeavours evidently succeeding, to judge from their current catalogues. On the other hand a growing number of works from well-established publishers and academic bodies are found to be treasure troves, with a wealth of detail, adequate notes, tables, appendices, and a good index. Inevitably these call for a high selling price, which adversely affects sales. Laptop publishing is therefore on the increase. Grahame Boyes, compiler of 'Ottley III', says in his introduction to that work (p. 11) that the best of the output since 1980 is superior to the best comparable works of earlier decades.

An event of great significance in the furtherance of railway and canal history research was the establishment in 1951 of the British Transport Commission's Historical Records Department. This occupied two floors of a large building in Porchester Road, London, flanking the Great Western line at Royal Oak station. All existing documents of over a thousand railway and canal undertakings were now made available to researchers. A supporting library to this vast archive made 'Porchester Road' a study-centre of immense value. Two regional archives were also opened, in Edinburgh and York. The Edinburgh collection was transferred to the

Scottish Record Office in 1969 and the one in York was amalgamated into the main body of documents in London. In 1972 ownership of this combined collection passed to the Public Record Office, and it is now at Kew.

'Porchester Road' became at once a source where study could lead to the publication of works of far greater authenticity and usefulness. Jack Simmons was among those who were clearly at one with this quite new and exciting facility. There was now free access to documents hitherto unavailable, and in 1953 he promoted two ventures which would, he believed, be of practical benefit to all students of transport history and would incidentally put Leicester University on the map for this subject. He founded, with the cooperation of Leicester University Press, the *Journal of Transport History* and this he edited with Michael Robbins from the start. It was published twice yearly and included, uniquely, an annual bibliography of essays on transport from non-transport periodicals and journals. Since 1980 the work has been published by Manchester University Press. He also made an offer of 250 books on transport history (mostly railways) to Leicester University library. In accepting, the Library Board expressed the hope that the gift would be the nucleus of a developing source of research. Thus was the Transport History Collection born. By 1973 it had grown to well over 20 000 items and was the largest collection on the subject in libraries of the British Isles. Sadly, the THC has always suffered from there being no Transport History Department to energize it, and it has been used only very little as an interdisciplinary collection.

A third event, in 1954, was the founding of the Railway & Canal Historical Society, unique among amateur transport history bodies in that it publishes in its quarterly *Journal* only essays which have been wrought from original research. One of its stated aims, from the outset, has been to provide bibliographical help and this it certainly has done by including an annual bibliography in its *Journal* from 1986 as a continuation of the two first volumes of Ottley's *Bibliography of British Railway History*, 1965, 1988. These annual bibliographies were amalgamated by Grahame Boyes who, with the help of Matthew Searle, Donald Steggles, and a team of R&CHS collaborators, produced in 1998, on behalf of the society, a Second Supplement. This is mainly for works published since 1985 (the cut-off date for the 1988 Supplement) with amendments and newly discovered items not described in the previous two volumes. Grahame's Second Supplement is an admirable achievement. The three volumes describe nearly 20 000 works.

But to return to the BM Reading Room in 1954. Jack Simmons as a frequent user of the Reading Room's facilities soon became aware of what I was doing (sometimes referred to by colleagues as 'Ottley's Folly') and his ready encouragement and friendliness was just what I needed at that time and throughout the 12 years spent in compiling the main work. This was published eventually in 1965, and when it went out of print in 1974, Jack urged me to consider the prospect of a reprint with corrections. This was done and it was he who paved the way for its appearance in 1983. Likewise the Supplement, in 1988. Both were published by the Science Museum and HMSO and both owe their being to his gift of communication and his widely acknowledged standing. Professor Jack Simmons was Ottley's good shepherd and mentor.

As a dedication in *British Transport: an Economic Survey from the Seventeenth*

Century to the Twentieth, the authors, H.J. Dyos and D.H. Aldcroft, chose these words:

> To Jack Simmons who has done more than anyone to advance the study and enjoyment of transport history.

'Amen' to that!

Appendix

Jack Simmons:
a Bibliography of his Published Writings[1]

Diana Dixon and Robert Peberdy

This bibliography is arranged chronologically, with material allocated to the following categories: editorships of series and journals, books, articles, and reviews, the last as a separate section. Within each category, entries are in alphabetical order (with reviews alphabetized by book author or title of article as appropriate). New editions and important reprints of books, and reprints of some articles, are noted under the year of first publication. The place of publication for books is London unless another place is stated. A journal's volume number covers the year under which it is entered in the bibliography unless specified otherwise.

Abbreviations

D&CN&Q	*Devon and Cornwall Notes and Queries*
EHR	*English Historical Review*
JTH	*Journal of Transport History*
RM	*Railway Magazine*
T&T	*Time and Tide*
TLAHS	*Transactions of the Leicestershire Archaeological and Historical Society*
TLAS	*Transactions of the Leicestershire Archaeological Society*
WCM	*West Country Magazine*

1   Acknowledgement: a list of Jack Simmons's publications from 1940 to 1975, compiled by Diana Dixon, appeared in *JTH*, n.s., 3 (1975–6), pp. 145–58 (Leicester University Press). It is incorporated in the present fuller bibliography by kind permission of the Continuum International Publishing Group Ltd, London and New York.

The Bibliography

Jack Simmons's first publications on railways appeared in the magazine *Locomotion*, which he, Michael Robbins, and Roger Kidner founded as schoolboys at Westminster (see above, pp. 2, 5). Jack Simmons contributed the following articles:

1931   'The Ivatt Locomotives of the G.N.R.', 2, p. 4.
        'The Ivatt Locomotives of the G.N.R., II, The Atlantics', 2, p. 12.
        'The Ivatt Locomotives of the G.N.R., III' [mistitled 'II'], 2, p. 20.
        'The Ivatt Locomotives of the G.N.R., IV', 2, p. 28.
1936   'Notes on the Naming of Locomotives', 7, p. 15.
1937   'The Rev. Dr. Dionysius Lardner, 1793–1859', 8, p. 24.
1938   'George Bradshaw, 1801–1853', 9, p. 26.
1939   'The Great Western Museum, Paddington', 10, p. 9.
        'North Staffordshire Locomotives', 10, p. 29.
        'The Rastrick Collection, University of London', 10, p. 33 (unsigned).

His literary interests were also represented in a school publication:

1934   'The Poetry of the Seventeenth Century', *The Westminster Chameleon*, No. 1, May 1934, pp. 15–20.

1940

'The Waller and Kirk Papers' (in 'Notes and News'), *Bodleian Library Record*, 1 (1938–41), pp. 145–7.

1941

'Some Letters from Bishop Ward of Exeter 1663–1667', *D&CN&Q*, 21 (1940–1), pp. 222–7, 282–8, 329–36, 359–68.
'A Suppressed Passage in *Livingstone's Last Journals* Relating to the Death of Baron von der Decken', *Journal of the Royal African Society*, 40, pp. 335–46.

1942

(With M.F. Perham) *African Discovery: An Anthology of Exploration*. Faber & Faber, 280 pp. Abridged version published by Penguin, 1948; 2nd edn, 1957.
'Some Letters from the Bishops of Exeter 1668–1688', *D&CN&Q*, 22 (1942–6), pp. 43–8, 72–8, 108–12, 143–4, 153–5, 166–8.
'The Thornton Papers' (in 'Notes and News'), *Bodleian Library Record*, 2 (1941–9), pp. 39–40.
'An Unpublished Letter from Abraham Cowley', *Modern Language Notes*, 57, pp. 194–5.

Reviews

C. Birkby, *It's a Long Way to Addis*, 1942, *Listener*, 28 (July–December 1942), p. 760.

A.H. Brodrick, *North Africa*, 1942, *Listener*, 28 (July–December 1942), p. 728.

P. Knaplund, *The British Empire, 1815–1939*, 1941, *Listener*, 28 (July–December 1942), p. 664.

T. Lever, *The Life and Times of Sir Robert Peel*, 1942, *Listener*, 27 (January–June 1942), p. 762.

M.E. Macmillan, *Savage Landor*, 1942, *Listener*, 27 (January–June 1942), p. 728.

G. Mattingly, *Catherine of Aragon*, 1942, *Listener*, 27 (January–June 1942), pp. 504, 507.

E. Rosenthal, *The Fall of Italian East Africa*, 1942, *Listener*, 27 (January–June 1942), p. 761.

1943

'A Frenchman's Impressions of Devon and Cornwall in 1706', *D&CN&Q*, 22 (1942–6), pp. 172–4, 183–6.

'Reflections on Southey's Centenary', *T&T*, 24, pp. 230, 232.

'A Representative Englishman: Robert Southey', *Country Life*, 93 (January–June 1943), pp. 532–3.

'Some Early Railway Bye-laws', *RM*, 89, p. 237.

Reviews

J.W. Blake (ed.), *Europeans in West Africa, 1450–1560*, 1942, *History*, 28, pp. 114–15.

'Critic's Commentary', *T&T*, 24, p. 911; review of A. Campbell, *Smuts and Swastika*, 1943.

S.E. Crowe, *The Berlin West Africa Conference, 1884–5*, 1942, *Listener*, 29 (January–June 1943), p. 278.

I.E. Edwards, *Towards Emancipation: A Study in South African Slavery*, 1942, *Oxford Magazine*, 61 (1942–3), pp. 367–8.

M.V. Jackson, *European Powers and South East Africa*, 1942, *EHR*, 58, pp. 372–3.

W. Oakeshott, *Founded Upon the Seas*, 1942, *Listener*, 29 (January–June 1943), p. 24.

E. Walker, *British Empire: Its Structure and Spirit*, 1943, *Africa*, 14 (1943–4), pp. 221–2.

1944

'Armada Dagger', *D&CN&Q*, 22 (1942–6), pp. 224–6.

'Edmund Elys: Rector of East Allington 1659–1689', *D&CN&Q*, 22 (1942–6), pp. 209–12, 219–21, 231–3 (reprinted in *Parish and Empire*, 1952).

'Kinglake and Eothen', *T&T*, 25, p. 1127.

'Reflections on Southey's Centenary', *T&T*, 25, pp. 230–2.

Reviews

'America, Britain and Africa', *T&T*, 25, p. 468; review of A.N. Cook, *British Enterprise in Nigeria*, 1943; Sir A. Burns, *History of Nigeria*, 1942; Lord Hailey, *The Future of Colonial Peoples*, 1943.

A.N. Cook, *British Enterprise in Nigeria*, 1943, *Africa*, 14 (1943–4), pp. 421–2.

'County History', *T&T*, 25, p. 514; review of A. Macdonald, *Worcestershire in English History*, 1943.

'The Essential Wordsworth', *T&T*, 25, p. 424; review of E. de Selincourt (ed.), *The Poetical Works of William Wordsworth*, vol. II, 1944; J.C. Smith, *A Study of Wordsworth*, 1944.

E. Huxley and M. Perham, *Race and Politics in Kenya*, 1944, *Africa*, 14 (1943–4), pp. 469–71.

'The Liberator', *New Statesman and Nation*, 27 (January–June 1944), pp. 408–9; review of C.V. Wedgwood, *William the Silent*, 1944.

'Our Cartography', *T&T*, 25, p. 1090; review of E. Lynam, *British Maps and Map-makers*, 1944.

'Plural Societies', *T&T*, 25, pp. 990–1; review of E.A. Walker, *Colonies*, 1944.

'Provincial History', *T&T*, 25, pp. 903–4; review of R.W. Ketton-Cremer, *Norfolk Portraits*, 1944.

'A Restoration Divine', *T&T*, 25, pp. 1012–13; review of P.H. Osmond, *Isaac Barrow: His Life and Times*, 1944.

'The Thunderer', *T&T*, 25, p. 78; review of D. Hudson and H. Child, *Thomas Barnes of the Times*, 1943.

1945

Southey, Collins, 256 pp., illus. US edition, New Haven, Connecticut: Yale University Press, 1948. Reprinted by Kennikat Press, 1968.

'A Devonshire Town' [Totnes], *T&T*, 26, p. 725.

'Sydney Smith', *T&T*, 26, pp. 166–7.

'Taunton Church', *T&T*, 26, pp. 270–2.

'William Jessop (1745–1814)', *D&CN&Q*, 22 (1942–6), pp. 267–70 (reprinted in *Parish and Empire*, 1952, as 'William Jessop: Civil Engineer').

Reviews

'Colonial Development', *T&T*, 26, p. 357; review of H.W. Foster, *Ourselves and Empire*, 1944.

'The Expansion of England', *T&T*, 26, p. 378; review of J.A. Williamson, *Great Britain and the Empire*, 1944.

'Explorer's Guide', *T&T*, 26, p. 694; review of F.K. Ward, *Modern Exploration*, 1945.

George Saintsbury: The Memorial Volume, 1945, *T&T*, 26, p. 881.

'The Georgian Church', *T&T*, 26, pp. 482–4; review of W.K.L. Clarke, *Eighteenth-Century Piety*, 1944.

'The Historical Imagination', *T&T*, 26, pp. 1059–60; review of A.L. Rowse, *West Country Stories*, 1945.

F.A. Lea, *Shelley and the Romantic Revolution*, 1945, *Listener*, 34 (July–December 1945), p. 189.

L.J. Ragatz, *A Bibliography for the Study of African History in the Nineteenth and Twentieth Centuries*, 1943, *Africa*, 15, pp. 167–8.

'Victorian Missionary', *T&T*, 26, p. 839; review of J.P.R. Wallis (ed.), *The Matabele Journals of Robert Moffat*, 1945.

1946

Reviews

E. Blunden, *Shelley*, 1946, *Listener*, 35 (January–June 1946), p. 589.

'Britain in Pictures', *T&T*, 27, p. 472; review of A. Elton, *British Railways*, 1946; C. Hadfield, *English Rivers and Canals*, 1946.

'The English Parish', *T&T*, 27, p. 1103; review of W.E. Tate, *The Parish Chest*, 1946.

'The Explorer's Mind', *T&T*, 27, p. 109; review of Sir R. Coupland, *Livingstone's Last Journey*, 1945.

'Macaulay in India', *Manchester Guardian*, 7 September 1946, p. 4; review of C.D. Dharker (ed.), *Lord Macaulay's Legislative Minutes*, 1946.

'A Traveller's Tales', *T&T*, 27, p. 87; review of P.B. du Chaillu, *Exploration and Adventures in Tropical Africa*, 1845.

'La Vie de Provence', *T&T*, 27, pp. 255–6; review of C.A. Oglander, *Nunwell Symphony*, 1946.

J.P.R. Wallis (ed.), *The Matabele Journals of Robert Moffat, 1829–1860*, 1945, *Africa*, 16, pp. 271–3.

J.P.R. Wallis (ed.), *The Matabele Mission*, 1945, *Africa*, 16, pp. 274–5.

M. Wight, *The Development of the Legislative Council*, 1946, *Listener*, 35 (January–June 1946), p. 553.

'Wordsworth's Art', *T&T*, 27, pp. 927–8; review of E. de Selincourt and H. Darbishire (eds), *The Poetical Works of William Wordsworth*, vol. III, 1946.

1947

The Maryport and Carlisle Railway (Oakwood Library of Railway History, no. 4). Chislehurst: Oakwood Press, 34 pp., illus.

'A Georgian Eccentric: Richard Coffin of Portledge', *D&CN&Q*, 23 (1947–9), pp. 33–4.

'In the Cloisters at Wells', *WCM*, 2 (1947–8), pp. 252–6 (reprinted in *Parish and Empire*, 1952).

'National Parks in South Devon', *Countrygoer* ('Countrygoer Books'), No. 10, pp. 45–50.

'Rajah Brooke at Burrator', *WCM*, 2 (1947–8), pp. 51–3.

'Three Unpublished Brunel Letters', *RM*, 93, p. 246.

Reviews

J.R. Crocker, *On Governing Colonies*, 1947, *Listener*, 37 (January–June 1947), p. 635.
A.M.D. Hughes, *The Nascent Mind of Shelley*, 1947, *Listener*, 37 (January–June 1947), p. 553.
'Scott's Last Year', *T&T*, 28, p. 362; review of J.G. Tait (ed.), *The Journal of Sir Walter Scott, 1829–32*, 1946.

1948

Editor (with W.G. Hoskins) of *The Leicestershire and Rutland Magazine*, 1, nos. 1–2 (December 1948, March 1949).
Editor, *TLAS*, 1948–61 (*TLAHS* from 1956).
The British Commonwealth. Bureau of Current Affairs, 52 pp.
'Devon', in C.E.M. Joad (ed.), *The English Counties Illustrated*, Odhams, pp. 255–68.
'An English Story', *Listener*, 39 (January–June 1948), p. 812.
'European Expansion in Africa and South-east Asia', in M. Beloff (ed.), *History: Mankind and his Story*. Odhams, pp. 275–94.
'The Great Western Passes', *WCM*, 3, pp. 46–8 (reprinted in J.C. Trewin (ed.), *West Country Book*, No. 1, 1949, pp. 139–41, and in *Parish and Empire*, 1952, as 'The End of the Great Western Railway').
'A Huguenot in Somerset', *WCM*, 3, pp. 208–11.
'The Proconsuls', *Listener*, 39 (January–June 1948), pp. 932–3, 936.

Reviews

'Leicester and England', *The Leicestershire and Rutland Magazine*, 1, no. 1 (December 1948), pp. 50–2; review of C.D.B. Ellis, *History in Leicester*, 1948.

1949

From Empire to Commonwealth: Principles of British Imperial Government. Odhams, 240 pp.
'The Civil War in the West', 'Monmouth's Rebellion', in A.L. Rowse (ed.), *The West in English History*. Hodder and Stoughton, pp. 83–91, 105–15.
'Falmouth, Truro, Fowey and Looe in 1759', *D&CN&Q*, 23 (1947–9), pp. 280–1.
'Murray's Berkshire', *Times Literary Supplement*, 26 August 1949, p. 553 [letter].
'Notes on a Leicester Architect: John Johnson, 1732–1814', *TLAS*, 25, pp. 144–58 (reprinted in *Parish and Empire*, 1952, as 'A Leicester Architect').
'The Proconsuls', in *Ideas and Beliefs of the Victorians: An Historic Revaluation of the Victorian Age* [talks broadcast on the Third Programme]. Sylvan Press, pp. 410–16 (reprinted in *Parish and Empire*, 1952, as 'The Victorian Proconsuls').
'Visitors at Burley', *The Leicestershire and Rutland Magazine*, vol. I, no. 2 (March 1949), pp. 63–7.

Reviews

'Eighteenth-Century Crisis', *Spectator*, 183 (July–December 1949), p. 748; review of
H. Butterfield, *George III, Lord North and the People, 1779–80*, 1949.
'The English Midlands', *Spectator*, 183 (July–December 1949), p. 86; review of
W.G. Hoskins, *Midland England*, 1949; D. Gray, *Nottingham Through Five
Hundred Years: A Short History of Town Government*, 1949.
H. Evans (ed.), *Men in the Tropics*, 1949, *Listener*, 42 (July–December 1949), p. 882.
'A Georgian Election', *Spectator*, 183 (July–December 1949), p. 270; review of
R.J. Robson, *The Oxfordshire Election of 1754*, 1949.
C. Jeffries, *Partners for Progress: The Men and Women of the Colonial Service*, 1949,
Listener, 42 (July–December 1949), p. 504.
'King's Friend', *Spectator*, 182 (January–June 1949), pp. 860–2; review of N.S.
Jucker (ed.), *The Jenkinson Papers, 1760–1766*, 1949.
D. Middleton, *Baker of the Nile*, 1949, *Listener*, 42 (July–December 1949), p. 734.
E.C.G. Weeks, *History of St. Nicholas' Parish Church, Leicester*, 1949, *TLAS*, 25,
pp. 166–8.
'Wordsworth and his Editors', *Spectator*, 183 (July–December 1949), pp. 578–80;
review of E. de Selincourt and H. Darbishire (eds.), *The Poetical Works of William
Wordsworth*, vol. V, 1949.

1950

*Local, National and Imperial History: An Inaugural Lecture Delivered at University
College Leicester* [16 February 1948]. Leicester: University College, 22 pp.
(reprinted in *Parish and Empire*, 1952).
'De Quincey', 'Gay', 'Hazlitt', 'Southey', entries in *Chamber's Encyclopaedia*, New
Edition, respectively vol. IV, pp. 466–7; vol. VI, pp. 192, 781; vol. XII, p. 771.
'Falkland, the Good Landlord of Great Tew', *Listener*, 43 (January–June 1950),
p. 373 [talk broadcast on the Midland Home Service].
'West-Country Guide-Books', *WCM*, 5, pp. 8–10 (reprinted in *Parish and Empire*,
1952).

Reviews

C.E. Carrington, *The British Overseas*, 1950, *Listener*, 44 (July–December 1950),
p. 655.
J. Coatman, *The British Family of Nations*, 1950, *Listener*, 44 (July–December 1950),
p. 156.
A.E. Eagar, *Cole Orton and the Beaumonts in Art, History and Literature*, 1949,
TLAS, 26, pp. 140–2.
'Guide Books', *Spectator*, 184 (January–June 1950), pp. 832–4; review of E.S.
Roscoe, *Buckinghamshire*, rev. edn, 1950; A.L. Salmon, *Cornwall*, rev. edn, 1950;
T.L. Tudor, *Derbyshire*, rev. edn, 1950; R.F. Jessup, *Kent*, rev. edn, 1950.
L.G.H. Horton-Smith, *For Them That Are Yet For To Come*, 1950, *TLAS*, 26, pp.
143–4.

'Settlement', *Manchester Guardian*, 29 August 1950, p. 4; review of A.F. Hattersley,
    The British Settlement of Natal, 1950.
'A Tour of Wessex', *Spectator*, 184 (January–June 1950), p. 588; review of R. Dutton,
    Wessex, 1950.

1951

The City of Leicester, Leicester: City of Leicester Publicity Dept, 64 pp., illus. 2nd
    edn, 1957; 3rd edn, 1960.
Editor, *Journeys in England: An Anthology*, Odhams, 288 pp., including Editor's
    'Introduction', pp. 13–34, illus. 2nd edn, David and Charles, Newton Abbot, 1969.
Editor, *Letters from England* by Robert Southey. Cresset Press, xxvi + 494 pp.,
    including 'Editor's Introduction', pp. ix–xxvi. Reprinted by Alan Sutton, Stroud,
    1984.
 'Drawings of Devon and Cornwall in the Sutherland Collection at Oxford',
    D&CN&Q, 24 (1950–1), pp. 170–4.
'Foreword' to G.R. Mellor, *British Imperial Trusteeship, 1783–1850*. Faber & Faber,
    pp. 7–8.

Reviews

W.G. Hoskins, *Essays in Leicestershire History*, 1950; idem, *The East Midlands and
    the Peak*, 1951, *TLAS*, 27, pp. 107–8.
L.H. Irvine (compiler), *TLAS: General Index to Volumes I–XX (1855–1939)*, 1951,
    TLAS, 27, p. 109.
'Mr. Rowse's Masterpiece', *National and English Review*, 136 (January–June 1951),
    pp. 44–5; review of A.L. Rowse, *The England of Elizabeth*, 1950.
E. Morris and L. Cox, *The Church of St. Mary Magdalene, Knighton*, 1951, *TLAS*, 27,
    p. 114.
G. Paget and L. Irvine, *Leicestershire*, 1950, *TLAS*, 27, pp. 112–13.
F.E. Skillington, *The Plain Man's History of Leicester*, 1950, *TLAS*, 27, pp. 113–14.

1952

Parish and Empire: Studies and Sketches. Collins, 256 pp., illus. (16 essays, most
    reprinted).
'African Nationalism', *Fortnightly*, New Series, 171 (January–June 1952), pp. 241–2
    (comment on 'African Nationalism' by Sir Charles Dundas, *ante*, pp. 147–51).
'Papers Relating to Leicestershire Published in *Reports and Papers of the Associated
    Architectural Societies, 1877–1931*', *TLAS*, 28, pp. 84–8.
'Transition in West Africa', in Sir D. Shiels, *The British Commonwealth*, Odhams,
    pp. 195–205.

Reviews

F. Burgess, *English Churchyard Sculpture*, n.d., *TLAS*, 28, p. 103.
C.D.B. Ellis, *Leicestershire and the Quorn Hunt*, 1951, *TLAS*, 28, pp. 96–7.
C.D. Hawley (ed.), *List of Antiquities in the Administrative County of Surrey*, 4th edn, 1951, *TLAS*, 28, pp. 102–3.
C. Howard (ed.), *West African Explorers*, 1951, *Listener*, 47 (January–June 1952), pp. 801, 803.
Leicester Museum and Art Gallery Bulletin (3rd ser., vol. I, no. 1), 1952, *TLAS*, 28, pp. 98–9.
The Preservation of Our Churches, 1952, *TLAS*, 28, pp. 100–2.

1953

Founder (with M. Robbins) of *JTH*; Editor, 1953–73 (with M. Robbins until 1965).
Compiler, 'Transport Bibliography' [annual survey], *JTH*, 1–7 (1953–65).
(With M. Robbins) 'Editorial', *JTH*, 1 (1953–4), pp. 1–2.
Editor of the series 'A New Survey of England' [historical and topographical guides to English counties], Collins, 1953–4, including 'Editor's Introduction', in M. Robbins, *Middlesex*, 1953, pp. xiii–xvii, and W.G. Hoskins, *Devon*, 1954, pp. xi–xv.
The Coronation: Some Historical and Religious Aspects of the Ceremony, Leicester: Leicester Corporation (City of Leicester official coronation souvenir, 2 June 1953).
'Sir Reginald Coupland', *Britain Today*, 202, pp. 6–10.

Reviews

E.L. Ahrons, *Locomotive and Train Working in the Latter Part of the Nineteenth Century*, vols. I, II, IV, 1951–2, *JTH* 1 (1953–4), pp. 62–3, 131–2.
'Colonial Documents', *Spectator*, 191 (July–December 1953), p. 708; review of V. Harlow and F. Madden, *British Colonial Developments, 1774–1834*, 1953.
T. Fuller (ed. J. Freeman), *The Worthies of England*, 1952, *TLAS*, 29, p. 87.
C.N. Hadfield, *Charnwood Forest*, 1952, *TLAS*, 29, p. 83.
Handlist of Leicestershire Parish Register Transcripts, 1953, *TLAS*, 29, p. 84.
D. Knowles and J.K. St Joseph, *Monastic Sites from the Air*, 1952, *TLAS*, 29, p. 85.
A Memorandum on the Ancient Monuments Acts, 1952; *The Preservation of Buildings of Historic Interest: A Note on the Town and Country Planning Act 1947*, 1953, *TLAS*, 29, p. 88.
J.E. Neale, *Elizabeth and her Parliaments*, 1953, *Apollo*, 58 (July–December 1953), pp. 196–7.
J. Nickalls (ed.), *The Journal of George Fox*, 1952, *TLAS*, 29, pp. 86–7.
J. Pudney, *The Thomas Cook Story*, 1953, *JTH* 1 (1953–4), p. 131.
R.A. Whitehead and F.D. Simpson, *The Story of the Colne Valley*, 1951, *JTH*, 1 (1953–4), pp. 130–1.

1954

(With M. Robbins) Editorial, *JTH*, 1 (1953–4), pp. 133–4.

Foreword to Sir Reginald Coupland, *Welsh and Scottish Nationalism*. Collins, pp. vii–xi.

'Railway History in English Local Records', *JTH*, 1 (1953–4), pp. 155–69.

Reviews

E.L. Ahrons, *Locomotive and Train Working in the Latter Part of the Nineteenth Century*, vol. VI, 1954, *JTH*, 1 (1953–4), pp. 255–6.

C.H. Ellis, *The Midland Railway*, 1953, *JTH*, 1 (1953–4), pp. 195–6.

R. Gunnis, *Dictionary of British Sculptors, 1660–1851*, 1953, *TLAS*, 30, pp. 137–9.

Handlist of Records of Leicester Archdeaconry, 1954, *TLAS*, 30, p. 132.

R.W. Kidner, *The South Eastern Railway and the S.E. and C.R.*, 1953; E.L. Ahrons, *Locomotive and Train Working in the Latter Part of the Nineteenth Century*, vol. V, 1953, *JTH*, 1 (1953–4), pp. 193–4.

R.C. Overton, *Gulf to Rockies*, 1953, *Business History Review*, 28, pp. 191–3.

'Readable History', *T&T*, pp. 53–4; review of J.A. Williamson, *The Tudor Age*, 1953.

The Victoria History of Leicestershire, vol. II, 1954, *TLAS*, 30, pp. 139–40.

C. Woodforde, *English Stained and Painted Glass*, 1954, *TLAS*, 30, p. 137.

1955

Leicester and Its University College. Leicester: University College, 50 pp. (reprinted as *Leicester and Its University*, Leicester: Leicester University Press, 1957).

Livingstone and Africa ['Men and their Times' series]. English Universities Press, x + 180 pp.

'Railways', in *The Victoria History of Leicestershire*, vol. III, pp. 108–27.

(With H.M. Colvin) 'Staunton Harold Chapel', 'King's Norton Church', 'Church Langton Church', *Archaeological Journal*, 112, pp. 173–6, 177–8, 191–2.

'A Valuable Local Society', *Listener*, 17 Feb. 1955, p. 277 (on the centenary of the Leicestershire Archaeological Society).

Reviews

'Historian's Mine', *T&T*, 36, pp. 1339–40; review of C. Read, *Mr Secretary Cecil and Queen Elizabeth*, 1955.

C.E. Lee, *The Swansea and Mumbles Railway*, 1954, *JTH*, 2 (1955–6), p. 64.

'Leviathan Awakes', *T&T*, 36, pp. 1295–6; review of A.L. Rowse, *The Expansion of Elizabeth England*, 1955.

P.J. Wexler, *La Formation du vocabulaire des chemins de fer en France (1772–1848)*, 1955, *JTH*, 2 (1955–6), pp. 125–6.

1956

(With M. Robbins) 'Editorial', *JTH*, 2 (1955–6), pp. 129–31.

'For and Against the Locomotive', *JTH*, 2 (1955–6), pp. 144–51 (revised version published in *The Express Train...*, 1994, as 'The Horse and the Locomotive').

'The Last 40 Years', *Twentieth Century*, 159, pp. 112–22 (contribution to a 'Redbrick Universities' issue).

Reviews

N.R.P. Bonsor, *North Atlantic Seaway*, 1955, *JTH*, 2 (1955–6), pp. 187–8.

T.C. Cochran, *Railroad Leaders, 1845–1890*, 1953, *JTH*, 2 (1955–6), p. 190.

K.O. Diké, *Trade and Politics in the Niger Delta, 1830–1885*, 1956, *West Africa*, 40, p. 133.

Eighteen Postcards [published by Leicester Museums and Art Gallery], 1956, *TLAHS*, 32, pp. 103–4.

G.O. Holt, *A Short History of the Liverpool and Manchester Railway*, 1955, *JTH*, 2 (1955–6), pp. 255–6.

Local Portraits [catalogue published by Leicester Museums and Art Gallery], 1956, *TLAHS*, 32, p. 103.

London General: The Story of the London Bus, 1856–1956, 1956; T.S. Lascelles, *The City and South London Railway*, 1955; C.E. Lee, *The Metropolitan District Railway*, 1956, *JTH*, 2 (1955–6), pp. 252–3.

W. Lord, *A Night to Remember*, 1956; R.M. Wilson, *The Big Ships*, 1956, *JTH*, 2 (1955–6), pp. 249–50.

B. Stewart, *The Library and the Picture Collection of the Port of London Authority*, 1956, *JTH*, 2 (1955–6), pp. 255–6.

1957

'Business Records and the Historian', *Aslib Proceedings*, 9, pp. 199–202.

'County was Challenge to Author', *Leicester Mercury*, 21 November 1957 [about W.G. Hoskins].

'Introduction' to *Leicester and the Midland Railway* [exhibition catalogue]. Leicester: Leicester City Museums and Art Gallery, pp. 4–8. Reprinted as 'Leicester and the Midland Railway' in *Journal of the Stephenson Locomotive Society*, 33 (1957), pp. 230–3.

(With J.L. Hunter) 'University of Leicester', *Nature*, 179, pp. 839–49.

'The University of Leicester', *Universities Review*, 30, pp. 5–10.

Reviews

G. Blake, *The Ben Line*, 1956, *JTH*, 3 (1957–8), pp. 59–60.

C.R. Clinker, *Birmingham and Derby Junction Railway*, 1956, *EHR*, 72, p. 384.

F.E. Hyde, *Blue Funnel*, 1956, *JTH*, 3 (1957–8), p. 128.

G.R. Taylor and I.D. Neu, *The American Railroad Network, 1861–1890*, 1956, *JTH*, 3 (1957–8), pp. 126–7.

1958

New University. Leicester: Leicester University Press, 234 pp., illus.
'Canada', 'Australia', 'Africa', in J. Bowle (ed.), *The Concise Encyclopaedia of World History.* Hutchinson, pp. 402–6, 414–18, 444–50.
'Leicestershire', in J. Betjeman (ed.), *Collins Guide to English Parish Churches.* Collins, pp. 229–33.
'The Scottish Records of the British Transport Commission', *JTH*, 3 (1957–8), pp. 158–67.

Reviews

D.S. Barrie, *The Brecon and Merthyr Railway*, 1957; J.M. Dunn, *The Wrexham, Mold and Connah's Quay Railway*, 1957; O.S. Nock, *Branch Lines*, 1957, *JTH*, 3 (1957–8), pp. 182–3.
British Railways Pre-Grouping Atlas and Gazetteer, 1958, *JTH*, 3 (1957–8), pp. 244–5.
R. Bucknall, *Boat Trains and Channel Packets*, 1957; R. Scott, *The Gateway of England*, n.d., *JTH*, 3 (1957–8), pp. 183–4.
A.W. Currie, *The Grand Trunk Railway of Canada*, 1958, *JTH*, 3 (1957–8), pp. 239–40.
C.H. Ellis, *A Picture History of Ships*, 1957, *JTH*, 3 (1957–8), pp. 175–6.
'Midland Life and Landscapes', *Economist*, 15 February 1958, pp. 574–5; review of W.G. Hoskins, *The Midland Peasant*, 1957; idem., *Leicestershire: An Illustrated Essay on the History of the Landscape*, 1957.
Postcards from Leicester City Museums and Art Gallery, 1958, *TLAHS*, 34, p. 93.
The Victoria History of Leicestershire, vol. IV, 1958, *TLAHS*, 34, pp. 90–3.

1959

'Brooke Church, Rutland: With Notes on Elizabethan Church-Building', *TLAHS*, 35, pp. 36–55.
'The Opening of Tropical Africa, 1870–85', in *Cambridge History of the British Empire.* Cambridge: Cambridge University Press, vol. III, pp. 65–94.
'Reference and Special Libraries: A Reader's View', in *Report of a Conference of the Reference Special Libraries and Information Section of the Library Association* (held 7 April 1959), pp. 5–14.
'Sir Reginald Coupland, 1884–1952', *Proceedings of the British Academy*, 45, pp. 287–95.
'South Western versus Great Western: Railway Competition in Devon and Cornwall', *JTH*, 4 (1959–60), pp. 13–36.

Reviews

P.B. Chatwin, *Incidents in the Life of Matthew Holbeche Bloxam*, 1959, *TLAHS*, 35, pp. 90–1.
R.S. Fletcher, *Eureka: From Cleveland by Ship to California, 1849–50*, 1959, *JTH*, 4 (1959–60), p. 132.
P.E. Hunt, *The Story of Melton Mowbray*, 1959, *TLAHS*, 35, pp. 89–90.
R.W. Kidner, *The Southern Railway*, 1958, *JTH*, 4 (1959–60), pp. 126–7.
J.M. Lee, *Leicestershire History: A Handlist to Printed Sources in the Libraries of Leicester*, 1958, *TLAHS*, 35, p. 90.
E. Melling (ed.), *Kentish Sources I: Some Roads and Bridges*, 1959, *JTH*, 4 (1959–60), pp. 131–2.
E.M. Patterson, *The Belfast and County Down Railway*, 1958, *JTH*, 4 (1959–60), pp. 127–8.
H. Rees, *British Ports and Shipping*, 1958, *JTH*, 4 (1959–60), pp. 62–3.

1960

(With D.T.-D. Clarke) 'Old Leicester: An Illustrated Record of Change in the City', *TLAHS*, 36, pp. 45–8.
'Photographic Records for Local History', *Library Association Record*, 62, pp. 328–33.

Reviews

E.W. Bovill, *The England of Nimrod and Surtees, 1815–54*, 1959, *JTH*, 4 (1959–60), pp. 190–1.
G. Carnall, *Robert Southey and his Age*, 1960; R. Southey (ed. A. Cabral), *Journals of a Residence in Portugal 1800–1801, and a Visit to France 1838*, 1960, *Listener*, 63 (January–June 1960), pp. 463–4.
G. Dow, *Great Central*, vol. I, 1959, *JTH*, 4 (1959–60), pp. 252–3.
R.B. Martin, *The Dust of Combat: A Life of Charles Kingsley*, 1960, *Listener*, 63 (January–June 1960), pp. 717, 720.
P. Quennell (ed.), *Memoirs of William Hickey*, 1960, *Listener*, 64 (July–December 1960), p. 853.
T.B. Sands, *The Midland and South-Western Junction Railway*, 1959, *JTH*, 4 (1959–60), pp. 257–8.
The Victoria History of Wiltshire, vol. IV, 1959, *JTH*, 4 (1959–60), pp. 189–90.
P.A. Wright, *Traction Engines*, 1959, *Agricultural History Review*, 8, p. 123.

1961

Editor of the series 'A Visual History of Modern Britain'. Vista Books, 1961–8.
England in Colour, Batsford, 71 pp. (Introduction and notes).

The Railways of Britain: An Historical Introduction. Routledge and Kegan Paul, xii + 264 pp., illus. Reprinted with corrections, 1965; 2nd edn, Macmillan, 1968; 3rd edn, extensively revised, Macmillan, 1986; 4th edn, as *The Railways of Britain: A Journey through History*, Sheldrake/Uralia, 1991.

Reviews

C.H. Ellis, *The Beauty of Railways*, 1960, *JTH*, 5 (1961–2), p. 132.
C. Highet, *The Wirral Railway*, 1961, *JTH*, 5 (1961–2), pp. 129–30.
D. St J. Thomas, *A Regional History of the Railways of Great Britain*, vol. I, 1960; L.H. Ruegg, *The Salisbury and Yeovil Railway*, 1960, *JTH*, 5 (1961–2), pp. 62–3.
R.J. Woodfin, *The Centenary of the Cornwall Railway*, 1960, *JTH*, 5 (1961–2), p. 66.

1962

Transport ['Visual History of Modern Britain' series]. Vista Books, x + 206 pp., illus.
'In Memoriam: Harry Percy Gee', *Leicester University Gazette*, April 1962, p. 106.
'Sir Robert Martin' [obituary], *TLAHS*, 37 (1961–2), pp. 60–3.

Reviews

C. Colles, *A Survey of the Roads of the United States of America*, 1961, *JTH*, 5 (1961–2), pp. 194–5.
The Victoria History of Yorkshire: The City of York, 1961, *JTH*, 5 (1961–2), pp. 258–9.

1963

'The English County Historians: A Lecture Given on 6 May 1961 to the Hunter Archaeological Society and Others', *Transactions of the Hunter Archaeological Society*, 8, pp. 272–87.
(With M. Robbins) 'Ten Years of Transport History', *JTH*, 6 (1963–4), pp. 1–2.
'Three Midland Towns' [Northampton, Leicester, Nottingham], *Northamptonshire Past and Present*, 3, pp. 136–40.

Reviews

D.S. Barrie, *The Barry Railway*, 1962, *JTH*, 6 (1963–4), pp. 122–3.
W.G. Hoskins, *Provincial England: Essays in Social and Economic History*, 1963, *Economist*, 16 November 1963, p. 677.
W.T. Jackman, *Transportation in Modern England*, 1962, *JTH*, 6 (1963–4), pp. 63–4.
'100 Years Underground', *Yorkshire Post*, 5 September 1963; review of T.C. Barker and M. Robbins, *A History of London Transport*, vol. I, *The Nineteenth Century*, 1963.
R. Sellick, *The West Somerset Mineral Railway*, 1962, *JTH*, 6 (1963–4), p. 126.

1964

Leicester: A Brief Guide to the City. Leicester: Leicester Corporation, 32 pp. Frequently revised.

Editorial Notes, *JTH*, 6 (1963–4), pp. 197–8.

'The Swiss Museum of Transport', *JTH*, 6 (1963–4), pp. 142–9.

Reviews

G. Dow, *Great Central*, vol. II, 1962, *JTH*, 6 (1963–4), p. 195.

C.R.V. Gibbs, *British Passenger Liners of the Five Oceans*, 1963, *JTH*, 6 (1963–4), pp. 259–60.

O.S. Nock, *Continental Main Lines*, 1963, *JTH*, 6 (1963–4), p. 196.

1965

Britain and the World: A Study of Power and Influence ['Visual History of Modern Britain' series]. Studio Vista, 286 pp., illus.

Foreword to G. Ottley, *A Bibliography of British Railway History.* Allen and Unwin, pp. 11–12 (pp. 9–10 of 2nd edn, 1983).

'Industrial Archaeology', *Listener*, 74, pp. 695–7.

'Introduction' to B. Blake (ed.), *Industrial Archaeology: A Guide to the Technological Revolution of Britain.* BBC, pp. 1–2.

Reviews

C.H. Ellis, *The Splendour of Steam*, 1965, *JTH*, 7 (1965–6), pp. 61–2.

H. Robinson, *Carrying British Mails Overseas*, 1964, *JTH*, 7 (1965–6), pp. 62–3.

1966

'Mid-Victorian Leicester', *TLAHS*, 41 (1965–6), pp. 41–56.

Reviews

G. Dow, *Great Central*, vol. III, 1965, *JTH*, 7 (1965–6), pp. 186–8.

1967

'Introduction' to S. Watts, *A Walk Through Leicester*, Leicester: Leicester University Press, pp. vii–xiii.

1968

Four Locomotives, Leicester: Leicester City Museums, 10 pp., illus.
Radio Leicester: The First Year: A Review, Leicester: Leicester Local Broadcasting Co., 11 pp.
St Pancras Station, Allen and Unwin, 120 pp., illus.
Introduction to Sir R. Coupland, *The Exploitation of East Africa 1856–1890: The Slave Trade and the Scramble*, 2nd edn. Faber & Faber, pp. v–viii.
Preface to A. Thornley, *Going into Print: Notes for the Guidance of Authors in Preparing Copy for the Printer, Correcting Proofs, etc.* Leicester: Armstrong-Thornley Printers, p. 3.

1969

Rugby Junction. Stratford-upon-Avon: Dugdale Society (Occasional Papers, No. 19), 25 pp. (revised version published in *The Express Train...*, 1994).
'Edwardian Elegance', *Clio*, 1, pp. 21–7.
'Introduction' to Sir A. Helps, *Life and Labours of Mr Brassey*. Evelyn, Adams and Mackay, pp. v–xvii.
'Twentieth Century Leicester', in C.D.B. Ellis, *History in Leicester*, 2nd edn. Leicester: City of Leicester Publicity Dept, 129–38. Revised version published in 3rd edn, 1976.

1970

Transport Museums in Britain and Western Europe. Allen and Unwin, 300 pp., illus.
Introduction to a reprint of *Murray's Handbook for Travellers in Switzerland, 1838* ['Victorian Library' series]. Leicester: Leicester University Press, pp. 9–29.
Introduction to a reprint of C. Torr, *Small Talk at Wreyland*. Bath: Adams and Dart, pp. v–xiv. Reprinted in paperback by Oxford University Press, 1979.
Introduction to J. Wilshere, *William Gardiner of Leicester, 1770–1853: Hosiery Manufacturer, Musician and Dilettante*. Leicester: Leicester Research Services, p. 6.
'Introductory Statement: Industrial Conservation', in N.A.F. Smith, *Victorian Technology and its Preservation in Modern Britain: A Report Submitted to the Leverhulme Trust*. Leicester: Leicester University Press, pp. ix–xi.
'Leicester Past and Present', in A.E. Brown (ed.), *The Growth of Leicester*. Leicester: Leicester University Press, pp. 87–92.
'The Pattern of Tube Railways in London: A Note on the Joint Select Committee of 1892', *JTH*, 7 (1965–6, published 1970), pp. 234–40.

1971

Editor of the series 'Classical County Histories'. Wakefield: EP Publishing, 1971–8.

A Devon Anthology. Macmillan, 285 pp. Reprinted in paperback, as No. 1 in the 'Devon Library', by Anthony Mott, 1983.

Life in Victorian Leicester. Leicester: Leicester Museums, ii + 82 pp.

'A Holograph Letter from George Stephenson', *JTH*, n.s., 1 (1971–2), pp. 108–15.

Introduction to a reprint of W.B. Adams, *English Pleasure Carriages*. Bath: Adams and Dart, pp. v–xviii.

Introduction to a reprint of J. Nichols, *History and Antiquities of the County of Leicester* ['Classical County Histories' series]. Wakefield: EP Publishers, vol. I, pp. v–xiii.

'Introduction' to a reprint of *The Railway Travellers' Handy Book of Hints, Suggestions and Advice before the Journey, on the Journey, after the Journey.* Bath: Adams and Dart, pp. 7–13.

'Mr Colin Ellis' [obituary], *TLAHS*, 45 (1969–70; published 1971), pp. 68–73.

'On Writing the History of Leicester', *Clio*, 3, pp. 5–10.

Preface and Introduction to *The Birth of the Great Western Railway: Extracts from the Diary and Correspondence of George Henry Gibbs*. Bath: Adams and Dart, pp. vii–viii, 1–14.

1972

'Communications and Transport', in N. Pye (ed.), *Leicester and Its Region*. Leicester: Leicester University Press for the Local Committee of the British Association, pp. 311–24.

'A Victorian Social Worker: Joseph Dare and the Leicester Domestic Mission', *TLAHS*, 46 (1970–1; published 1972), pp. 65–80.

Reviews

C.F.D. Marshall, *A History of British Railways down to the Year 1830*, 2nd edn, 1971, *JTH*, n.s., 1 (1971–2), pp. 193–4.

N.W. Webster, *Joseph Locke: Railway Revolutionary*, 1970, *JTH*, n.s., 1 (1971–2), pp. 188–9.

1973

Editor of the series 'Man on the Move'. Bath: Adams and Dart (Moonraker Press).

Introduction to a reprint of J. Wright, *The History and Antiquities of the County of Rutland* ['Classical County Histories' series]. Wakefield: EP Publishing, pp. v–xi. (Reprinted in *English County Historians*, 1978.) A special limited edition produced for Rutland County Council also contains 'Rutland County Council: A Brief History', 6 pp.

'The Power of the Railway', in H.J. Dyos and M. Wolff (eds.), *The Victorian City*. Routledge and Kegan Paul, pp. 277–310.

(With R.L. Goodstein et al.) 'Professor E.A. Stewardson' [obituary], *Times*, 11 Sept. 1973, p. 18.

Reviews

J. Marshall, *The Lancashire and Yorkshire Railway*, vol. III, 1972, *JTH* n.s., 2 (1973–4), p. 122.

G. Wilson, *London United Tramways – A History 1894–1939*, 1971, *EHR*, 88, pp. 236–7.

1974

Leicester Past and Present. Eyre Methuen. Vol. I, *Ancient Borough to 1860*, xvi + 192 pp., illus.; vol. II, *Modern City, 1860–1974*, xv + 159 pp., illus. Vol. I reprinted in paperback by Alan Sutton, Stroud, 1983.

A View of Leicester [Harry Peach Memorial Lecture, delivered on 12 March 1974]. Leicester: Leicester University Press, 23 pp.

Introduction to *Memoirs of a Stationmaster*, a reprint of H.A. Simmons, *Ernest Struggles*, vol. I, 1879. Bath: Adams and Dart, pp. v–x.

Introduction to a reprint of G.P. Neele, *Railway Reminiscences: Notes and Reminiscences of Half a Century's Progress in Railway Working, and of a Railway Superintendent's Life, Particularly on the London and North Western Railway.* Wakefield: EP Publishing, pp. x–xi.

Introductory Note to a reprint of O. Manning and W. Bray, *The History and Antiquities of the County of Surrey* ['Classical County Histories' series]. Wakefield: EP Publishing, vol. I, pp. v–vii.

'J.A. Hone, William Hone (1780–1842), Publisher and Bookseller: An Approach to early 19th century Radicalism', *Historical Studies*, 16, pp. 55–70.

'Transport Museums – A Visitor's View', *Transport Museums* [*Yearbook* of the International Association of Transport Museums], 1. Gdansk: Centraine Muzeum Morskie, pp. 52–7.

1975

Editor, *Rail 150: The Stockton and Darlington Railway and What Followed.* Eyre Methuen, 198 pp., illus., including Editor's 'The Railway in Britain 1825–1947', pp. 47–117.

Introduction to a reprint of Sir Richard Colt, Bt, *The Ancient History of Wiltshire* ['Classical County Histories' series]. Wakefield: EP Publishing, vol. I, pp. 7–11.

'Thomas Cook of Leicester', *TLAHS*, 49 (1973–4; published 1975), pp. 18–32.

Reviews

P. Howard Anderson, *Forgotten Railways: The East Midlands*, 1973; R. Keys and L. Porter, *The Manifold Valley and its Light Railway*, 1972, *Bulletin of Local History: East Midland Region*, 10, pp. 60–1.

W. Kidd, *Leicester Old and New*, 1975, *TLAHS*, 49 (1973–4; published 1975), p. 65.

'The Town in Urban History', *Literature and History*, No. 2, pp. 96–9; review of C.W. Chalklin, *The Provincial Towns of Georgian England: A Study of the Building Process, 1740–1820*, 1974; A. Armstrong, *Stability and Change in an English Country Town: A Social Study of York, 1801–51*, 1974; F.H. Hill, *Victorian Lincoln*, 1974.

1976

Great British Locomotives. John Pinches, xxii + 114 pp., illus. (private publication for purchasers of ingots depicting locomotives).

Introduction to a reprint of John P. Anderson, *The Book of British Topography: A Classified Catalogue of the Topographical Works in the Library of the British Museum Relating to Great Britain and Ireland*, 1881. Wakefield: EP Publishing, pp. vii–ix.

Introduction to a reprint of Sir George Findlay, *The Working and Management of an English Railway*, 6th edn, 1899. Wakefield: EP Publishing, pp. i–x.

'Museums in London', *The London Journal*, 2, pp. 250–65 ('Viewpoint' article).

Reviews

D. Stuart, *Edwardian Burton* (pt 1 of *County Borough: The History of Burton-upon-Trent 1901–74*), 1975, *Midland History*, 3, pp. 306–7.

1977

'History Observed: A Lecture given at the University of Leicester, 9 December 1975', *University of Leicester Convocation Review*, pp. 12–22.

'The Museums of London', *Museums Journal*, 77, pp. 15–18.

1978

Editor, *English County Historians*, First Series. Wakefield: EP Publishing, viii + 246, illus., including Editor's 'Preface', pp. vii–viii; 'The Writing of English County History', pp. 1–21; 'James Wright', pp. 44–55.

The Railway in England and Wales, 1830–1914, vol. I, *The System and its Working*. Leicester: Leicester University Press, 295 pp., illus. (Due to publishing difficulties other volumes in this projected series were published separately. See *The Railway in Town and Country*, 1986; *The Victorian Railway*, 1991.)

Introduction to a reprint of William Upcott, *A Bibliographical Account of the Principal Works Relating to English Topography*, 1818. Wakefield: EP Publishers, pp. v–x.

'The Lysons Brothers and their Work', '*Magna Britannia: Bedfordshire*', introductions to a reprint of D. and S. Lysons, *Magna Britannia: Bedfordshire*

['Classical County Histories' series]. Wakefield: EP Publishers, pp. v–xxiv, xxv–xxvii.

'The Lysons Brothers and their Work', '*Magna Britannia: Berkshire*', introductions to a reprint of D. and S. Lysons, *Magna Britannia: Berkshire* ['Classical County Histories' series]. Wakefield: EP Publishers, pp. v–xxiv, xxv–xxvii.

'The Lysons Brothers and their Work', '*Magna Britannia: Cambridgeshire*', introductions to a reprint of D. and S. Lysons, *Magna Britannia: Cambridgeshire* ['Classical County Histories' series]. Wakefield: EP Publishers, pp. v–xxiv, xxv–xxvii.

'Technology in History', *History of Technology*, 3, pp. 1–12 (paper read at Imperial College, London, 19 March 1977).

1979

A City and A County: Leicester and Leicestershire from the Ninth Century to the Twentieth. Leicester: Leicestershire Archaeological and Historical Society in Association with Leicester City Council, 16 pp.

A Selective Guide to England. Edinburgh and London: John Bartholomew and Son, 1979, xiv + 521 pp., illus.

Reviews

'Leopold II's Furious Ambitions', *Books and Bookmen*, 24, no. 11 (August 1979), pp. 13–15; review of B. Emerson, *Leopold II of the Belgians*, 1979; T. Gould, *In Limbo: The Story of Stanley's Rear Column*, 1979.

1980

Foreword to F. McKenna, *The Railway Workers, 1840–1970*. Faber & Faber, pp. 15–16.

Foreword to R.H.G. Thomas, *The Liverpool and Manchester Railway*. Batsford, pp. 7–8.

'Rail 150: 1975 or 1980?' *JTH*, 3rd ser., 1, pp. 1–8.

Reviews

'True Autobiography', *Books and Bookmen*, 25, no. 5 (February 1980), pp. 24–5; review of A.L. Rowse, *A Man of the Thirties*, 1979.

1981

Dandy-cart to Diesel: The National Railway Museum. H.M.S.O. for Science Museum and National Railway Museum, vii + 68 pp., illus.

1982

'The Idea of a University in Leicester', *University of Leicester Convocation Review*, pp. 13–22.

'The Railway in Victorian Cornwall, 1835–1914', *Journal of the Royal Institution of Cornwall*, n.s., 9, pt 1, pp. 11–29 (reprinted in *Back Track*, 1 (1987), pp. 148–57; revised version published in *The Express Train...*, 1994).

1983

Editor, *The Men Who Built Railways: A Reprint of F.R. Conder's 'Personal Recollections of English Engineers'*. Thomas Telford, xvi + 204 pp., illus., including 'Editor's Introduction', pp. i–ix (revised version published in *The Express Train...*, 1994, as 'Engineer, Contractor and Writer').

1984

The Victorian Hotel [H.J. Dyos Memorial Lecture, delivered on 15 May 1984]. Leicester: Victorian Studies Centre, University of Leicester, 30 pp. (revised version published in *The Express Train...*, 1994, as 'Railways and Hotels in Britain, 1839–1914').

'Cook, John Mason', 'Cook, Thomas', 'Fay, Sir Samuel', 'Fowler, Sir John', in D.J. Jeremy (ed.), *Dictionary of Business Biography*. Butterworths, vol. I, pp. 763–5, 766–9; vol. II, pp. 328–31, 412–14.

Reviews

R.L. Hills and D. Patrick, *Beyer Peacock: Locomotive Builders to the World*, 1982; B. Reed *Crew Locomotive Works and its Men*, 1982; C.H. Hewison, *Locomotive Boiler Explosions*, 1983, *JTH*, 3rd ser., 5, pp. 76–8.

1985

Image of the Train: A 10th Birthday Present for the National Railway Museum. Bradford: National Museum of Photography, Film and Television, 16 pp., illus. (accompanied exhibition in 1985–6 at the National Museum in Bradford, National Railway Museum in York, and Science Museum in London).

'Suburban Traffic at King's Cross, 1852–1914', *JTH*, 3rd ser., 6, pp. 71–8 (revised version published in *The Express Train...*, 1994).

1986

The Railway in Town and Country, 1830–1914. Newton Abbot: David and Charles, 400 pp., illus.
'Stroudley, William', in D.J. Jeremy (ed.), *Dictionary of Business Biography.* Butterworths, vol. III, pp. 392–6.

1987

Editor, *The Pattern of English Building* by Alec Clifton-Taylor, 4th edn. Faber and Faber, 480 pp., illus., including Editorial Note, pp. 7–8; Editor's Foreword, pp. 11–14.

Reviews

C. Wilson, *First with the News: The History of W.H. Smith, 1792–1872*, 1985, *JTH*, 3rd ser., 8, pp. 228–9.

1988

'The Diary of a London and North Western Engineman, 1855–62', *Back Track*, 2, pp. 66–8 (revised version published in *The Express Train...*, 1994).
Foreword to G. Ottley, *A Bibliography of British Railway History, Supplement: 7951–12956.* H.M.S.O., pp. 9–10.

Reviews

T.R. Gourvish, *British Railways, 1948–73: A Business History*, 1987, *EHR*, 103, pp. 1018–20.

1989

Reviews

J. Richards and J.M. MacKenzie, *The Railway Station: A Social History*, 1986, *EHR*, 104, pp. 247–8.

1991

Compiler, *Railways: An Anthology.* Collins, xii + 260 pp., including Compiler's 'Introduction', pp. 1–6.
The Victorian Railway. Thames and Hudson, 416 pp., illus. Reprinted in paperback, 1995.

1992

Reviews

G. Briwnant-Jones, *Welsh Steam*, 1991, *JTH*, 3rd ser., 13, pp. 202–3.

J. Freeman and D.H. Aldcroft (eds), *Transport in Victorian Britain*, 1988, *EHR*, 107, pp. 233–4.

M. Hunter and R. Thorne (eds), *Change at King's Cross*, 1990, *Economic History Review*, 2nd series, 45, pp. 430–1.

1993

Image of the Train: The Victorian Era. Bradford: National Museum of Photography, Film and Television, 31 pp., illus. (accompanied exhibition held 23 February–23 May 1993).

(With M. Robbins) 'Forty Years On: A Message from the Founding Editors', *JTH*, 3rd ser., 14, pp. iv–vi.

'Treffry, Joseph Thomas', in C.S. Nicholls (ed.), *The Dictionary of National Biography: Missing Persons*. Oxford: Oxford University Press, pp. 680–1.

1994

The Express Train and Other Railway Studies. Nairn: Thomas and Lochar, 240 pp. (Collection of 14 essays, including seven previously unpublished: 'The Origins and Early Development of the Express Train', 'The Removal of a North Eastern General Manager, 1871', 'A Powerful Critic of Railways: John Tenniel in Punch', 'Railway Prospectuses', 'Bradshaw', 'Working Timetables', 'Accident Reports, 1840–90').

'Railways in their Context', in R. Shorland-Ball (ed.), *Common Roots – Separate Branches: Railway History and Preservation*. Science Museum for National Railway Museum, York, pp. 103–12.

1996

Foreword to G. Briwnant-Jones and D. Dunstone, *The Vale of Neath Line: From Neath to Pontypool Road*. Llandysul: Gomer, p. 7.

Foreword to B. Buchanan and G. Hulme, *St Mark's Church, Leicester: An Architectural and Historical Study*. Leicester: Leicester Group of the Victorian Society, pp. 1–2.

Introduction [new text] to a reprint of C. Torr, *Small Talk at Wreyland*. Liverton: Forest Publishing, pp. vii–xvi.

'Public Transport in Leicestershire, 1814–80', *TLAHS*, 70, pp. 105–27.

'The Writing of Railway History in Britain', *Revue Française de civilisation Britannique*, 8, No. 4 (June 1996), pp. 105–9 (contribution to issue on 'Le Voyage').

1997

Editor (with Gordon Biddle), *The Oxford Companion to British Railway History.* Oxford: Oxford University Press, xvi + 591 pp., including 316 entries by Professor Simmons. Reprinted in paperback, 1999.

'A.L. Rowse' [obituary], *The Independent*, 6 Oct. 1997, p. 16.

'This Junk', in C. Bruce (ed.), *Proceedings of a Seminar on Railway Records Held at the Public Record Office on 25 April 1997*. Railway Heritage Committee, pp. 31–40.

1998

Introduction to a reprint of L.T.C. Rolt, *Red for Danger* (1966 edn.). Stroud: Sutton Publishing, pp. 7–9.

1999

Foreword to M.A. Vanns, *The Railways of Newark-on-Trent*, Usk: Oakwood Press, p. 5.

'Where Do We Go From Here?', *Back Track*, 13, p. 515 (guest editorial).

List of Sponsors

Professor Derek Aldcroft
Maurice E. Bann
Margaret E. Bennett
Professor Michael Biddiss
Gordon Biddle
Dr J.M. Bourne
Gwyn Briwnant-Jones
Professor W. H. Brock
Dr David Brooke
Ivor John Bufton
Evan Bumford
Brian Burch
Philip Burkett
Burlington Hotel (Leicester) Staff
Cynthia G.B. Burt
David Bushell
Esther I. Buss
David Butler
Jean A. Chennell
Dr Peter Clark
Duncan Cloud
Dr Henry J. Cohn
Professor Philip Collins
Ann Conolly
David Conway
D. Madeleine Cooke
Professor Philip L. Cottrell
Graham Cousins
Dr Roger Craik
Dr James Crighton
Professor J. Mordaunt Crook
Professor I.M.T. Davidson
Dr Paul Davies
Derek Day
John Dean
Professor Colin Divall

Diana Dixon
Professor Michael Dockrill
Timothy F. Doust
Sylvia Dowling
Clifford Dunkley
Dr Alastair Durie
Dr Kenneth Edwards
Rupert Evans
Professor Alan Everitt
Paul Fincham
Margaret Findlay
Deborah J. Fisher
The Reverend Michael Fisher
Wilfred Flemming
Richard Float
Colin Ford
Dorothy Forster
Mary Forsyth
Hilary Fry
Dr Howard and Mrs Giana Fry
Terry and Sylvia Garfield
Rosemary Gascoyne
Herald and Joan Goddard
Dr John Gough
Dr Terry Gourvish
Olivia Grant
Professor J.K. Grodecki
George M. Gunson
Alan J. Harding
Michael Harris
Effie Harvie
Dr Alfred Ch. Heinimann
Professor Luke Herrman
Dr Hilary P. Higgins
History Department, University of
   Leicester

Professor John Hoffman
The Very Reverend Derek Hole
Dieter Hopkin
Norah Howe-Smith
Robert Hull
Jean Humphreys
Bruce Hunter
Alan A. Jackson
David Jeffreys
Colin Jennings
E. Marjorie Johnson
Stephen H. Johnson
The Venerable T. Hughie Jones
Christine Jordan
Leicester Group of the Victorian Society
Leicestershire Archaeological &
    Historical Society
Leicestershire Industrial History Society
Robin Leleux
Elizabeth Leon
Professor J.H.W.G. Liebeschuetz
Elizabeth Lowry
David W. Lyne
Rosemary McCrum
Professor John McManners
Dr A.D. McWhirr
Dr Lionel Madden
Professor Geoffrey Martin
Air Vice-Marshall A.R. Martindale
Professor A. J. Meadows
Stephen Medhurst
Roy Millward
Joyce Moody
Ruth Moon
Christine Morton-Thorpe
Eileen Mosson
Peter Neaverson
Professor Aubrey Newman
Veronica Nicholas
Sir Fraser Noble
Joan F.M. North
Richard Ollard
George Ottley

Professor Marilyn Palmer
Stuart Paterson
Dr Robert Peberdy
Dr John Pemble
Dr Dennis Petch
Professor Charles Phythian-Adams
Chris Pickford
Dr G.B. Pyrah
Dr D.A. Reeder
Professor R.C. Richardson
R.C. Riley
Michael Robbins
Angela Rogers
Norman Rosenthal
Leopold de Rothschild
John Runcie
Robert and Sybil Rutland
Peter Saunders
Norman Scarfe
Harold E. Scarlett
Andrew J. Scott
Sir Maurice Shock
Professor Brian Simon
Geoffrey Smith
Mike Smith
I.G. Stevens
Heather Stewart
Dr Joan Thirsk
Robin Titley
Tony and Rosemary Trollope
Dr David Turnock
Michael Vanns
Professor J.S. Wacher
Roger Warren
Professor Bernard Wasserstein
Margaret Watson
Dame Margaret Weston
Mollie Whitworth
Dr Lawrence Williams
Elizabeth Williamson
John Willis
Jan and Margaret Zientek

Index

Compiled by Professor R.C.Richardson

Abbott, Robert, 13
Aberystwyth, coming of the railway to, 221
Aberystwyth & Pwllheli line, 220
Accurate Description of the Liverpool and Manchester Railway (1830), 228
'Adlesdrop' (Edward Thomas), 153
Advanced Passenger Train (APT), 274
Advertising, 234
Aird, Sir John, 90
Allport, James, 120
American tourists, 208
Anatolian Railway, 81
APEX fares, 274
Arnold, Dr Thomas, 226
Arts and Crafts Movement, 184
Ashbee, C.R., 184
Ashby de la Zouch, canals and railways in, 18–27
ASLEF, 270, 277
Atholl Palace Hydro (Pitlochry), 207
Atkinson, J.C., 182, 185, 196
Attenborough, F.L., xix, 3
Aubrey, John, 184
Auden W.H., 136
Aytoun, W.E., 204

Baghdad railway, ch.6 *passim*
Baldwin, Henry, 151
Balfour, A.J., 79, 95
Banbury, railway church patronage in, 105
Bankes, Meyricke, 37
Barings, ch.6 *passim*
Barmouth, development of, 221, 223
Barnett, Henrietta, 178
Barrett, D.W., 101, 107, 108
Barry, 216, 217
Barry Island, ferries to, 215, 217
Barth, Boris, 78

Bath, coming of the railway to, 123
Baxter, Bertram, 14
Beaumont, Sir George, 17
Beaumont, Huntingdon, 11
Beeching, Dr Richard, 270
Bell, Robert, 244–5
Belvoir Castle railway, 28–9
Besant, Annie, 114
Betjeman, John, 136, 150
Bird, Revd F.J., 111
birds, wild (in Surrey), 192–3
Birmingham, 183
    industrial scene in, 237;
    market, 184
Blackbrook reservoir, 17
Bloomsbury Chapel, 105
Boer War, second, 79
Booth, Lawrence, 214
Booth, William, 113
Borrow, George, 182, 184
Boultbee, Joseph, 17
Bowle, John, 2
Bowles, Gibson, 91, 94
Bradlaugh, Charles (funeral), 114
Bradshaw's Railway Guide, 228
Braithwaite, John, 154
Brand, E., 13
Brassey, Thomas, 49
Brideshead Revisited, 153
Briggs, Asa, 117
British Library, 279
British Rail Council, 270
British Transport Commission, 270
Brock, Revd William, 105
Bruff, Peter, 154
Brunel, I.K., 119, 152, 215, 222
Buddicom, William, 49
Burdett, Sir Francis, 21

burial grounds, railway displacement of, 114n
Burlinson, N.J., 152
Bute, 200
Bywell, E.M., 249, 250, 253

Calke estate, limeworks, 24
Camden, smoke nuisance in, 126
Camden, William, 184
canal mania, 12
Carlisle, coming of the railway to, 123
carriage building costs, 52
Cassel, Sir Ernest, 79, 80, 81, 90, 96, 98
Chadwick, Edwin, 106
Chadwick, Jeremiah, 103
Chadwick, Owen, 102
Chadwick, Thomas, 214
Chamberlain, Joseph, 78
Chandler, Alfred Jr., 271
Charnwood Forest, canal and railways in, 14–17
Chester, coming of the railway to, 124
Christian Excavators Union, 108
'church interval', 103
Church Pastoral Aid Society, 107
Clare, John, 121
Close, Revd Francis, 101
club carriages, 51
Coalbrookdale, 12
coal trade, railways and, 11, 12, 215–16, 217
Cobbett, William, 184
cocktail bars, on trains, 54
Columbian Exposition and World Fair (Chicago, 1893), 208
Colwyn Bay, development of, 211, 214
Commuter's Charter (1979/80), 270
commuting, ch.13 *passim*, 201
Constantinople, British Chamber of Commerce at, 95
Conway, coming of the railway to, 124
Cook, John Mason, 108–9
Cook, Thomas, 108–9, 203
Corbett, Archibald Cameron, 175–6
Cornish Riviera Express, 152
corridor trains, 52
Coupland, Reginald, 3
Crewe, railway churches in, 104
Crieff Hydro, 209
Crossley, J.S., 119
Croydon, 237

Crystal Palace, 231

Davies, David, 163
Dawkins, Sir Charles, 79
Dean, Waddington & Company, canal carriers, 43
Defoe, Daniel, 184
Derby, coming of the railway to, 123
Deutsche Bank, 77, 78, 85, 87, 89
Devey, George, 184
Dewes, Simon (pseud. John Muriel), 158–9
dining cars, railway, 50
Dinsdale, Sir Joseph, 206
diocesan organisation, impact of railway on, 102
Dodgson, Charles, 211
Dolgellau, coming of the railway to, 222
Doncaster, railway church in, 104
Douglas Navigation, 36, 42
Dovey Junction, 163–5
'Dreadnought' rolling stock (GWR), 53
Dublin, industrial scene in, 238
Dunoon, 201
Dyos, H.J., xiii, 4n

Ealing, 171–3
'Ealing Express', 173
Eastleigh, 105
ecclesiastical patronage of railway companies, 104–5
Edinburgh, coming of the railway to, 123, 200
Edmonton, 173–5
Edmunds, John, 277
Ellis, Hamilton, 279, 280
enclosure, 192
Enfield, 174
English Local History, Jack Simmons and, xix–xxi
excursion agents, railway, 203

Farey, John, 12, 16, 18
Fayers, Thomas, 107
Fiennes, Celia, 184
Finberg, H.P.R., xix
Fishguard, 214, 215
Flying Scotsman, 53
Follett, Sir David, xxiii
footwarmers, 49
Ford, Colin, xxiii
Fort William, 204

Fowler, W. Warde, 122
Franklin, W.E., 203
freight wagon, design changes, 58–60
Furness Abbey, railway and, 124, 125

Garnett, Mrs Elizabeth, 107, 108
Gilbert, Dr John, 277
Gimson, Ernest, 184
Gladstone, W.E., 101
Glasson Dock, 43
Golders Green, coming of the railway to, 176–8
golfing (Scotland), 207
Goodmayes, 176
Gooch, Daniel, 152
Gorton railway works, Manchester, 102
Gray, Thomas, 118
Greenwood, James, 196
Grenfell, E.C., 97
Grey, Sir Edward, 97
Griggs, F.L., 184
grouse shooting, 206–7
Gwinner, Arthur von, 78, 82, 84, 87, 89, 91, 92, 93, 94, 96

Haida Pasha Port Company, 91
hairdressing salons on trains, 54
Haiton, Thomas, xix
Hampstead Garden Suburb, 178–9
Hamud, Sultan Abdul, 93, 96
Handbook for Travellers in Scotland, 199
Hardy, Thomas, 138, 153
Harper, J.B., 244, 245, 246
Harrison, Brian, 108
Harrison, T.E., 50
harvesting, 193–4
Hastings, Charles, 25
Hatton, Thomas, xix
Hawkshaw, Sir John, 125
Hawthorne, Nathaniel, 125
heating, in railway carriages, 48
Helfferich, Karl, 96
herb gathering, 195
Hindley–Liverpool canal project, 36
Hill, A.J., 58
Hillingdon, Lord, 83
History and Topographical Survey of Kent (Hasted), xx
History of the Gothic Revival (Eastlake), xxii
Hodgson, Richard, 120

Holyhead, 212
Hongkong & Shanghai Banking Corporation, 80
Hope, John, 202
Hoskins, W.G., xix, 3
Hudson, George, 118–19
Hudson, W.H., 181, 185, 196, 197n
Huish, Mark, 120
Hull, Philip Larkin on, 144
Hungerford Bridge, 126
Hurcomb, Sir Cyril, 253
Huxley, T.H., 121

Ilford, 175–6
Illustrations of the Scenery on the Line of the Whitby and Pickering Railway, 228
Imperial Bank of Persia, 80
Imperial Ottoman Bank, 78n
Ince Hall Coal & Canal Company, 43
inns, decline of, 108
International Mercantile Marine Company (NJ), 80
Irish Mail, 211

Jackson, Sir Thomas, 184
James, Henry, 121
James, William, 118, 119
Jefferies, Richard, 182, 185–96
Jekyll, Gertrude, 184
Jennings, Louis, 185, 196
Jessop, William, 16, 17, 18
Johnson, Leonard, 255
Journal of Transport History, 4, 281

Kingston upon Thames, 191
Kipling, Rudyard, 181
Kuwait, as supplier of horses, 90

labour migration, 190
ladies' retiring rooms, on trains, 54
Lake District railways, 132–3
Lancaster Canal, 37
landscape, impact of railways on, 120–2
Lansdowne, Lord, ch. 6 *passim*
Larkin, Philip, and railways, ch.10 *passim*
    as novelist, 138
    'Whitsun Weddings', 146–9
lavatories, on trains, 48
Law, Sir Edward Fitzgerald, 93
Leamington Spa, coming of the railway to, 125

Lecount, Lt. Peter, 48
Lee, Charles E., 13
Leicester Navigation Company, 14, 15
Leigh, Alexander, 36
Leland, John, 33, 184
Lenny & Company, coachbuilders, 237
Lewes, coming of the railway to, 124
Lewis, Michael, 14
Leyton, 173–5
Lichfield, Earl of, 120
Liddell, Dean, H.G., 211
Liddington, 186–90
Light from the East, 228
lighting, in railway carriages, 49
Lincoln, Bishop of, 107
Liverpool, coming of the railway to, 123
Llandudno, development of, 211–12
Llangollen, coming of the railway to, 222
Lloyd, Leslie, 277
Loch Ness, 204
Locke, Joseph, 119, 157
Locomotion No.1, 244
London County Council, 170
London railway suburbs, ch.13 *passim*
Love on a Branch Line (John Hadfield),
    158
Lowenthal, David, 262
Lucas & Aird, railway contractors, 107
luncheon baskets, on railways, 48

M'Cree, George, 105
Mackay, Charles (poet), 114
Maitland, F.W., 185
Major, John, 270
Manchester, 101–3
Manchester, Bishop of, Right Revd James
    Fraser, 101–3
Manchester, Bolton, and Bury Canal, 36
Manchester Mission, 103
Manchester Ship Canal, 41
Marples, Ernest, 271
Marsh, Richard, 270, 272
Marshall, William, 194n
Martineau, Harriet, 121
Matthews, Sir Ronald, 253
Maumbury Rings, Dorchester, 125
Measham Colliery, 12
Measom, Sir George Samuel, ch.17 *passim*
Mewburn, William, 105
Midnight on the Great Western (Thomas
    Hardy), 153

milk production, railway impact on, 186n,
    190
Mill, J.S., 121
Mitchell, Joseph, 119, 205
Mitford, Mary Russell, 184
Moffat (Scottish spa), 205
Moffat, Hugh, 157
Moira, 26
Moody, D.L., 109, 113
Morris, William, 133, 153
Mostyn, Edward, 212
Motion, Andrew, 141
motor car, and tourism, 209
Murray's *Handbooks*, 227
museums, Jack Simmons and, xxiii–xxiv, 6
music, on trains, 54

Nall, George, bookseller, 234
narrow gauge railways (Wales), 223
National Museum of Photography, Film and
    Television, xxiii
National Trust, 184
navvies, missions to, 106, 107, 108
Navvy Mission Society, 108
Newcastle, coming of the railway to, 125
Newquay, 212
New Railway Guide, The, 111
noise, railway, 127
Northern Belle, 54
Nottingham, 11
NUR, 277

Oakwood Press, 2n–3n
Oban, 200, 204
open coaches, 52
operating literature, railway, ch. 5 *passim*
Ordnance Survey, 183
Overend & Gurney, bankers, 106n
Oxford–Hereford line, 274
Oxford, Bishop of, Right Revd Samuel
    Wilberforce, 102n
Outram, Benjamin, 18
Oxford University Railway Society, 2

Paddle steamers, Clyde, 200
Parker, Sir Peter, 270, 271, 275
Paterson, Frank, 277
Patterson, A.T., 13
Paxton, Sir William, 220
Peach, Samuel (railway poet), 109
Peak District railways, 132

Pease, Edward, 118
Pease, Joseph, 118
Peebles, 200
Pemberton railway, 34, 37
Pembroke, 220
Penarth, 216–17
Pender, Sir John, 214
Perham, Margery, 3
Perrins, Mrs Dyson, 223
Peterborough, Bishop of, 107
Peto, Sir Samuel Morton, 105–6, 107
Pick, Frank, 179
Piercey, Benjamin, 163
Pinto, Vivian de Sola, 146
Plot, Robert, 184
poetry, railways and, 106, 109, 110, 121, 124, 136, ch.10 *passim*, 153, 226
Porter, J.M., 214
Porthcawl, 216, 218–19
Portland, Duke of, 118
Port Talbot, 218
Potter, Beatrix, 122
Preservation of Relics and Records, The (BTC Report, 1951), 254
Prestatyn, 211
Preston, 108
Price's Candle Company, 235
Prideaux, Dr John, 277
promenades, seaside, 226
Public Record Office, 257
Pugin, A.W.N., 122
pullman cars, 48, 49

rail strategy, 'business-led', 273
Railroaders Memorial Museum (Pennsylvania), 266
Railton, Herbert, 184
railway accidents, 74, 276
Railway & Canal Historical Society, 281
railway archives, 5, ch.5, 280–1
railway coaches, steel-framed, 55
railway companies
  Bolton & Leigh, 39
  Cambrian Railways, 220
  Chester & Holyhead, 211
  Coleorton, 25
  Cromford & High Peak, 30
  Dingwall & Skye, 204, 205
  Garnkirk & Glasgow, 201–2
  Glasgow, Paisley, & Greenock, 202
  Grand Junction, 212

  Great North of Scotland, 207
  Great Western, 55, 190
  Invergarry & Fort Augustus, 204, 209
  Kendal & Windermere, 121
  Kenyon & Leigh Junction, 39
  Kilmarnock & Troon, 118
  Lancashire & Yorkshire, 39
  Lancashire Union, 40
  Leicester & Swannington, 17, 25, 29–30, 118
  Liverpool & Manchester, 39, 118, 225
  London & Birmingham, 212
  London & Greenwich, 104
  London & North Western, 39, 213
  London, Tilbury & Southend, 65n
  Metropolitan District, 172–3
  Midland, 25, 66n, ch.5 *passim*, 120
  Mid Suffolk, 156–7
  Necropolis, 113–14, 235, 237
  North Union, 39
  Oystermouth, 225
  Pembroke & Tenby, 220
  Portpatrick & Wigtownshire Joint, 204
  Scottish Central, 205
  Southampton & Dorchester, 121
  South Eastern, 182
  Southwold, 160–1
  Stockton & Darlington, 118, 225, 243
  Vale of Llangollen, 222
  Wigan Branch, 39
Railway Companion. By a Tourist (1833), 228
railway guidebooks, ch. 17 *passim*
Railway Magazine, 63
Railway Mission, London, 109
railway primers, 228
Railway Signal, 101, 109
railway stations
  Birmingham New St., 126
  Exeter, St David's, 122, 124
  Glasgow, St Enoch's, 104
  Leeds, 124
  London, Euston, 122, 124
  London, Liverpool St., 17
  London, St Pancras, 104, 115
rail-users, canvassed, 57
'Railway Sunday', 103
Raven, Vincent, 59
Rayner, Peter, 275
Reid, Sir Robert (Bob Reid I), 271
Reid, Sir Bob (Bob Reid II), 271, 275

religion and railways, ch. 7 *passim*
Report on the Agriculture of Suffolk (1849), 151
resorts, Scottish holiday, 200
Revelstoke, Lord, 79, 85, 86, 88, 90, 94, 95, 98
Rhyl, 211
Richardson, Robert, 15
ritual year, 190
Robbins, Michael, 122, 280
Robertson, General Sir Brian, 270
rolling stock, railway, ch.4 *passim*
Rothesay, 200
Rothschilds, 79, 91
Routh, Dr, 151
Royal Commission on Transport (1928), 54
'Royal Palaces on Wheels' (NRM exhibition), 265–6
Royal Scot, 53
Rudgard, H.J., 244
rule books, railway, 63n
rural tradition, railways and, ch.14 *passim*
Ruskin, John, 121, 122, ch.9 *passim*
    on Derbyshire railways, 132;
    on Lake District railways, 132;
    on morality, 130
    on the steam engine, 129
    on tourists, 133
    opposed to laissez-faire, 131
    Tory convictions, 130
Russell, Charles, 152, 158

sabbatarianism, 103, 105, 223
Sackville-West, Victoria, 184n
safety, rail, ch. 5 *passim*, 276
Salvation Army, 113
Sanders, Joseph, 118
Sankey, Ira D., 109, 113
Sankey Navigation, 36
Saunders, C.A., 152
Savin, Thomas, 222
Scarfe, Samuel, 158
Scholes, John, 255
seaside resorts, 114
Season at Harwich, A (1851), 227
season tickets, 171, 174, 176, 213
seating, railway carriage, 49
senior citizens railcards, 270
Serpell Report (1983), 271, 272, 275
Settle & Carlisle Railway, 75, 119
Seven Kings, 176

Severn Tunnel, 214, 217
Shaw, Norman, 184
Sheffield, 183
Sheppard, Edward, 158
shower cubicles in railway sleeping cars, 54
Shrewsbury, coming of the railway to, 124
Shugborough, 120
Simmons, Jack, xiii, xv–xvii, xix–xxiv, 1–7, 11, 62, 69, 77n, 115, 151, 153–4, 181, 225n, 259, 263, 269, 280, 281, 283–306
sleeping cars, 50
slums, railways and, 125
Smiles, Samuel, 106
Smith, G.R., 254
Smith, Sir Henry Babington, 92, 94
Smith, Samuel, 106, 108
Smith, W.H., 112
smoke nuisance of railways, 126
smoking cars, 49, 55
Soar Navigation, 14
Somerleyton, Suffolk, 106
Southampton, coming of the railways to, 123
South Sea Bubble, 119
SPCK, 111, 112
Spettisbury Rings, Dorset, 125
'Spiritual Railway' (poem), 110
Spurgeon, Charles Haddon, 112
Standing Committee on Mineral Transport (1927), 59
Stantonbury, railway church in, 104
Staveley, Christopher, 14, 16
Stephenson, George, 30, 119
Stephenson, Robert, 30, 119
Strathpeffer (Scottish spa), 205, 206
Stretton, Clement E., 12–13
Sturt, George, 182, 190
Suez Canal, 96
supplementary fare expresses, 56
Swindon, railway church in, 104

Taine, Hyppolyte, 103, 112
Tanner, Lawrence, 1–2
tarmac, 181n
Temperance movement, 106, 108–9
Tenby, 220
Thatcher, Margaret, 276
Thompson, Flora, 182, 190
Thorne, James, 196
threshing machines, 187n
timetables, railway working, 62–74

Topographical Dictionary of Wales (Samuel Lewis), 221
Torr, Cecil, xxii, 181
Tottenham, 173–5
tourism, 133, ch.15 *passim*, ch.16 *passim*
transport museums, xxiii, ch.19 *passim*
travelling charts, railway, 182, 228
Trevelyan, G.M., 263, 264

urban environment, railway impact on, 123–7

ventilation and air-conditioning, on trains, 54, 58
vernacular tradition, 184
Vicinus, Martha, 111n
Victorian Library (Leicester University Press), xxii
Victorian Studies, Jack Simmons and, xxi–xxii
visionaries, railway, 118–19
visitors to transport museums, 261–3, 267–8
Vitali, Count, 81

Waddington, David, 105
wakes holidays, 213
Walker, Thomas Andrew, 217
Walters, Alan, 270
Walthamstow, 173–5
Warner, Thomas Courtenay Theydon, 175

Watkin, Edward, 120
Webb, Francis, 104
West Coast Main Line, 119
West Highland line, 204
whistle signals, 70
White, Gilbert, 184
Whitelaw, William, 209
Whitemore Bay, 218
Wigan Coal & Iron Company, 35
Wigan coalfield, canals and railways in, ch.3 *passim*
Wigan pier, 37
Wilkes, Joseph, 12
Williams, Owen, 212
Williamson, Oliver, 271
Willmer, Henry, 105
Wilson, Harold, 179
Winchester, Bishop of, Right Revd Charles Sumner, 113
Wolverton, railway church in, 104
Wooley, William, 184
Wordsworth, William, 106, 121, 124, 226
workmen's fares, 170–1
Wright, Patrick, 262

Yerkes, Charles Tyson, 176
Yolland, Colonel, 74
York, coming of the railway to, 124
York, Railway Museum, ch.18 *passim*

Zuleika Dobson, 153